Biological Thermodynamics

Biological Thermodynamics provides an introduction to the study of energy transformation in the biological sciences. Don Haynie uses an informal writing style to discuss this core subject in a way which will appeal to the interests and needs of undergraduate students of biology and biochemistry. The emphasis throughout the text is on understanding basic concepts and developing problem-solving skills, but mathematical difficulty is kept to a minimum. Each chapter comprises numerous examples taken from different areas of biochemistry, as well as a broad range of exercises and list of references for further study. Topics covered include energy and its transformation, the First Law of Thermodynamics, the Second Law of Thermodynamics, Gibbs free energy, statistical thermodynamics, binding equilibria, reaction kinetics, and a survey of the most exciting areas of biological thermodynamics today, particularly the origin of life.

Don Haynie has a B.S. in physics (South Florida) and a Ph.D. in biophysics (Biocalorimetry Center, The Johns Hopkins University). Following post-doctoral research in physical biochemistry at the University of Oxford, he took up a lectureship in biochemistry in the Department of Biomolecular Sciences at the University of Manchester Institute of Science and Technology (UMIST), UK. He has taught thermodynamics, to biology, biochemistry, and engineering students at Johns Hopkins and UMIST, and at Louisiana Tech University, where he is now Associate Professor in the Institute for Micromanufacturing.

Biological Thermodynamics

Donald T. Haynie

CAMBRIDGE
UNIVERSITY PRESS

PUBLISHED BY THE PRESS SYNDICATE OF THE UNIVERSITY OF CAMBRIDGE
The Pitt Building, Trumpington Street, Cambridge, United Kingdom

CAMBRIDGE UNIVERSITY PRESS
The Edinburgh Building, Cambridge CB2 2RU, UK
40 West 20th Street, New York, NY 10011-4211, USA
477 Williamstown Road, Port Melbourne, VIC 3207, Australia
Ruiz de Alarcón 13, 28014 Madrid, Spain
Dock House, The Waterfront, Cape Town 8001, South Africa

http://www.cambridge.org

First published 2001
Reprinted 2003

Printed in the United Kingdom at the University Press, Cambridge

Typeface Swift 9.5/12.25pt.   System QuarkXPress™   [SE]

A catalogue record for this book is available from the British Library

ISBN 0 521 79165 0 hardback
ISBN 0 521 79549 4 paperback

—

For my great-grandfather
JOSEPH RAFFEL

my grandmother
HELENE MINDER

and my friend
BUD HERSCHEL

—

The trouble with simple things is that one must understand them very well

ANONYMOUS

Contents

Chapter 5 | Gibbs free energy – applications

Chapter 6 | Statistical thermodynamics

Preface

Interest in the biological sciences has never been greater. At the dawn of twenty-first century, biology and biochemistry are captivating the minds of young people in the way that physics did 40–50 years ago. There has been a massive shift in public opinion and in the allocation of resources for university-based research. Breakthroughs in genetics, cell biology, and medicine are transforming the way we live, from improving the quality of produce to eradicating disease; they are also stimulating pointed thinking about the origin and meaning of life. Growing awareness of the geometry of life, on length scales extending from an individual organism to a structural element of an individual macromolecule, has led to a reassessment of the principles of design in all the engineering disciplines, including computation. And a few decades after the first determination at atomic resolution of the structures of double-stranded DNA and proteins, it is becoming increasingly apparent that both thermodynamic and structural information are needed to gain a deep sense of the functional properties of biological macromolecules.

This book is about the thermodynamics of living organisms. It was written primarily for undergraduate university students of the biological, biochemical and medical sciences. It could also serve as an introductory text for undergraduate students of chemistry or physics who are interested in biology, and for graduate students of biology or biochemistry who did their first degree in a different subject. The style and depth of presentation reflect my experience learning thermodynamics as an undergraduate student, doing graduate-level research on protein thermodynamics at the Biocalorimetry Center at Johns Hopkins University, teaching thermodynamics to biochemistry undergraduates in the Department of Biomolecular Sciences at the University of Manchester Institute of Science and Technology and to pre-meds at Johns Hopkins, and discussing thermodynamic properties of proteins with colleagues in Oxford Centre for Molecular Sciences.

My sense is that an integrated approach to teaching this subject, where the principles of physical chemistry are presented not as a stand-alone course but as an aspect of biology, has both strengths and weaknesses. On the one hand, most biological science students prefer to encounter physical chemistry in the context of learning about living organisms, not in lectures designed for physical chemists. On the other hand, applications-only courses tend to obscure fundamental concepts. The treatment of thermodynamics one finds in general biochemistry textbooks can compound the difficulties, as the subject is usually treated separately, in a single chapter, with applications being touched on only here and there in the remainder of the text. Moreover, most general biochemistry texts are written by scientists who have little or no special training in thermodynamics, making a coherent and integrated presentation of the subject that much more difficult. A result of this is that many students of the biological sciences complete their undergraduate study with either a shallow or fragmented knowledge of thermodynamics, arguably the most basic area of all the sciences and engineering. Indeed, many scientists would say that the Second Law of Thermodynamics is the most general idea in science and that energy is its most important concept.

It is not difficult to find compelling statements in support of this view. According to Albert Einstein, for example, 'Classical thermodynamics . . . is the only physical theory of universal content concerning which I am convinced that, within the framework of applicability of its basic concepts, will never be overthrown.' Einstein, a German–American physicist, lived from 1879 to 1955. He was awarded the Nobel Prize in Physics in 1921 and described as 'Man of the Century' by *Time* magazine in late 1999. Sir Arthur S. Eddington (1882–1944), the eminent British astronomer and physicist, has said, 'If your theory is found to be against the second law of Thermodynamics I can give you no hope; there is nothing for it but to collapse in deepest humiliation.' Sir Charles P. Snow, another British physicist, likened lack of knowledge of the Second Law of Thermodynamics to ignorance of Shakespeare, to underscore the importance of thermodynamics to basic awareness of the character of the physical world. Finally, M. V. Volkenstein, member of the Institute of Molecular Biology and the Academy of Sciences of the USSR, has written, 'A physical consideration of any kind of system, including a living one, starts with its phenomenological, thermodynamic description. Further study adds a molecular content to such a description.'

The composition and style of this book reflect my own approach to teaching thermodynamics. Much of the presentation is informal and qualitative. This is because knowing high-powered mathematics is often quite different from knowing what one would like to use mathematics to describe. At the same time, however, a firm grasp of thermodynamics and how it can be used can really only be acquired through quantitative problem solving. The text therefore does not avoid expressing ideas in the form of equations where it seems fitting. Each chapter is imbued with *l'esprit de géométrie* as well as with *l'esprit de finesse*. In general, the mathematical difficulty of the material increases from beginning to end. Worked examples are provided to illustrate how to use and appreciate the mathematics, and a long list of references and suggestions for further reading are given at the end of each chapter. In addition, each chapter is accompanied by a broad set of study questions. These fall into several categories: brief calculation, extended calculation, multiple choice, analysis of experimental data, short answer, and 'essay.' A few of the end-of-chapter questions are open-ended, and it would be difficult to say that a 'correct' answer could be given to them. This will, I hope, be seen as more of a strength of the text than a weakness. For the nature of the biological sciences is such that some very 'important' aspects of research are only poorly defined or understood. Moreover, every path to a discovery of lasting significance has had its fair share of woolly thinking to cut through. Password-protected access to solutions to problems is available on line at http://chem.nich.edu/homework

Several themes run throughout the book, helping to link the various chapters into a unified whole. Among these are the central role of ATP and glucose in life processes, the proteins lysozyme and hemoglobin, the relationship between energy and biological information, and the human dimension of science. The thermodynamics of protein folding/unfolding is used to illustrate a number of points because it is well known to me and because it is of more general significance than some people might think. After all, about 50% of the dry mass of the human body is protein, no cell could function without it, and a logical next step to knowing the amino acid sequence encoded by a gene is predicting the three-dimensional structure of the corresponding functional protein. We also draw attention to how thermodynamics has developed over the past several hundred years from contributions from researchers of many different countries and backgrounds.

The principal aim of this text is to help students of the biological sciences gain a clearer understanding of the basic principles of energy transformation as they apply to living organisms. Like a physiologically meaningful assembly of biological macromolecules, the organization of the text is hierarchical. For students with little or no preparation in thermodynamics, the first four chapters are essential and may in some cases suffice for undergraduate course content. Chapter 1 is introductory. Certain topics of considerable complexity are dealt with only in broad outline here; further details are provided at appropriate points in later chapters. This follows the basic plan of the book, which highlights the unity and primacy of thermodynamic concepts in biological systems and processes and not simply the consistency of specific biological processes with the laws of thermodynamics. The second and third chapters discuss the First and Second Laws of Thermodynamics, respectively. This context provides a natural introduction to two important thermodynamic state functions, enthalpy and entropy. Chapter 4 discusses how these functions are combined in the Gibbs free energy, a sort of hybrid of the First and Second Laws and the main thermodynamic potential function of interest to biological scientists. Chapter 4 also elaborates several basic areas of physical chemistry relevant to biology. In Chapter 5, the concepts developed in Chapter 4 are applied to a wide range of topics in biology and biochemistry, an aim being to give students a good understanding of the physics behind the biochemical techniques they might use in an undergraduate laboratory. Chapters 4 and 5 are intended to allow maximum flexibility in course design, student ability, and instructor preferences. Chapters 6 and 7 focus on the molecular interpretation of thermodynamic quantities. Specifically, Chapter 6 introduces and discusses the statistical nature of thermodynamic quantities. In the next chapter these ideas are extended in a broad treatment of macromolecular binding, a very common and extremely important class of biochemical phenomenon. Chapter 8, on reaction kinetics, is included in this book for two main reasons: the equilibrium state can be defined as the one in which the forward and reverse rates of reaction are equal, and the rate of reaction, be it of the folding of a protein or the catalysis of a biochemical reaction, is determined by the free energy of the transition state. In this way inclusion of a chapter on reaction kinetics gives a more complete understanding of biological thermodynamics. Finally, Chapter 9 touches on a number of topics at the forefront of biochemical research where thermodynamic concepts are of considerable importance.

A note about units. Both joules and calories are used throughout this book. Unlike monetary exchange rates and shares on the stock exchange, the values of which fluctuate constantly, the conversion factor between joules and calories is constant. Moreover, though joules are now more common than calories, one still finds both types of unit in the contemporary literature, and calories predominate in older publications. Furthermore, the instrument one uses to make direct heat measurements is a called a calorimeter not a joulimeter! In view of this it seems fitting that today's student should be familiar with both types of unit.

Three books played a significant role in the preparation of the text: *Introduction to Biomolecular Energetics* by I. M. Klotz, *Foundations of Bioenergetics* by H. J. Morowitz, and *Energy and Life* by J. Wrigglesworth. My interest in biophysics was sparked by the work of Ephraim Katchalsky (not least by his reflections on art and science!) and Max Delbrück, which was brought to my attention by my good friend Bud Herschel. I can only hope that my predecessors will deem my approach to the subject a helpful contribution to thermodynamics education in the biological sciences.

The support of several other friends and colleagues proved invaluable to the project. Joe Marsh provided access to historical materials, lent volumes from his personal library, and encouraged the work from an early stage. Paul C. W. Davies offered me useful tips on science writing. Helpful information was provided by a number of persons of goodwill: Rufus Lumry, Richard Cone, Bertrand Garcia-Moreno Esteva, Alan Eddy, Klaus Bock, Mohan Chellani, Bob Ford, Andy Slade, and Ian Sherman. Van Bloch was an invaluable source of encouragement and good suggestions on writing, presenting, and publishing. I thank Chris Dobson. Norman Duffy, Mike Rao, Alan Cooper, Bertrand Garcia-Moreno Esteva, John Ladbury, Alison Roger, Terry Brown, and several anonymous reviewers read parts of the text and provided valuable comments. I wish to thank Katrina Halliday, Ward Cooper, Beverley Lawrence and Sue Tuck at Cambridge University Press for the energy and enthusiasm they brought to this project. I am pleased to acknowledge Tariq, Khalida, and Sarah Khan for hospitality and kindness during the late stages of manuscript preparation. I am especially grateful to Kathryn, Kathleen, and Bob Doran for constant encouragement and good-heartedness.

Every attempt has been made to produce a spotless textbook. I alone am responsible for any infelicities that may have escaped the watchful eyes of our red pens.

D. T. H.
15th August 2000
Oxford, England

References

Eddington, A. S. (1930). *The Nature of the Physical World*, p. 74. New York: MacMillan.

The Editor (2000). Resolutions to enhance confident creativity. *Nature*, **403**, 1.

Klein, M. J. (1967). Thermodynamics in Einstein's Universe. *Science*, **157**, 509.

Volkenstein, M. V. (1977). *Molecular Biophysics*. New York: Academic.

Chapter 1

Energy transformation

A. | Introduction

Beginning perhaps with Anaximenes of Miletus (fl. *c.* 2550 years before present), various ancient Greeks portrayed man as a microcosm of the universe. Each human being was made up of the same elements as the rest of the cosmos – earth, air, fire and water. Twenty-six centuries later, and several hundred years after the dawn of modern science, it is somewhat humbling to realize that our view of ourselves is fundamentally unchanged.

Our knowledge of the matter of which we are made, however, has become much more sophisticated. We now know that all living organisms are composed of hydrogen, the lightest element, and of heavier elements like carbon, nitrogen, oxygen, and phosphorus. Hydrogen was the first element to be formed after the Big Bang. Once the universe had cooled enough, hydrogen condensed to form stars. Then, still billions[1] of years ago, the heavier atoms were synthesized in the interiors of stars by nuclear fusion reactions. We are 'made of stardust,' to quote Allan Sandage (b. 1926), an American astronomer.

Our starry origin does not end there. For the Sun is the primary source of the energy used by organisms to satisfy the requirements of life (Fig. 1.1). (Recent discoveries have revealed exceptions to this generalization: see Chapter 9.) Some organisms acquire this energy (Greek, *en*, in + *ergon*, work) directly; most others, including humans, obtain it indirectly. Even the chemosynthetic bacteria that flourish a mile and a half beneath the surface of the sea require the energy of the Sun for life. They depend on plants and photosynthesis to produce oxygen needed for respiration, and they need the water of the sea to be in the liquid state in order for the plant-made oxygen to reach them by convection and diffusion. This is not necessarily true of bacteria *everywhere*. The recent discovery of blue-green algae beneath ice of frozen lakes in Antarctica has indicated that bacteria *can* thrive in such an environment. Blue-green algae, also known as cyanobacteria, are the most ancient photosynthetic, oxygen-producing organisms known. In order to thrive, however,

---

[1] 1 billion $= 10^9$.

A diagram of how mammals capture energy. The Sun generates radiant energy from nuclear fusion reactions. Only a tiny fraction of this energy actually reaches us, as we inhabit a relatively small planet and are far from the Sun. The energy that does reach us – $c. 5 \times 10^{18}$ MJ yr^{-1} $(1.7 \times 10^{17}$ J s$^{-1})$ – is captured by plants and photosynthetic bacteria, as well as the ocean. (J = joule. This unit of energy is named after British physicist James Prescott Joule (1818–1889)). The approximate intensity of direct sunlight at sea level is 5.4 J cm^{-2} min^{-1}. This energy input to the ocean plays an important role in determining its predominant phase (liquid and gas, not solid), while the energy captured by the photosynthetic organisms (only about 0.025% of the total; see Fig. 1.2) is used to convert carbon dioxide and water to glucose and oxygen. It is likely that all the oxygen in our atmosphere was generated by photosynthetic organisms. Glucose monomers are joined together in plants in a variety of polymers, including starch (shown), the plant analog of glycogen, and cellulose (not shown), the most abundant organic compound on Earth and the repository of over half of all the carbon in the biosphere. Animals, including grass eaters like sheep, do not metabolize cellulose, but they are able to utilize other plant-produced molecules. Although abstention from meat (muscle) has increased in popularity over the past few decades, in most cultures humans consume a wide variety of animal species. Muscle tissue is the primary site of conversion from chemical energy to mechanical energy in the animal world. There is a continual flow of energy and matter between micro-organisms (not shown), plants (shown), and animals (shown) and their environment. The sum total of the organisms and the physical environment participating in these energy transformations is known as an **ecosystem**.

polar bacteria must be close to the surface of the ice and near dark, heat absorbing particles. Solar heating during summer months liquifies the ice in the immediate vicinity of the particles, so that liquid water, necessary to life as we know it, is present. During the winter months, when all the water is frozen, the bacteria are 'dormant.' **Irrespective of form, complexity, time or place, all known organisms are alike in that they must capture, transduce, store and use energy in order to live.** This is a profound statement, not least because **the concept of energy is the most basic one of all of science and engineering**.

How does human life in particular depend on the energy output of the Sun? Green plants flourish only where they have access to light. Considering how green our planet is, it is amazing that much less than 1% of the Sun's energy that penetrates the protective ozone layer, water vapor and carbon dioxide of the atmosphere, is actually absorbed by plants (Fig. 1.2). The chlorophyll and other pigment molecules of plants act as antennas that enable them to absorb photons of a relatively limited range of energies (Fig. 1.3). On a more detailed level, a pigment molecule, made of atomic nuclei and electrons, has a certain electronic *bound* state that can interact with a photon (a *free* particle) in the visible range of the electromagnetic spectrum (Fig. 1.4). When a photon is absorbed, the bound electron makes a transition to a higher energy but less stable 'excited' state. Energy captured in this way is then transformed by a very complex chain of events (Chapter 5). The mathematical relationship between wavelength of light, λ, photon frequency, ν, and photon energy, E, is

$$E = hc/\lambda = h\nu \tag{1.1}$$

where h is Planck's constant[2] $(6.63 \times 10^{-34}$ J s) and c is the speed of light *in vacuo* $(2.998 \times 10^8$ m s$^{-1})$. Both h and c are fundamental constants of nature. Plants combine trapped energy from sunlight with carbon dioxide and water to give $C_6H_{12}O_6$ (glucose), oxygen and heat. In this way solar energy is turned into chemical energy and stored in the form of chemical bonds, for instance the $\beta(1 \rightarrow 4)$ glycosidic bonds between the glucose monomers of cellulose and the chemical bonds of glucose itself (Fig. 1.1).

---

[2] Named after the German physicist Max Karl Ernst Ludwig Planck (1858–1947). Planck was awarded the Nobel Prize in Physics in 1918.

Animals feed on plants, using the energy of digested and metabolized plant material to manufacture the biological macromolecules they need to maintain existing cells, the morphological units on which life is based, or to make new ones. The protein hemoglobin, which is found in red blood cells, plays a key role in this process in humans, transporting oxygen from the lungs to cells throughout the body and carbon dioxide from the cells to the lungs. Animals also use the energy of digested foodstuffs for locomotion, maintaining body heat, generating light (e.g. fireflies), fighting off infection by microbial organisms, growth, and reproduction (Fig. 1.5). These biological processes involve a huge number of exquisitely specific biochemical reactions, each of which requires energy in order to proceed.

To summarize in somewhat different terms. The excited electrons of photosynthetic reaction centers are reductants. The electrons are transferred to carbon dioxide and water, permitting (*via* a long chain of events) the synthesis of organic molecules like glucose and cellulose. The energy of organic molecules is released in animals in a series of reactions in which glucose, fats, and other organic compounds are oxidized (burned) to carbon dioxide and water (the starting materials) and heat. This chain of events is generally 'thermodynamically favorable' because we live in a highly oxidizing environment: 23% of our atmosphere is oxygen. Don't worry if talk of oxidation and reduction seems a bit mystifying at this stage: we shall return to it treat it in due depth in Chapter 4.

Two of the several **requirements for life** as we know it can be inferred from these energy transformations: *mechanisms* **to control** *energy* **flow**, for example the membrane-bound protein 'machines' involved in photosynthesis; and *mechanisms* **for the storage and transmission of biological** *information*, namely polyribonucleic acids. The essential role of *mechanisms* in life processes implies that *order* **is a basic characteristic of living organisms**. A most remarkable and puzzling aspect of life is that the structures of the protein enzymes that regulate the flow of energy and information in a cell are encoded by nucleic acid within the cell. We can also see from the preceding discussion that energy flow in nature resembles the movement of currency in an

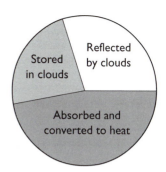

Fig. 1.2 Pie plot showing the destiny of the Sun's energy that reaches Earth. About one-fourth is reflected by clouds, another one-fourth is absorbed by clouds, and about half is absorbed and converted into heat. Only a very small amount ($\ll 1\%$) is fixed by photosynthesis (not shown).

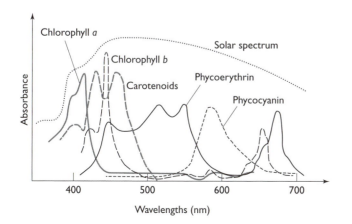

Fig. 1.3 Absorption spectra of various photosynthetic pigments. The chlorophylls absorb most strongly in the red and blue regions of the spectrum. Chlorophyll *a* is found in all photosynthetic organisms; chlorophyll *b* is produced in vascular plants. Plants and photosynthetic bacteria contain carotenoids, which absorb light at different wavelengths from the chlorophylls. The relationship between photon wavelength and energy is given by Eqn. 1.1 and illustrated in Fig. 1.4.

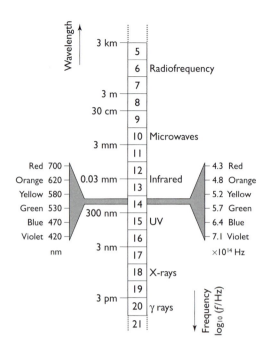

Fig. 1.4 The electromagnetic spectrum. The visible region, the range of the spectrum to which the unaided human eye is sensitive, is expanded. As photon wavelength increases (or frequency decreases), energy decreases. The precise relationship between photon energy and wavelength is given by Eqn. 1.1. Photon frequency is shown on a $\log_{10}$ scale. Redrawn from Fig. 2.15 in Lawrence *et al.* (1996).

economy: energy 'changes hands' (moves from the Sun to plants to animals...) and is 'converted into different kinds of currency' (stored as chemical energy, electrical energy, etc.).

A deeper sense of the nature of energy flow in biology can be gained from a bird's-eye view of the biochemical roles of adenosine triphosphate (ATP), a small organic compound. This molecule is synthesized from photonic energy in plants and chemical energy in animals. The mechanisms involved in this energy conversion are very complicated, and there is no need to discuss them in detail until Chapter 5. The important point here is that, once it has been synthesized, **ATP plays the role of the main energy 'currency' of biochemical processes in all known organisms**. For instance, ATP is a component of great importance in chemical communication between and within cells, and it is the source of a building block of deoxyribonucleic acid (DNA), the molecules of storage and transmission of genetic information from bacteria to humans (Fig. 1.6). We can see from

Fig. 1.5 Log plot of energy transformation on Earth. Only a small amount of the Sun's light that reaches Earth is used to make cereal. Only a fraction of this energy is transformed into livestock tissue. And only part of this energy is transformed into human tissue. (What happens to the rest of the energy?) A calorie is a unit of energy that one often encounters in older textbooks and scientific articles and in food science. A calorie is the heat required to increase the temperature of 1 g of pure water from 14.5 °C to 15.5 °C. 1 calorie = 1 cal = 4.184 J *exactly*. Based on Fig. 1-2 of Peusner (1974).

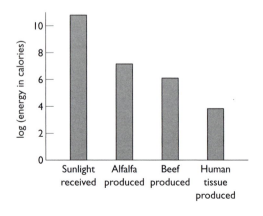

this that ATP is of very basic and central importance to life as we know it, and we shall have a good deal to say about it throughout the book.

Let's return to the money analogy and develop it further. Just as there is neither an increase nor a decrease in the money supply when money changes hands: so in the course of its being transformed, energy is neither created nor destroyed. **The total energy is *always* constant.** As we shall see in the next chapter, this is a statement of the First Law of Thermodynamics. However, unlike the money analogy, energy transformations certainly can and do indeed affect the relative proportion of energy that is available in a form that is useful to living organisms. This situation arises not from defects inherent in the biomolecules involved in energy transformation, but from the structure of our universe itself. We shall cover this aspect of energy transformation in Chapter 3.

Thus far we have been talking about energy as though we knew what it was. After all, each of us has at least a vague sense of what energy transformation involves. For instance, we know that it takes energy to heat a house in winter (natural gas or combustion of wood); we know that energy is required to cool a refrigerator (electricity); we know that energy is used to start an automobile engine (electrochemistry) and to keep it running (gasoline). But we still have not given a precise definition of *energy*. What *is* energy? **Being able to say what energy is with regard to living organisms is what this book is about.**

B. Distribution of energy

Above we said that throughout its transformations energy is conserved. The idea that something can change and remain the same may seem strange, but we should be careful not to think that the idea is therefore untrue. We should be open to the possibility that some aspects of physical reality might differ from our day-to-day macroscopic experience of the world. In the present context, the something that stays the same is a quantity called the total **energy**, and the something that changes is

Fig. 1.6 ATP fuels an amazing variety of cellular processes. In the so-called ATP cycle, ATP is formed from adenosine diphosphate (ADP) and inorganic phosphate (P_i) by photosynthesis in plants and by metabolism of 'energy rich' compounds in most cells. Hydrolysis of ATP to ADP and P_i releases energy that is trapped as usable energy. This form of energy expenditure is integral to many key cellular functions and is a central theme of biochemistry. Redrawn from Fig. 2-23 of Lodish *et al.* (1995).

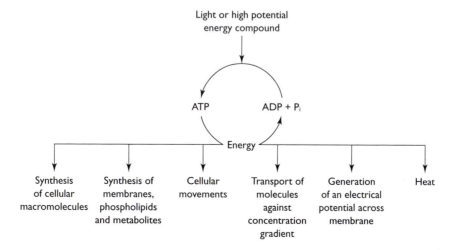

how that energy is *distributed* – where it is found and in which form and at which time. A crude analog of this would be a wad of chewing gum. Neglecting the change in flavor with time, the way in which the gum molecules are distributed in space depends, first of all, on whether the gum is in your mouth or still in the wrapper! Once you've begun to work your jaw and tongue, the gum changes shape a bit at a time, though it can change quite dramatically when you blow a bubble. Regardless of shape and the presence or absence of bubbles, however, the *total amount* of gum is *constant*. But one should not infer from this that energy is a material particle.

Elaboration of the money–energy analogy will help to illustrate several other important points. Consider the way a distrustful owner of a busy store might check on the honesty of a certain cashier. The owner knows that m_b dollars were in the till at the beginning of the day, and, from the cash register tape, that m_e dollars should be in the till at the end of trading. So, of course, the owner knows that the net change of money must be $m_e - m_b = \Delta m$, where 'Δ,' the upper case Greek letter *delta*, means 'difference.' This, however, says nothing at all about the way the cash is distributed. Some might be in rolls of coins, some loose in the till, and some in the form of dollar bills of different denomination. (*bill = banknote*.) Nevertheless, when all the accounting is done, the pennies, nickels, dimes and so on should add up to Δm, if the clerk is careful and honest. A simple formula can be used to do the accounting:

$$\Delta m = \$0.01 \times (\text{number of pennies}) + \$0.05 \times (\text{number of nickels}) + \ldots + \$10.00 \times (\text{number of ten dollar bills}) + \$20.00 \times (\text{number of twenty dollar bills}) + \ldots \quad (1.2)$$

This formula can be modified to include terms corresponding to coins in rolls:

$$\Delta m = \$0.01 \times (\text{number of pennies}) + \$0.50 \times (\text{number of rolls of pennies}) + \$0.05 \times (\text{number of nickels}) + \$2.00 \times (\text{number of rolls of nickels}) + \ldots + \$10.00 \times (\text{number of ten dollar bills}) + \$20.00 \times (\text{number of twenty dollar bills}) + \ldots \quad (1.3)$$

A time-saving approach to counting coins would be to weigh them. The formula might then look like this:

$$\Delta m = \$0.01 \times (\text{weight of unrolled pennies})/(\text{weight of one penny}) + \$0.50 \times (\text{number of rolls of pennies}) + \$0.05 \times (\text{weight of unrolled nickels})/(\text{weight of one nickel}) + \$2.00 \times (\text{number of rolls of nickels}) + \ldots + 10.00 \times (\text{number of ten dollar bills}) + 20.00 \times (\text{number of twenty dollar bills}) + \ldots \quad (1.4)$$

There are several points we can make by means of the money analogy. One, the number of each type of coin and bill is but one possible distribution of Δm dollars. A different distribution would be found if a wiseacre paid for a $21.95 item with a box full of unrolled nickels instead of a twenty and two ones (Fig. 1.7)! One might even consider measuring the distribution of the Δm dollars in terms of the proportion in pennies, nickles, dimes, and so on. We shall find out more about this in Chapter

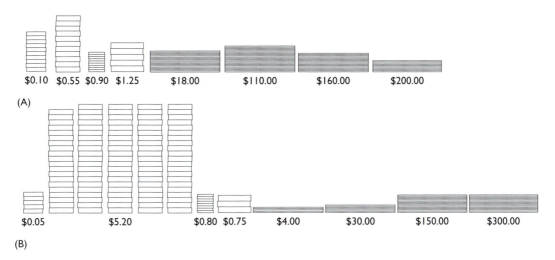

$0.10 $0.55 $0.90 $1.25 $18.00 $110.00 $160.00 $200.00

(A)

$0.05 $5.20 $0.80 $0.75 $4.00 $30.00 $150.00 $300.00

(B)

3. Two, given a distribution of Δm dollars into so many pennies, nickles, dimes, and so forth, there are many different ways of arranging the coins and bills. For example, there are many different possible orderings of the fifty pennies in a roll. The complexity of the situation would increase even further if we counted coins of the same type but different date as 'distinguishable' and ones of the same type and same date as 'indistinguishable.' Three, the more we remove ourselves from counting and examining individual coins, the more abstract and theoretical our formula becomes. (As Aristotle[3] recognized, the basic nature of scientific study is to proceed from observations to theories; theories are then used to explain observations and make predictions about what has not yet been observed. Theories can be more or less abstract, depending on how much they have been developed and how well they work.) And four, although measurement of an abstract quantity like Δm might not be very hard (the manager could just rely on the tape if the clerk were known to be perfectly honest and careful), determination of the contribution of each relevant component to the total energy could be a time-consuming and difficult business – if not impossible, given current technology and definitions of thermodynamic quantities. We shall have more to say about this in Chapter 2.

So, how does the money simile illustrate the nature of the physical world? **A given quantity of energy can be distributed in a multitude of ways.** Some of the different forms it might take are chemical energy, elastic energy, electrical energy, gravitational energy, heat energy, mass energy, nuclear energy, radiant energy, and the energy of intermolecular interactions. But **no matter what the form, the total amount of energy is constant.** All of the mentioned forms of energy are of interest to the biological scientist, though some clearly are more important to

Fig. 1.7 Two different distributions of the same amount of money. The columns from left to right are: pennies ($0.01), nickels ($0.05), dimes ($0.10), quarters ($0.25), one dollar bills ($1.00), five dollar bills ($5.00), ten dollar bills ($10.00) and twenty dollar bills ($20.00). Panel (A) differs from panel (B) in that the latter distribution involves a relatively large number of nickels. Both distributions correspond to the same total amount of money. The world's most valuable commodity, oil, is the key energy source for the form of information flow known as domestic and international travel.

---

[3] Aristotle (384–322 BC) was born in northern Greece. He was Plato's most famous student at the Academy in Athens. Aristotle established the Peripatetic School in the Lyceum at Athens, where he lectured on logic, epistemology, physics, biology, ethics, politics, and aesthetics. According to Aristotle, minerals, plants and animals are three distinct categories of being. He was the first philosopher of science.

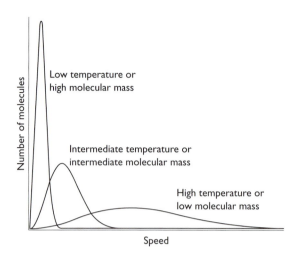

Fig. 1.8 The Maxwell distribution of molecular speeds. The distribution depends on particle mass and temperature. The distribution becomes broader as the speed at which the peak occurs increases. Low, intermediate, and high temperatures correspond to the solid, liquid, and gaseous states, respectively. James Clerk Maxwell, a Scot, lived 1831–1879. He is regarded as the nineteenth-century scientist who had the greatest influence on twentieth-century physics and is ranked with Isaac Newton and Albert Einstein for the fundamental nature of his contributions. He did important work in thermodynamics and the kinetic theory of gases. Based on Fig. 0.8 of Atkins (1998).

us than others; some are relevant only in somewhat specialized situations. The terms denoting the different types of energy will be defined below as we go along. In living organisms the main repositories of energy are macromolecules, which store energy in the form of covalent and non-covalent chemical bonds, and unequal concentrations of solutes, principally ions, on opposite sides of a cell membrane. In Fig. 1.3 we can see another type of energy distribution, the solar spectrum. For a given amount of solar energy that actually reaches the surface of our planet, more of the photons have a wavelength of 500 nm than 250 or 750 nm. According to the kinetic theory of gases, a subject we shall discuss at several points in this book, the speeds of gas molecules are distributed in a certain way, with some speeds being much more common than others (Fig. 1.8). In general, slow speeds and high speeds are rare, near-average speeds are common, and the average speed is directly related to the temperature. A summary of the chief forms of energy of interest to biological scientists is given in Table 1.1.

C. System, boundary, and surroundings

Before getting too far underway, we need to define some important terms. This is perhaps done most easily by way of example. Consider a biochemical reaction that is carried out in aqueous solution in a test tube (Fig. 1.9A). The **system** consists of the solvent, water, and all chemicals dissolved in it, including buffer salts, enzyme molecules, the substrate recognized by the enzyme and the product of the enzymatic reaction. The system is that part of the universe chosen for study. The **surroundings** are simply the entire universe excluding the system. The system and surroundings are separated by a **boundary**, in this case the test tube.

At any time, the system is in a given thermodynamic **state** or condition of existence (which types of molecule are present and the amount of each, the temperature, the pressure, etc.). The system is said to be closed if it can exchange *heat* with the surroundings but not *matter*. That is, the boundary of a **closed system** is *impermeable* to matter. A dialysis bag that is permeable

Table 1.1 | Energy distribution in cells. Contributions to the total energy can be categorized in two ways: kinetic energy and potential energy. Each category can be subdivided in several ways

Kinetic energy (motion)	Potential energy (position)
Heat or thermal energy – energy of molecular motion in organisms. At 25 °C this is about 0.5 kcal mol^{-1}.	*Bond energy* – energy of covalent and non-covalent bonds, for example a σ bond between two carbon atoms or van der Waals interactions. These interactions range in energy from as much as 14 kcal mol^{-1} for ion–ion interactions to as little as 0.01 kcal mol^{-1} for dispersion interactions; they can also be negative, as in the case of ion–dipole interactions and dipole–dipole interactions.
Radiant energy – energy of photons, for example in photosynthesis. The energy of such photons is about 40 kJ mol^{-1}.	*Chemical energy* – energy of a difference in concentration across a permeable barrier, for instance the lipid bilayer membrane surrounding a cell of a substance which can pass through the membrane. The magnitude of this energy depends on the difference in concentration across the membrane. The greater the difference, the greater the energy.
Electrical energy – energy of moving charged particles, for instance electrons in reactions involving electron transfer. The magnitude of this energy depends on how quickly the charged particle is moving. The higher the speed, the greater the energy.	*Electrical energy* – energy of charge separation, for example the electric field across the two lipid bilayer membranes surrounding a mitochondrion. The electrical work required to transfer monovalent ions from one side of a membrane to the other is about 20 kJ mol^{-1}.

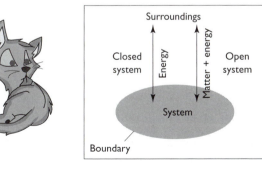

Fig. 1.9 Different types of system. (A) A closed system. The stopper inhibits evaporation of the solvent, so essentially no matter is exchanged between the test tube and its surroundings (the air surrounding the test tube). Energy, however, can be exchanged with the surroundings, through the glass. (B) An open system. All living organisms are open systems. A cat is a particularly complex open system. A simplified view of a cat as a system is given in Fig. 1.10. (C) A schematic diagram of a system.

to small molecules but not to large ones is not a closed system! As long as no matter is added to the test tube in Fig. 1.9A during the period of observation, and as long as evaporation of the solvent does not contribute significantly to any effects we might observe, the system can be considered closed. This is true even if the biochemical reaction we are studying results in the release or absorption of heat energy; as we have said, energy transfer between system and surroundings is possible in a closed system. Another example of a closed system is Earth itself: our planet continually receives radiant energy from the Sun and continually gives off heat, but because Earth is neither very heavy nor very light it exchanges practically no matter with its surroundings (Earth is not so massive that its gravitational field pulls nearby bodies like the Moon into itself, as a black hole would do, but there is enough of a gravitational pull on air to prevent it going off into space, which is why asteroids have no atmosphere).

If matter can be exchanged between system and surroundings, the system is open. An example of an **open system** is a cat (Fig. 1.9B). It breathes in and exhales matter (air) continually, and it eats, drinks, defecates and urinates periodically. In barely-sufferable technospeak, a cat is an open, self-regulating and self-reproducing heterogeneous system. The system takes in food from the environment and uses it to maintain body temperature, power all the biochemical pathways of its body, including those of its reproductive organs, and to run, jump and play. The system requires nothing more for reproduction than a suitable feline of the opposite sex. And the molecular composition of the brain of the system is certainly very different from that of its bone marrow. In the course of all the material changes of this open system, heat energy is exchanged between it and the surroundings, the amount depending on the system's size and the difference in temperature between its body and its environment. A schematic diagram of a cat is shown in Fig. 1.10. **Without exception, all living organisms that have ever existed are open systems**.

Finally, in an **isolated system**, the boundary permits neither matter nor energy to enter or exit. A schematic diagram of a system, surroundings and boundary are shown in Fig. 1.9C.

Fig. 1.10 The 'plumbing' of a higher animal. Once inside the body, energy gets moved around a lot (arrows). Following digestion, solid food particles are absorbed into the circulatory system (liquid), which delivers the particles to all cells of the body. The respiratory system enables an organism to acquire the oxygen gas it needs to burn the fuel it obtains from food. If the energy input is higher than the output (excretion + heat), there is a net increase in body weight. In humans, the ideal time rate of change of body weight, and therefore food intake and exercise, varies with age and physical condition. Based on Fig. 1–5 of Peusner (1974).

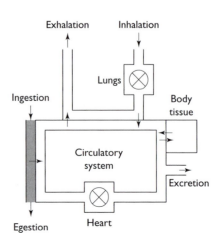

D. | Animal energy consumption

Now let's turn briefly to a more in-depth view of the relationship between food, energy, and life than we have seen so far. We wish to form a clear idea of how the energy requirements of carrying out various activities, for instance walking or sitting, relate to the energy available from the food we eat. The comparison will be largely qualitative. The discussion will involve the concept of heat, and a formal definition of the term will be given.

Energy measurements can be made using a calorimeter. Calorimetry has made a huge contribution to our understanding of the energetics of chemical reactions, and there is a long tradition of using calorimeters in biological research. In the mid-seventeenth century, pioneering experiments by Robert Boyle (1627–1691) in Oxford demonstrated the necessary role of air in combustion and respiration. About 120 years later, in 1780, Antoine Laurent Lavoisier (1743–1794) and Pierre Simon de Laplace (1749–1827) extended this work by using a calorimeter to measure the *heat* given off by a live guinea pig. On comparing this heat with the amount of oxygen consumed, the Frenchmen correctly concluded that respiration is a form of combustion. Nowadays, a so-called bomb calorimeter (Fig. 1.11) is used to measure the heat given off in the oxidation of a combustible substance like food, and nutritionists refer to tables of combustion heats in planning a diet. There are many different kinds of calorimeter. For instance, the instrument used to measure the energy given off in an atom smasher is called a calorimeter. In this book we discuss three of them: bomb calorimeter, isothermal titration calorimeter and differential scanning calorimeter.

Thermodynamics is the study of energy transformations. It is a hierarchical science. This means that the more advanced concepts assume knowledge of the basics. To be ready to tackle the more difficult but more interesting topics in later chapters, we had better take time to develop a good understanding of *what* is being measured in a

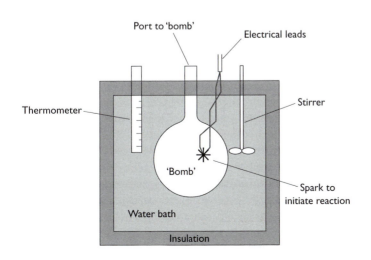

Port to 'bomb'

Electrical leads

Thermometer

Stirrer

'Bomb'

Spark to initiate reaction

Water bath

Insulation

Fig. 1.11 Schematic diagram of a bomb calorimeter. A sample is placed in the reaction chamber. The chamber is then filled with oxygen at high pressure (>20 atm) to ensure that the reaction is fast and complete. Electrical heating of a wire initiates the reaction. The increase in the temperature of the water is recorded, and the temperature change is converted into an energy increase. The energy change is divided by the total amount of substance oxidized, giving units of $J\,g^{-1}$ or $J\,mol^{-1}$. Insulation helps to prevent the escape of the heat of combustion. Based on diagram on p. 36 of Lawrence *et al.* (1996).

Table 1.2 | Heat released upon oxidation to CO_2 and H_2O

Substance	Energy yield			
	kJ (mol^{-1})	kJ (g^{-1})	kcal (g^{-1})	kcal $(g^{-1}$ wet wt)
Glucose	2817	15.6	3.7	—
Lactate	1364	15.2	3.6	—
Palmitic acid	10040	39.2	9.4	—
Glycine	979	13.1	3.1	—
Carbohydrate	—	16	3.8	1.5
Fat	—	37	8.8	8.8
Protein	—	23	5.5	1.5
Protein to urea	—	19	4.6	—
Ethyl alcohol	—	29	6.9	—
Lignin	—	26	6.2	—
Coal	—	28	6.7	—
Oil	—	48	11	—

Note:

D-glucose is the principal source of energy for most cells in higher organisms. It is converted to lactate in anaerobic homolactic fermentation (e.g. in muscle), to ethyl alcohol in anaerobic alcoholic fermentation (e.g. in yeast), and to carbon dioxide and water in aerobic oxidation. Palmitic acid is a fatty acid. Glycine, a constituent of protein, is the smallest amino acid. Carbohydrate, fat and protein are different types of biological macromolecule and sources of energy in food. Metabolism in animals leaves a residue of nitrogenous excretory products, including urea in urine and methane produced in the gastrointestinal tract. Ethyl alcohol is a major component of alcoholic beverages. Lignin is a plasticlike phenolic polymer that is found in the cell walls of plants; it is not metabolized directly by higher eukaryotes. Coal and oil are fossil fuels that are produced from decaying organic matter, primarily plants, on a timescale of millions of years. The data are from Table 2.1 of Wrigglesworth (1997) or Table 3.1 of Burton (1998). See also Table A in Appendix C.

bomb calorimeter. We know from experience that the oxidation (burning) of wood gives off *heat*. Some types of wood are useful for building fires because they ignite easily (e.g. splinters of dry pine); others are useful because they burn slowly and give off a lot of heat (e.g. oak). The amount of heat transferred to the air per unit volume of burning wood depends on the density of the wood and its *structure*, and the same is true of food. Fine, but this does not tell us what heat is.

It is the nature of science to tend towards formality and defining terms as precisely as possible. With accepted definitions in hand, there will be relatively little ambiguity about what is meant. What we need now is a good definition of heat. **Heat**, q, or thermal energy, is a form of kinetic energy, energy arising from motion. Heat is the change in energy of a system that results from its temperature differing from that of the surroundings. For instance, when a warm can of Coke is placed in a refrigerator, it gives off heat continuously until it has reached the same temperature as all other objects inside, including the air. The heat *transferred* from the Coke can to the air is absorbed by the other items in the fridge. Heat is said to *flow* from a region of higher temperature (greater molecular motion) to one of lower temperature (lesser molecular motion). The flow of heat resembles a basic property of a liquid. But this does not mean that heat is a material particle.

Heat is rather a type of energy transfer. Heat makes use of *random molecular motion*. Particles that exhibit such motion (*all* particles!) do

Table 1.3	Energy expenditure in humans		
Activity	Time (min)	Energy cost (kJ min^{-1})	Total energy expenditure (kJ)
Lying	540	5.0	2700
Sitting	600	5.9	3540
Standing	150	8.0	1200
Walking	150	13.4	2010
Total	1440	—	9450

Note:
The measurements were made by indirect calorimetry. Digestion increases the rate of metabolism by as much as 30% over the basal rate. During sleep the metabolic rate is about 10% lower than the basal rate. The data are from Table 2.2 of Wrigglesworth (1997).

so according to the laws of (quantum) mechanics. A familiar example of heat being transferred is the boiling of water in a saucepan. The more the water is heated, the faster the water molecules move around. The bubbles that form on the bottom of the pan give some indication of just how fast the water molecules are moving. This is about as close as we can get to 'see' heat being transferred, apart from watching something burn. But if you've ever been in the middle of a shower when the hot water has run out, you will know very well what it is to *feel* heat being transferred! By convention, $q > 0$ if energy is transferred *to* a system as heat. In the case of a cold shower, and considering the body to be the system, q is negative.

Now we are in a position to have a reasonably good quantitative grasp of the oxidation of materials in a bomb calorimeter and animal nutrition. The heat released or absorbed in a reaction is ordinarily measured as a change in temperature; calibration of an instrument using known quantities of heat can be used to relate heats of reaction to changes in temperature. One can plot a standard curve of temperature *versus* heat, and the heat of oxidation of an unknown material can then be determined experimentally. Table 1.2 shows the heats of oxidation of different foodstuffs. Importantly, different types of biological molecule give off more heat per unit mass than do others. Some idea of the extent to which the energy obtained from food is utilized in various human activities is given in Table 1.3.

It also seems fitting to mention here that animals, particularly humans, 'consume' energy in a variety of ways, not just by eating, digesting and metabolizing food. For instance, automobiles require gasoline to run, and in order to use electrical appliances we first have to generate electricity! The point is that we can think about energy transformation and consumption on many different levels. As our telescopic examining lens becomes more powerful, the considerations range from one person to a family, a neighborhood, city, county, state, country, continent, surface of the Earth, biosphere, solar system, galaxy . . .. As the length scale decreases, the considerations extend from a whole person to an organ, tissue, cell, organelle, macromolecular

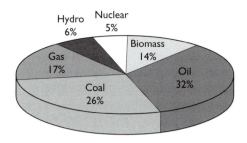

Fig. 1.12 Global human energy use. As of 1987, this totaled about 4×10^{20} J yr^{-1}. Energy production has increased substantially since then, but the distribution has remained about the same. Note that the rate of energy consumption is about four orders of magnitude smaller than the amount of radiant energy that is incident on Earth each year (cf. Fig. 1.1). Note also that c. 90% of energy consumption depends on the products of photosynthesis, assuming that fossil fuels are the remains of ancient organisms. Redrawn from Fig. 8.12 in Wrigglesworth (1997).

assembly, protein, atom, nucleus, proton or neutron . . . . Fig. 1.12 gives some idea of mankind's global energy use per sector. It will come as no surprise that a comprehensive treatment of all these levels of energy transformation is well beyond the scope of this undergraduate text-book. Instead, our focus here is on the basic principles of energy trans-formation and their application in the biological sciences.

E. | Carbon, energy, and life

We close this chapter with a brief look at the *energetic* and *structural* role of carbon in living organisms. The elemental composition of the dry mass of the adult human body is roughly 3/5 carbon, 1/10 nitrogen, 1/10 oxygen, 1/20 hydrogen, 1/20 calcium, 1/40 phosphorus, 1/100 potassium, 1/100 sulfur, 1/100 chlorine and 1/100 sodium (Fig. 1.13). We shall see all of these elements at work in this book. The message here is that carbon is the biggest contributor to the weight of the body. Is there is an energetic 'explanation' for this?

Yes! Apart from its predominant structural feature – extraordinary chemical versatility and ability to form asymmetric molecules – carbon forms especially stable single bonds. N—N bonds and O—O bonds have an energy of about 160 kJ mol^{-1} and 140 kJ mol^{-1}, respectively, while the energy of a C—C bond is about twice as great (345 kJ mol^{-1}). The C—C bond energy is moreover nearly as great as that of a Si—O bond.

Fig. 1.13 Composition of the human body. Protein accounts for about half of the dry mass of the body. On the level of individual elements, carbon is by far the largest component, followed by nitrogen, oxygen, hydrogen and other elements. It is interesting that the elements contributing the most to the dry mass of the body are also the major components of air, earth, water, and carbon-based combustible matter. Based on data from Freiden (1972).

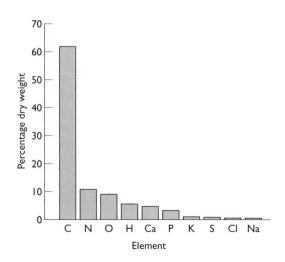

Chains of Si—O are found in tremendous abundance in the silicate minerals that form the crust of our planet, and one might guess therefore that silicates could support life in distant solar systems, if not elsewhere in our own. Although this possibility cannot be ruled out, Si—O is unlikely to be as useful for life as C—C because it is practically inert. The predominant importance of carbon in the molecules of life is likely to be the rule throughout the universe rather than the exception here on Earth.

F. | References and further reading

Atkins, P. W. (1998). Physical Chemistry, 6th edn, ch. 0. Oxford: Oxford University Press.

Atkinson, D. E. (1977). *Cellular Energy Metabolism and its Regulation*. New York: Academic Press.

Atwater, W. A. & Benedict, F. G. (1903). Experiments on the metabolism of matter and energy in the human body. *Experiment Station Record United States Department of Agriculture*. **136**.

Berdahl, P. (1993). Energy Conversion. In *Encyclopedia of Applied Physics*, vol. 6, ed. G. L. Trigg, pp. 229–243. New York: VCH.

Blaxter, K. (1989). *Energy Metabolism in Animals and Man*. Cambridge: Cambridge University Press.

Bridger, W. A. & Henderson, J. F. (1983). *Cell ATP*. New York: John Wiley.

Brunner, B. (1998). *The Time Almanac 1999*, International Energy & Basic Planetary Data. Boston, Massachusetts: Information Please LLC.

Burton, R. F. (1998). *Biology by Numbers: an Encouragement to Quantitative Thinking*, ch. 3. Cambridge: Cambridge University Press.

Cork, J. M. (1942). *Heat*. New York: John Wiley.

Encyclopædia Britannica CD 98, 'Calorie,' 'Earth,' 'Energy,' 'Heat,' 'Mole,' 'Nutrition,' and 'Principles of Thermodynamics.'

Feynman, R. P., Leighton, R. B. & Sands, M. (1963). *Lectures on Physics*, vol. I, cc. 1, 3 & 4. Reading, Massachusetts: Addison-Wesley.

Fraústo da Silva, J. R. & Williams, R. J. P. (1991). *The Biological Chemistry of the Elements*. Oxford: Oxford University Press.

Frieden, E. (1972). The chemical elements of life. *Scientific American*, **227**, no. 1, 52–60.

Fruton, J. S. (1999). *Proteins, Enzymes, Genes: the Interplay of Chemistry and Biology*. New Haven: Yale University Press.

Gates, D. M. (1963). The energy environment in which we live. *American Scientist*, **51**, 327–348.

Gillispie, Charles C. (ed.) (1970). *Dictionary of Scientific Biography*. New York: Charles Scribner.

Gislason, E. A. & Craig, N. C. (1987). General definitions of work and heat in thermodynamic processes. *Journal of Chemical Education*, **64**, 660–668.

Gold, T. (1992). The deep, hot biosphere. *Proceedings of the National Academy of Sciences of the United States of America*. **89**, 6045–6049.

Goodsell, D. S. (1993). *The Machinery of Life*. New York: Springer-Verlag.

Harold, F. M. (1986). *The Vital Force: a Study of Bioenergetics*, cc. 1 & 2. New York: W. H. Freeman.

Harris, D. A. (1985). *Bioenergetics at a Glance*, ch. 1. Oxford: Blackwell Science.

Kildahl, N. K. (1995). Bond energy data summarized. *Journal of Chemical Education*, **72**, 423–424.

Kondepudi, D. & Prigogine, I. (1998). *Modern Thermodynamics: from Heat Engines to Dissipative Structures*, ch. 2.6. Chichester: John Wiley.

Krebs, H. A. & Kornberg, H. L. (1957). *Energy Transformations in Living Matter*. New York: Springer-Verlag.

Lawrence, C., Roger, A. & Compton, R. (1996). *Foundations of Physical Chemistry*. Oxford: Oxford University Press.

Lodish, H. Baltimore, D., Berk, A., Zipursky, S. L. Madsudaira, P. & Darnell, J. (1995). *Molecular Cell Biologies*, 3rd edn, ch. 2. New York: W. H. Freeman.

Losee, J. (1993). *A Historical Introduction to the Philosophy of Science*, 3rd edn. Oxford: Oxford University Press.

Morowitz, H. J. (1968). *Energy Flow in Biology*. New York: Academic Press.

Peusner, L. (1974). *Concepts in Bioenergetics*, ch. 1. Englewood Cliffs: Prentice-Hall.

Price, G. (1998). *Thermodynamics of Chemical Processes*, ch. 1. Oxford: Oxford University Press.

Priscu, J. C., Fritsen, C. H., Adams, E. E., Giovannoni, S. J., Paerl, H. W., McKay, C. P., Doran, P. T., Gordon, D. A., Lanoil, B. D. & Pinckney, J. L. (1998). Perennial Antarctic lake ice: an oasis for life in a polar desert. *Science*, **280**, 2095–2098.

Shaw, D. J. & Avery, H. E. (1989). *Physical Chemistry*, ch. 2.7. London: MacMillan.

Skulachev, V. P. (1992). The laws of cell energetics. *European Journal of Biochemistry*, **208**, 203–209.

Smith, C. A. & Wood, E. J. (1991). *Energy in Biological Systems*, cc. 1.1, 1.2 & 2.2. London: Chapman & Hall.

Voet, D. & Voet, J. G. (1995). *Biochemistry*, 2nd edn. New York: John Wiley.

Walter, D. K. (1993). Biomass Energy. In *Encyclopedia of Applied Physics*, vol. 2, ed. G. L. Trigg, pp. 473–487. New York: VCH.

Watt, B. K. & Merill, A. L. (1963). *Composition of Foods*. Washington, D. C.: United States Department of Agriculture.

Wiegert, R. G. (1976). *Ecological Energetics*. Stroudsburg: Dowden, Hutchinson and Ross.

Wink, D. (1992). The conversion of chemical energy. Biochemical examples. *Journal of Chemical Education*, **69**, 264–267.

Williams, T. I. (ed.) (1969). *A Biographical Dictionary of Scientists*. London: Adam & Charles Black.

Wrigglesworth, J. (1997). *Energy and Life*, cc. 1.1 & 2.1. London: Taylor & Francis.

Youvan, D. C. & Marrs, B. L. (1987). Molecular mechanisms of photosynthesis. *Scientific American*, **256**, no. 6, 42–48.

G. Exercises

1. What *is* energy? What is the etymology of *energy*? When did *energy* acquire its present scientific meaning? (Hint: consult the *Oxford English Dictionary* and any good encyclopedia of physics.)

2. Some primitive religions teach that the celestial bodies we call stars (or planets) are gods. This view was common in the ancient Greek world, and it was espoused by Thales of Miletus (fl. 6th century BC), one of the greatest thinkers of all time. Needless to say, the Greeks knew nothing about nuclear fusion in stars, though they were certainly aware that the Sun is much larger than it appears to the

unaided eye and that plants need light and water to grow. Explain briefly how the belief that stars are gods was remarkably insightful, even if polytheism and animism are rejected on other grounds.

3. According to Eqn. 1.1, E is a *continuous* and *linear* function of λ^{-1}; the energy spectrum of a *free* particle is *not* characterized by discrete, step-like energy levels. A continuous function is one that changes value smoothly; a linear function is a straight line. Consider Eqn. 1.1. Is there a fundamental limit to the magnitude of the energy of a photon? In contrast, the electronic *bound* state with which a photon interacts in photosynthesis is restricted to certain energy *levels*, and these are determined by the structure of the pigment molecule and its electronic environment; electromagnetic radiation interacts with *matter* as though it existed in small packets (photons) with *discrete* values. All of the energy levels are the bound electron are below a certain threshold, and when this energy level is exceeded, the electron becomes a free particle. What effect does exceeding the energy thershold have on the plant? What part of the biosphere prevents high-energy photons from the Sun from doing this to plants?

4. Chlorophylls absorb blue light and red light relatively well, but not green light (Fig. 1.3). Explain why tree leaves are green in summer and brown in late autumn.

5. The wavelength of blue light is about 4700 Å, and that of red light is about 7000 Å. (1 Å $= 10^{-10}$ m; the Ångström is named in honor of the Swedish physicist Anders Jonas Ångström (1814–1874).) Calculate the energy of a photon at these wavelengths. About 7 kcal mol^{-1} is released when ATP is hydrolyzed to ADP and inorganic phosphate (under 'standard state conditions'). Compare the energy of the photons absorbed by plants to the energy of ATP hydrolysis (1 mole $= 6.02 \times 10^{23}$).

6. In the anabolic (biosynthetic) reduction–oxidation reactions of plant photosynthesis, 8 photons are required to reduce one molecule of CO_2. 1 mol of CO_2 gives 1 mol of carbohydrate (CH_2O). What is the maximum possible biomass (in g of carbohydrate) that can be produced in 1 hour by plants receiving 1000 μE s^{-1} of photons of a suitable wavelength for absorption? Assume that 40% of the photons are absorbed. (1 E $=$ 1 einstein $=$ 1 mol of photons. The einstein is named in honor of Albert Einstein.) The atomic masses of H, C and O are 1, 12 and 16, respectively.

7. The energy of oxidation of glucose to H_2O and CO_2 is -2870 kJ mol^{-1}. Therefore, at least 2870 kJ mol^{-1} are needed to synthesize glucose from H_2O and CO_2. How many 700 nm photons must be absorbed to fix one mole of CO_2? If the actual number needed is 3 to 4 times the minimum number, what is the efficiency of the process?

8. Devise your own analogy for energy conservation and distribution. Explain how the analog resembles nature and where the similarity begins to break down.

9. Give examples of a spatial distribution, a temporal distribution, and a spatio-temporal distribution.

10. Give three examples of a closed system. Give three examples of an open system.

11. Describe the preparation of a cup of tea with milk in terms of energy transformation.

12. Describe an astronaut in a spaceship in terms of open and closed systems.

13. The growth temperatures of almost all organisms are between the freezing and boiling points of water. Notable exceptions are marine organisms that live in seas a few degrees below 0 °C. Homeothermic (Greek, *homoios*, similar + *therme*, heat) organisms maintain an almost constant body temperature, independent of the temperature of the environment. Human beings are an example, as are horses and cats. Fluctuations about the average temperature of these organisms are generally less than 1 °C. All such organisms have an average temperature between 35 and 45 °C; a narrow range. Most birds strictly regulate their body temperatures at points between 39 and 44 °C. In some species, however, body temperature can vary by about 10 degrees centigrade. Poikilotherms, which include reptiles, plants, microorganisms, show much less temperature regulation. Eubacteria and archaebacteria exhibit the greatest range of growth temperatures of all known organisms. Suggest how a reptile might regulate its temperature. What about a plant?

14. Calculate the heat energy released by complete burning of an 11 g spoonful of sugar to carbon dioxide and water (Table 1.2).

15. Banana skins turn brown much more rapidly after the fruit has been peeled than before. Why?

16. Human daily energy requirement. A metabolic rate is a measure of energy consumption per unit time. Basal metabolic rate (BMR) is measured after a 12 h fast and corresponds to complete physical and mental rest. A 70 kg man might have a BMR of 80 W (1 W $= 1$ watt $= 1$ J s^{-1}. The watt is named after Scottish inventor James Watt (1736–1819)). A very active man might have a BMR three times as large. Calculate the minimal daily energy requirement of a man who has a BMR of 135 W.

17. The energy of protein catabolism (degradation) in living organisms is different from the energy of protein combustion in a calorimeter. Which energy is larger? Why?

18. Consider a 55 kg woman. Suppose she contains 8 kg of fat. How much heavier would she be if she stored the same amount of energy as carbohydrate?

19. Student A spends 15 hr day^{-1} sitting in the classroom, library, student cafeteria or dormitory. Another half-hour is spent walking between the dorm and lecture halls, and an hour is used for walking in the morning. Using Table 1.3, calculate Student A's daily energy requirement. Student B's routine is identical to Student A's except that his hour of exercise is spent watching television. Calculate the difference in energy requirements for these two students. Referring to Table 1.2, calculate the mass of fat, protein or carbohydrate Student A would have to ingest in order to satisfy her energy needs.

How much glucose does Student A need for daily exercise? List the underlying assumptions of your calculations.

20. In nuclear fusion, two deuterium atoms ^{2}H combine to form helium and a neutron

$$^2H + ^2H \rightarrow ^3He + n$$

The mass of ^{2}H is 2.0141 a.m.u. (atomic mass units), the mass of ^{3}He is 3.0160 a.m.u., and the mass of a neutron is 1.0087 a.m.u. 1 a.m.u. $= 1.6605 \times 10^{-27}$ kg. Perhaps the most famous mathematical formula in the history of civilization on Earth is $E = mc^2$, where m is mass in kg, c is the speed of light, and E is heat energy. Show that the heat released on formation of one mole of helium atoms and one mole of neutrons from two moles of deuterium atoms is about 3.14×10^8 kJ.

21. Worldwide energy production (WEP) of 320 quadrillion (320×10^{15}) Btu (British thermal units; 1 Btu $= 1.055$ kJ) in 1987 increased by 55 quadrillion Btu by 1996. Give the magnitude of energy production in 1996 in joules and the percentage increase ($[(\text{WEP}_{1996} - \text{WEP}_{1987})/\text{WEP}_{1987}] \times 100$). Calculate the average annual rate of increase in WEP between 1987 and 1996. In 1996, the U.S. produced 73 quadrillion Btu, more than any other country. Compute the contribution of the U.S. to WEP in 1996. Only about 0.025% of the Sun's energy that reaches Earth is captured by photosynthetic organisms. Using the data in the legend of Fig. 1.1, calculate the magnitude of this energy in kJ s^{-1}. Find the ratio of WEP$_{1996}$ to the Sun's energy captured by photosynthetic organisms. Assuming that 173 000 $\times 10^{12}$ W of the Sun's energy reaches Earth and is then either reflected or absorbed, calculate the total energy output of the Sun. Diameter of Earth $= 12756$ km; area of a circle $= \pi \times (\text{diameter}/2)^2$; surface area of a sphere $= 4 \times \pi \times \text{radius}^2$; mean distance of Earth from Sun $= 149.6 \times 10^6$ km.) Using your result from the previous problem, calculate the number of moles of ^{2}H consumed when a heat this large is released. Calculate the energy equivalent of the Earth (mass $= 5.976 \times 10^{27}$ g). Compare the mass energy of Earth to the energy of the Sun that reaches Earth in one year.

22. It is said that energy is to biology what money is to economics. Explain.

For solutions, see http://chem.nich.edu/homework

Chapter 2

The First Law of Thermodynamics

A. | Introduction

In order to have a good understanding of the laws of thermodynamics, it would be helpful to have a deep appreciation of the meaning of the words *law* and *thermodynamics*. So let's take a moment to think about words before launching into a detailed discussion of the First Law. We are aided in this by the nature of science itself, which unlike ordinary prose and poetry aims to give words a more or less precise meaning.

We are familiar with the concept of law from our everyday experience. Laws are rules that we are not supposed to break; they exist to protect someone's interests, possibly our own, and there may be a penalty to pay if the one who breaks a law gets caught. Such are civil and criminal laws. Physical laws are similar but different. They are similar in that they regulate something, namely how matter behaves under given circumstances. They are different in that violations are not known to have occurred and they describe what is considered to be a basic property of nature. If a violation of a physical law should ever seem to occur, one would think first that the experiment had gone wrong at some stage, and only second that maybe the 'law' wasn't a law after all. For instance, Galileo,[1] like Copernicus,[2] believed that the orbits of the planets were circles, for the circle is a perfect shape and perfection is of the heavens. Galileo also thought that the motion of celestial objects like planets was qualitatively different from the motion of terrestrial objects like cannonballs and feathers. But in fact, the orbits of planets are ellipses, not circles,[3] and the mechanical laws of planetary motion are the same as those of a missile flying through the air on a battlefield,

---

[1] Galileo Galilei, Italian astronomer and physicist, lived 1564–1642. His model of the Earth, the Moon, the Sun and planets was based on that of Copernicus, who had proposed a Sun-centered planetary system in his *De Revolutionibus Orbium Coelestium* (1543). Galileo is widely considered the father of modern science, because he emphasized the role of observations and experimentation in the discovery of new aspects of nature.

[2] Nicolaus Copernicus (1473–1543) was a Polish Catholic priest who was fascinated by astronomy.

[3] This was demonstrated by the German astronomer Johannes Kepler (1571–1630).

an object rolling down an inclined plane, and an apple falling the ground in an orchard.[4] The point is not that Galileo was poor at science, for his contributions have played an extremely important role in its development. Rather, the point is that what was considered a 'law' was later shown not to be a law (and that sometimes a great gets it wrong). There are many other examples one could cite from the history of science. It is the nature of human awareness of the physical world to develop in this way.

Whereas a human can break a law intentionally or unwittingly, a basic assumption of physical science is that a particle cannot break a law of physics. Particle motion is *governed* by the laws of physics (even if we don't know what those laws are). One very important fact as far as we are concerned is that no violation of any law of thermodynamics is known to have occurred in over 200 years of research in this area. Because of this many scientists consider the laws of thermodynamics to be the laws of physics least likely to be overturned or superseded by further research. For reasons to be discussed in Chapter 9, the laws of thermodynamics are generally described as the most general concepts of all of modern science. It behoves the biologist to be familiar with the principles of thermodynamics because they are of such basic importance. In view of all this, we might begin to suspect that the concepts we shall discuss are very deep and that considerable study and thought will be the price to pay for mastery of them. Thus has it ever been with basic things.

The word *thermodynamics* was coined around 1840 from two Greek roots: *therme*, heat, and *dynamis*, power. The same roots appear in *thermometer* (a device to measure temperature, or heat) and *dynamite* (a powerful explosive). Based on the meaning of the word alone, we should expect that thermodynamics will have to do with heat and power or movement. Indeed, this branch of physics is concerned with energy and its storage, transformation and dissipation. Thermodynamics aims to describe and relate – in relatively simple mathematical terms – the physical properties of systems of energy and matter. Thermodynamics has very much to do with molecular motion and heat. Though you might not think so, you certainly know something about thermodynamics, if not from having studied physics before starting university, then from having seen what happens when a pan of water is heated on the stove! At first, when the temperature of the water is about 25 °C, nothing seems to be happening; the eye does not detect any motion, or at least not much. As things heat up, however, motion becomes more readily apparent, so that by the time the boiling point is reached the water is moving about rather violently. *Everyone* knows something about thermodynamics, even if we don't normally think about it in terms of physics, and we can see that a lot has been known about thermodynam-

---

[4] As shown by the English mathematician, natural philosopher, and alchemist Isaac Newton (1642–1727). Sir Isaac is perhaps the greatest scientist of all time. His voluminous writings show that he was as interested in theology and alchemy as natural philosophy, i.e. science.

ics since well before the word was invented! We do not have time to say much about the history of thermodynamics here, but is worth mentioning that the general principles of this science grew out of attempts in the nineteenth century to understand something as specific and practical as how to make a steam engine work as efficiently as possible and why heat is generated when one drills the bore of a cannon. In view of this, it might be a good idea not to be too prescriptive about how science should develop.

Thermodynamics is mainly a matter of discovering and then using the principles, or laws, that govern energy transformations and working out the relationships between those principles. The First Law is one of several laws of thermodynamics. Like Kepler's laws of planetary motion and Newton's laws of mechanics, there are three laws of thermodynamics (plus one). We shall have a good deal to say about the first two laws here and in Chapter 3, as they form the core of classical thermodynamics. Discussion of the First and Second Laws also provides a good context for introducing concepts that underlie the Gibbs free energy, an extremely useful concept in the biological sciences that we shall study in Chapters 4 and 5. The Third Law of Thermodynamics is of less immediate importance to biologists, but we shall nevertheless cover it briefly at the end of Chapter 3, showing how it is raises some very interesting questions about the nature of living organisms. The central practical significance of the laws of thermodynamics for us is that they can provide insight into how a biological system may be working and serve as a logical framework for designing experiments and testing theories.

Before investigating the First Law, let's take one minute to go over the *Zeroth* Law of Thermodynamics. The function of this Law is to *justify* the concept of temperature and the use of thermometers (two things most of us are accustomed to take for granted!), and it is included here to provide a broader conceptual foundation to our subject. In outline, the Zeroth Law is identical in form to a famous logical argument known at least as early as the ancient Greeks. It goes like this: if $\alpha = \beta$ (one premise), and $\beta = \gamma$ (another premise), then $\gamma = \alpha$ (conclusion). The Zeroth Law is built on this *syllogism*, or logical argument consisting of three propositions. It involves the concept of **thermal equilibrium**, which means that two objects **A** and **B** are in contact and at the same temperature.[5] The Zeroth Law states: if **A** is in thermal equilibrium with **B**, and **B** is in equilibrium with object **C**, then **C** is also in thermal equilibrium with **A** (Fig. 2.1). Simple! Now we are free to use the concept of temperature as much as we like.

Armed with the Zeroth Law, we are ready to focus on the main topic of this chapter. **The First Law of Thermodynamics is a statement of the conservation of energy; though it can be changed from one form to another, energy can neither be created nor destroyed** (Fig. 2.2). (Note

Fig. 2.1 The Zeroth Law of Thermodynamics. If three systems, **A**, **B** and **C**, are in physical contact, at equilibrium all three will have the same temperature. The concept of equilibrium is discussed in depth in Chapter 4.

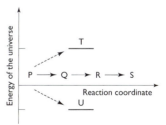

Fig. 2.2 The First Law of Thermodynamics. The total energy of the universe is constant, no matter what changes might occur within it. This principle also applies to an isolated system. Moreover, it is no less applicable to an open system or a closed system, as long as a complete account can be made of energy exchanged with the surroundings.

---

[5] That thermal equilibrium is characterized by the equality of a single parameter (temperature) for all systems was first stated Joseph Black (1728–1799), a Scottish chemist and physician.

how this resembles the law of conservation of matter, according to which the total amount of matter is constant in the course of a chemical reaction.) **The First Law of Thermodynamics is empirical in nature; it cannot be derived mathematically from more basic principles.** Unlike the Pythagorean Theorem[6], for example, which can be derived from the most basic principles of Euclidean geometry[7], there is no *mathematical* proof that the First Law of Thermodynamics is right. Then why should we believe it? Because the author of a textbook says so? Some people might consider that a fairly unconvincing argument! We accept the First Law on a number of different grounds, a necessary and the most important one being that it is based on the experience of many, many researchers. The First Law has been tested many times, and as far as anyone knows it has not been violated even once. It works. It is simple. It makes sense. For all these reasons and others besides, we believe in the First Law of the Thermodynamics. End of sermon.

Despite its lack of a rigorous mathematical foundation, the First Law is the basis for *all* quantitative accounts of energy, regardless of form. **The First Law makes energy the most important concept in physics.** As we saw in Chapter 1, the energy of a system can be converted from one form to another and distributed in a myriad of ways, and now we know that it also cannot be created or destroyed. **The energy of a system plus surroundings is constant in time.** In other words, you can change, say, the chemical energy of coal into heat, and use the heat to boil water and make steam, and the have steam push a piston or turn a turbine. But, the amazing thing is that throughout all these changes of form, the total energy remains the same.

B. | Internal energy

To see more clearly how the First Law operates, we need to add to our conceptual toolkit and equip ourselves with the concepts of internal energy and work. As with heat, both internal energy and work are measured in units of joules (or calories). **The internal energy is the energy *within* the system**. We shall represent the internal energy as U, and for our purposes U will refer only to those kinds of energy that can be modified by chemical processes – translational, vibrational, rotational, bonding, and non-bonding energies. A particle in a system may translate from point A to point B, a bond may rotate and vibrate, a bond may break and reform, and one particle may form a covalent bond with another particle. In contrast, nuclear energy, which is always present and certainly important in things like fusion reactions in the Sun, does not play much of a role in the typical biochemical reaction! Whenever

---

[6] Named after Pythagoras (*c.* 580–500 BC), a mathematically inclined religious philosopher of ancient Greece. The Pythagorean Theorem is $a^2 = b^2 + c^2$, where a, b and c are the lengths of the sides of a right triangle.

[7] The Greek mathematician Euclid lived *c.* 300 BC. His *Elements of Geometry* was the standard work on the subject until other types of geometry were invented in the nineteenth century.

we think about a particular biochemical reaction, we take the atoms involved as given and we do not consider their history. In other words, we can leave nuclear energy out of any calculation of the internal energy because the nuclear energy does not change in the typical biochemical reaction. Life's complicated enough as it is!

The internal energy defines the energy of a substance in the absence of *external* effects, for instance those due to capillarity, electric fields and magnetic fields. The internal energy is an **extensive** property of a substance, meaning that its value depends on the amount of the sample. For example, the internal energy of 2 g of fat is twice as great as the internal energy of 1 g of fat under the same conditions. In contrast, an **intensive** property, for example the concentration of sodium in a solution of sodium bicarbonate, is independent of the amount of solution. The internal energy is a special kind of thermodynamic function called a **state function**. This means that U, a physical property of a system, can be expressed in a mathematical form, and that the value of U depends only on the current **state of the system** (i.e. temperature, pressure and number of particles) and not at all on how that state was prepared. An example will help to illustrate the point. The internal energy of an aqueous buffer solution depends only on its current state and not on whether it was made directly by mixing some chemicals with water or was prepared from as a concentrated stock solution that had been frozen at $-20\,^\circ\mathrm{C}$ for an indefinite period of time. Other general characteristics of state functions will be introduced as we go along.

The internal energy cannot be measured *directly*; it must be calculated from other measured properties of the system. When this is done, it is not U that is measured but a change in U (ΔU). But this presents no problems, because normally we are normally interested in changes in U and not U itself. When a **process** brings about a change in a system from state 1 to state 2, its internal energy changes from U_1 to U_2 ($\Delta U = U_2 - U_1$). For example, when salt is dissolved in water a large amount of heat is released, and solvation of the ions can be measured as a change in temperature. State 1 is the crystalline form of the salt and pure water, and state 2 is the salt completely dissociated into ions and solvated. It does not matter whether we think of dissolution occurring in several steps (e.g. separation of ions in vacuum followed by solvation) or all in one go (Fig. 2.3); the energy difference between states 1 and 2 is the same. This implies that a change in U for a complete cycle, say a change from state 1 to state 2 and back again, must be zero, regardless of the **path** of the process – the succession of states through which the system passes. Many experiments have shown that U is independent of path and that ΔU for a complete cycle is zero, and no exception to the rule has been documented. This is the empirical basis on which U is considered a state function. In fact **all state functions are path-independent**. If the path-independence of U sounds like what we have said about the First Law, it's because changes in U are what the First Law is about! In the money analogy of Chapter 1, the total amount of money at the end of the day did not depend at all on whether payment was made in coins or bills, or

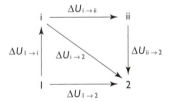

Fig. 2.3 Thermodynamic cycle. The difference in internal energy between state 2 and state 1 is $\Delta U_{1\to2}$. Because U is a state function, the energy difference is independent of path; the internal energy of a system is determined by the specific physical properties of the state of the system and not at all on how the state was prepared. It follows that if the system begins in state 1 and is brought back to this state, $\Delta U = 0$. In symbols, $\Sigma_{\text{loop}}\Delta X = 0$. This holds not just for the internal energy but for any state function, X.

on the order in which the customers made their purchases; it depended only on which purchases were made on a particular day. There are many, many ways in which money could exchange hands and still compute to a net change of Δm. The situation with ΔU is clearly similar to the money analogy. Now let's set the internal energy aside for a moment and return to it after a short discussion of work.

C. | Work

Work, w, is an important physical quantity in thermodynamics, and we had therefore better take time to know something about it. Work is similar to heat. When heat is added to a system, the internal energy changes (it increases). When work is done on a system, for example by compressing the volume of a gas, the internal energy changes (it increases). Both **heat and work are forms of energy transfer across the boundary of a system**; they are 'boundary phenomena'. You may have noticed that, unlike the internal energy, both heat and work are represented as lower case symbols. This is because internal energy is a state function but neither heat nor work is a state function. Rather, q and w are **path functions**. What this means in practical terms is that, unlike internal energy, which is stored or possessed by a system, both heat and work are *transient* quantities. So we see that heat and work are similar. But surely they must also be different or they would not go by different names! Work and heat differ in that **work is the equivalent of a force (e.g. gravity) acting through the displacement of an object**, while **heat is energy transfer due to a temperature difference**. Work involves the *non-random* movement of particles, heat the *random* movement of particles.

There are many different types of work. Here are just a few of them: lifting of a weight against the force of gravity, expansion of a gas against an external pressure, movement of a charge against an electric field gradient (voltage), rotation of a shaft driving a fan, driving of an electric generator, expansion of a liquid film against its surface tension. In each case a force acts through a displacement, and this results in work being done by or on the system. In rough terms, the system does work when it pushes on the surroundings, and work is done on the system when the surroundings push on the system. The heart does work on blood when pumping it through the vasculature.

The technical definition of *work* is similar to the one we are familiar with from everyday life. If someone works hard, they put a lot of effort into accomplishing a task or reaching a goal. Similarly, in physics work is done when an object is moved against an opposing force. For example, when a heavy box is lifted, it is moved against the opposing force of gravity; the atoms of the box are involved in a transfer of energy that occurs in an *organized* way, because all the atoms move together in the same direction. Here we make a crucial clarification of what we mean by *work*: although it might seem like work (non-technical meaning) to carry a heavy box on a level surface, if its distance from the ground does not change no work (technical meaning) is done. This is because the box is not being displaced against an opposing force. In this illustration, however, we have assumed that the box alone is the system. If the person carrying the box is considered explicitly as part of the system, then work is done in holding the box. That's because 'physiological' work is done in maintaining muscle tension. This work results from chemical energy being converted to mechanical energy in striated muscle (red meat), a process that involves a lot of nerve impulses, the energy molecule ATP (Chapter 5), polymerized actin (Chapter 8) and myosin (Chapter 8).

Another example will help to illustrate how work is done by moving an object in a coordinated way against an opposing force. Pumping air into a bicycle tire is easy if the tire is flat; there is not much stuff in the tube to resist the flow of air in. Once the tube starts to fill, however, and the pressure begins to build, it gets harder and harder to force air in. Here air in the pump is being moved in an organized way (in) against the opposing force of the compressed air in the tire.

In the course of being inflated, the tire expands somewhat but not much. This is because tires are made to adopt a certain shape, one that is well-suited to their function. The volume of the tire is approximately constant. Why does the pressure increase as more air is pumped into the tire? An analogy might help us towards an answer. If you have been to a nightclub or concert and found the music unbearably loud, it is probably because the music made your ears hurt. Such pain is not merely a psychological effect; it comes from huge waves of air pressure smashing into your eardrum! Even in the middle of a grassy plain on a still day, where there is not a sound to be heard anywhere, the eardrum is in contact with the air. Nitrogen, oxygen and carbon dioxide bombard it continually. We usually take no notice of this, because the same types of molecules are bombarding the eardrum from the opposite side with *equal* force. We detect a disturbance of the air as sound only when there is a pressure *difference* across the eardrum (a type of membrane) and the pressure difference is large enough. So, when a sound is so loud that it makes your ears hurt, it's because air molecules on the outside of your eardrum are banging on your eardrum with a much greater force than air molecules on the inside are banging on your eardrum, and you would be well advised to plug your ears immediately! The more air molecules in a container of fixed volume, the greater the pressure.

Now, at thermal equilibrium, all the particles of a system have the same (average) **thermal energy. The thermal energy of a particle is proportional to T, the absolute temperature**; temperature is a measure of the thermal energy of the particles of a system. Suppose our system is a gas at thermal equilibrium. The particles move about freely within the system in all directions; particle movement is *random*. From time to time a gas molecule collides with a wall of the container. Such collisions are the origin of pressure. If we keep the volume the same but increase the number of particles, the number of collisions goes up and the pressure is increased. If we keep the volume and number of particles the same but add heat, the temperature increases and so does the speed of the particles. As particle movement increases, the particles strike the walls of the system more often and the pressure is increased. (One can see this easily from the form of the ideal gas law, in which the pressure times the volume equals the number of moles of gas times the temperature times the gas constant: $pV = nRT$. We'll come back to the ideal gas law a bit later.)

What if we have a mixture of two different kinds of gas particles, a light gas like hydrogen and a 'heavy' gas like nitrogen? At thermal equilibrium, all the particles will have the same thermal energy. But will the helium molecule and nitrogen molecules bombard the walls of the container with equal force? No! Why not? Their masses are different. The kinetic energy of a particle (K. E.) is proportional to its mass times its velocity squared, or K. E. $\propto mv^2$. (The symbol '$\propto$' means 'proportional to.' A proportionality becomes an equality when a multiplicative factor is included. In the equation '$2x = 6$,' for example, the proportionality constant is the symbol 'x' and is equal to 3. The precise value of the proportionality constant in the relationship of kinetic energy to momentum is irrelevant to the discussion and therefore omitted.) If the velocity of an object doubles, say from 1 m s^{-1} to 2 m s^{-1}, its kinetic energy quadruples. In terms of particle momentum, $\mathbf{p}$, which is defined as mass $\times$ velocity, K. E. $\propto \mathbf{p}^2/m$. Unlike K. E., which varies as the square of the velocity, $\mathbf{p}$ is a linear function of v: doubling the velocity doubles the momentum, not quadruples it. We include momentum in this discussion because it's the change in momentum per unit time that gives rise to pressure: a change in momentum per unit time is a force, and a force per unit area is a pressure. In symbols, $\Delta\mathbf{p}/\Delta t = F/A = p$. At thermal equilibrium, the (average) kinetic energy of a particle is equal to the (average) thermal energy, so $\mathbf{p}^2/m \propto T$. Solving this relationship for $\mathbf{p}$ gives $\mathbf{p} \propto (Tm)^{1/2}$. Thus, in our mixture of gases at the same temperature, a nitrogen molecule will have a greater average momentum than a hydrogen molecule for the simple reason that nitrogen is heavier (more massive) than hydrogen. Such ideas underlie a good deal of physical biochemistry, and they should be kept running continually in the background of our thinking as we make our way through this book.

Now let's exchange our bicycle tire as an airtight cylinder with a piston. The number of air molecules in the system is constant. The piston is holding the air in at a certain pressure. If we apply greater force to the piston, the air within the system is compressed, and the number of collisions the molecules make with the walls of the piston increases. We have done work on the system by moving an object (the piston) from outside the system against an opposing force (the molecules pushing against the piston inside the system). Great! But what on Earth does a piston filled with air have to do with biology? Quite a lot, in fact, as we shall see in due course! Our main object at the moment, though, is to be sure we have a good idea of the meaning of *work*, because that's what's needed to appreciate the First Law of Thermodynamics, the subject of this chapter.

Before concluding this section, we want to introduce the concept of a heat engine. We'll give the heat engine the once-over-lightly treatment

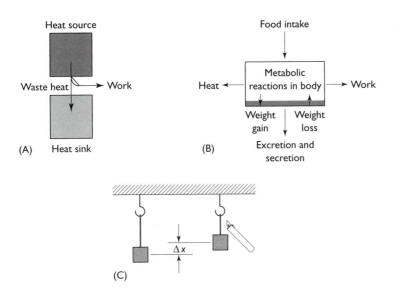

(A) Heat source / Waste heat → Work / Heat sink

(B) Food intake / Metabolic reactions in body / Heat ← → Work / Weight gain / Weight loss / Excretion and secretion

(C) Δx

Fig. 2.4 Heat transfer. (A) Heat is transferred from a source (a warm body) to a sink (a cold body). Some, but not all, of this heat can be used to do work. (B) Schematic representation of energy transformations within the body. The energy input is the food we eat. There are several 'outputs.' Aside from heat and work, which are mentioned explicitly in the First Law, there is excretion and change in body weight. Food intake − waste excreted = change in body weight + heat + work, according to the First Law of Thermodynamics. (C) A very simple heat engine. The rubber band contracts when heated, lifting a weight. Because the weight is translated against the force of gravity, work is done. Thus, heat is turned into work. The efficiency of this engine is very low! Panel (C) based on Fig. 44-1 of Feynman et al. (1963).

here but come back to it with vigor in Chapter 3. As depicted in Fig. 2.4A, a heat engine does works by transferring heat from a heat source (e.g. a radiator) to a heat sink (e.g. a cold room). Only some of the heat transferred can be used to do work, because there is a fundamental limit on engine efficiency. (This limitation is a statement of the Second Law of Thermodynamics, as we shall see in Chapter 3.) The heat energy that is not used to do work enters the heat sink as randomized molecular motion. As before, work is energy transfer by ordered motion, heat is energy transfer by random motion. Note how the First Law applies here: the energy lost by the heat source (ΔU) is either converted into work (w) or transferred to the heat sink (q), and w and q must sum to ΔU. Fig. 2.4B shows a diagram of a cell, though it could just as well represent a tissue, organ or entire organism. The system of inputs and outputs resembles the situation in panel (A), but now *everything is at the same temperature*; Fig. 2.4C shows how the heat energy of a candle can be used to do work. A rubber band dangles from a horizontal support, and attached to the rubber band is a weight. When heat from the candle is absorbed by the molecules of the rubber band, it contracts. The attached weight is translated a distance Δx against the opposing force of gravity, and work w is done. Some of the heat of the candle will of course be lost to the surrounding air (this heat engine is not very efficient), and only if sufficient care is taken will the rubber not melt before our eyes, leaving us with no engine at all! With all this in mind, let's take a closer look at how the First Law works.

D. | The First Law in operation

By convention, the internal energy of a system will *increase* either by transferring heat *to* it or by doing work *on* it (Table 2.1). Knowing this, we can write down a mathematical statement of the First Law of Thermodynamics as follows:

Table 2.1	Sign conventions for heat and work
Heat is transferred to the system	$q > 0$
Heat is transferred to the surroundings	$q < 0$
The system expands against an external pressure	$w < 0$
The system is compressed by an external pressure	$w > 0$

$$\Delta U = q + w \qquad (2.1)$$

Note that, in keeping with our earlier comments on measuring energy, the First Law defines only changes in U. The conceptual background to Eqn. 2.1 was formulated in 1847 by the German physicist Heinrich Helmholtz (1821–1894) (See *Ueber die Erhalting der Kraft* (Berlin: Reimer, 1847)), though the concept had been proposed in 1842 by the German physiologist Julius Robert von Mayer (1814–1878). It is interesting that a biological scientist played such an important role in establishing one of the most important concepts of thermodynamics. The conservation of mechanical energy (kinetic energy + potential energy = constant) had in fact been proposed much earlier, by the German philosopher and mathematician Gottfried Wilhelm Leibniz (1646–1716), son of a professor of moral philosophy, and was an accepted principle of mechanics. Mayer's statement of what is now called the First Law of Thermodynamics was based, curiously, on an analysis of the color of human blood. This is particularly amusing in view of the role of blood in *distributing food energy* throughout the body. When a system does work on its surroundings, w makes a negative contribution to ΔU because the system loses energy. Similarly, if heat is lost from the system, q makes a negative contribution to ΔU. In other words, and in very practical terms, ΔU measures the *net* amount of energy change in the system; it is the difference between the energy gained from the surroundings and the energy lost to the surroundings.

Let's see some examples of how Eqn. 2.1 works. James Prescott Joule, son of a beer brewer in Manchester, England, is famous for his studies on the conservation of thermal energy understood as the mechanical equivalent of heat (1843). Perhaps the best-known experiment Joule did was to monitor the temperature of a vat of water during stirring. In this experiment, increases in the temperature of the water represent positive increments in q, the heat transferred to the system. A motor turns a wheel in contact with water. The system is the water plus the wheel. As the wheel turns, mechanical energy is converted into increased motion of the water, and as we have seen, the motion of water is related to its temperature. Individual water molecules collide with the wheel and with each other. This increases their average speed, or temperature. In principle, vigorous and protracted stirring would eventually bring the vat of water to the boil. Mechanical energy is being substituted for the gas or electricity (or wood!) we would usually use to boil water. Note that in the course of this experiment, the *system* does no work; it does not expand against an opposing pressure or anything like that, so $w = 0$ and $\Delta U = q$.

Here's another example. Suppose it has been determined that a motor generates 30 kJ of mechanical work per second, and that 9 kJ is lost to the surroundings as heat in the same amount of time. The change in internal energy of the motor per second is $-9\,kJ - 30\,kJ = -39\,kJ$. The energy produced by the motor is negative because work is done by the system on the surroundings and heat is lost to the surroundings. In Chapter 8 we shall study some of the biochemical properties of the protein myosin, a molecular motor that is involved in the force generation of muscle contraction.

Now we want to see how these ideas can be applied to a biological system. Above we saw that no work is done in holding up a heavy box when the person is excluded from the system. If fact, though, physiological work is done as we described, and unless that energy is replenished by food, there is a net decrease in the internal energy of the body. Our bodies do work even when we're sleeping! When a person touches a metallic door handle on a wintry day, unless they have gloves on they can feel the heat being rapidly transferred from their flesh to the metal. Pain! This heat transfer makes a negative contribution to the internal energy of the body. When you walk up a flight of stairs, you do work against the force of gravity. If there are many steps to climb, as there are for instance on the way up to the cupola of the Duomo, the cathedral in Florence, Italy, or the top of Liberty's torch in New York Harbor, by the time you've reached the top you might well be out of breath and dripping with perspiration. Not only will you have lost energy moving against Earth's gravitational pull on your body, you will be losing a lot of heat to the surroundings in order to keep your body temperature from heating up too much. But, you'll probably also be awestruck by the view and not mind too much. In any case, the energy used to climb stairs and the energy lost as heat comes from food – biological thermodynamics.

We can be more quantitative about work with relatively little additional effort. From physics, the work done when an object is displaced a distance Δx ($x_{final} - x_{initial}$, where x refers to position) against an *opposing* force (hence the minus sign) of constant magnitude F is calculated as

$$w = -F\Delta x \qquad (2.2)$$

We see that work is the product of an 'intensity factor' that is independent of the size of the system (the force) and a 'capacity factor' (the change in the position of the object on which the force acts). For instance, the work done against gravity by a 50 kg woman in climbing to a point on a ladder 4 m above the ground is $- (50\,kg \times 9.8\,m\,s^{-2}) \times 4\,m = -1960\,kg\,m^2\,s^{-2} = -1.96\,kJ$. (Note: $1\,J = 1\,kg\,m^2\,s^{-2}$. Oddly enough, the dimensions of energy are [mass][length]^2[time]$^{-2}$.) The minus sign indicates that energy has been expended by the system, i.e. the woman. Diagrams will help visualize the situation. The work done in Eqn. 2.2 can be represented graphically as an *area*, as shown in Fig. 2.5A. Fig. 2.5B shows that the work done during a process depends on the path, because the shaded area need not be the same for all processes. This tells us that w is *not* a state function; it is a path function.

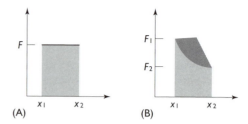

Fig. 2.5 Graphical representation of work. $|w| = F\Delta x$, where '$|w|$' means 'the absolute value or magnitude of w.' For example, $|-3| = 3$. (A) For a constant force, the magnitude of work is the area $F \times \Delta x$. B. If the force is variable, $|w|$ can no longer be calculated as $F \times \Delta x$. This illustrates one reason why w cannot be considered a state function: its value depends on the path. See Fig. 2.3. The graphical method of calculating the work done by a system was introduced by James Watt.

When a piston in a cylinder moves against a *constant* external pressure p_{ex} (as for instance in Fig. 2.6), the work done is

$$w = -p_{ex}\Delta V \qquad (2.3)$$

where ΔV represents the change in the volume of the *system*; and $p_{ex} = nRT/V$, to the extent that whatever is pushing on the system resembles an ideal gas. This type of work is called pV-work. Again, the work done is the product of an 'intensity factor' (p_{ex}) and a 'capacity factor' (ΔV). If the volume *of the system* increases, $\Delta V > 0$; the energy for expansion against an opposing force comes from the system itself, so the work done is negative. If there is no external pressure (if the surroundings are vacuum), then $p_{ex} = 0$; there is no force to oppose expansion of the system, and no work is done as V increases. Both p and V are known as **state variables**: they specify the state of the system.

E. Enthalpy

Now things are going to become a little less concrete. Another thermodynamic state function we need to know about is the **enthalpy**, H. It is covered here for several reasons, the most important one being that H is a component of the Gibbs free energy, the central topic of this book from a practical point of view. The term *enthalpy* comes from Greek elements *en + thalpe* and was coined around 1850 by the German physicist Rudolf Julius Emanuel Clausius (1822–1888), son of a pastor and schoolmaster. **The enthalpy is the heat absorbed by a system at constant pressure** (subscript 'p'). To see this, we rewrite the First Law as

$$q_p = \Delta U - w \qquad (2.4)$$

When the pressure is constant and the system expands from state 1 to state 2, doing work on the surroundings, and the only type of work is pV-work, we have

$$q_p = U_2 - U_1 + p(V_2 - V_1) \qquad (2.5)$$

We can rearrange Eqn. 2.5 as

$$q_p = (U_2 + pV_2) - (U_1 + pV_1) = \Delta U + p\Delta V \qquad (2.6)$$

The quantity $\Delta U + p\Delta V$ is thus the heat exchanged at constant pressure. There are two components to the heat, ΔU and $p\Delta V$. The right hand side of Eqn. 2.6 can be written as ΔH, where H, the enthalpy, is defined as:

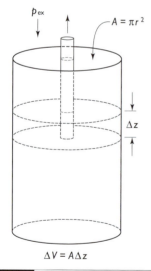

$A = \pi r^2$

$\Delta V = A\Delta z$

Fig. 2.6 Gas-filled cylinder. There is a constant external pressure, p_{ex}. For a change in position of the piston, Δz, there is a corresponding change in volume, ΔV. The work done *by the* system is $w = -p_{ex}\Delta V$. If ΔV is negative, i.e. if the gas in the cylinder is compressed, there is a positive contribution to ΔU.

$$H = U + pV \tag{2.7}$$

Eqn. 2.7 can seem confusing and abstract. Did we not say above that $w = -p_{ex}\Delta V$ is a *path* function? How can a state function plus a path function equal a state function? We need a way of making sense of this in terms of what we have already said and our everyday experience of the world. It is precisely because p and V specify the state of the system and they, like U, are independent of how that state was reached. H is therefore a state function, and it no less or more a state function than U. Eqn. 2.7 tells us that ΔH is the *total change* in the energy of a system, as it comprises both heat transfer and volume change. The development leading up to Eqn. 2.7 says that **if the pressure is constant, the amount of heat exchanged during a reaction is independent of whether the product is formed directly or indirectly, in one or in a series of steps**. This is the 'law of constant summation' of the Swiss–Russian chemist Germain Henri Hess (1802–1850), a physician and chemist. Hess's Law restates the First Law, though historically the former preceded the latter. Based on the results of experiments, Hess's Law says that **state functions and state function differences (e.g. ΔU or ΔH) are additive** (see Fig. 2.3).

Looked at another way, the enthalpy can be thought of as the amount of energy *possessed* by a thermodynamic system *for transfer* between itself and the environment. For example, in the bomb calorimetry experiments of Chapter 1, the change in enthalpy was (very close to) the heat of the oxidation. The heat of oxidation was the energy transferred from the oxidized compounds to the calorimeter. In a change of phase, say when liquid water becomes solid, the change in enthalpy of the system is the 'latent heat' of fusion, the heat given off to the environment by the process. And in a temperature change, for example as occurs when a warm can of Coke is placed in a fridge, the change in the enthalpy of the liquid with each degree is the **heat capacity** of the liquid at constant pressure. The heat capacity tells us how much energy is possessed by the system for transfer between itself and the environment *per degree*. This concept is so important that an entire section will be devoted to it shortly.

With all these ideas in mind, let's exercise our grasp of the mathematical representation of thermodynamic concepts to get a clearer view of things. When the enthalpy varies by a small but measurable amount, from Eqn. 2.7 we have

$$\Delta H = \Delta(U + pV) = \Delta U + \Delta(pV) = \Delta U + p\Delta V + V\Delta p \tag{2.8}$$

If the external pressure is constant, $\Delta p = 0$ and the last term vanishes. Substituting in Eqn. 2.1 and requiring pV-work only gives

$$\Delta H = q_p - p\Delta V + p\Delta V \tag{2.9}$$

The last two terms on the right hand side cancel, and we are left with

$$\Delta H = q_p \tag{2.10}$$

Although we have proved what we already knew, we can at least be a little more confident that we are following what's happening! **The heat transferred to a system at constant pressure measures the change in**

the enthalpy of the system. Why the emphasis on heat transfer at constant pressure in a book for biochemists? For one thing, most of the reactions we study are carried out at constant pressure (usually 1 atm). For another, as we shall see in Chapter 4, H is a component of a state function known as the Gibbs free energy, G. This is important because G predicts the direction of spontaneous change for a process at constant pressure and temperature, the biochemist's favorite experimental conditions.

It might seem from Eqn. 2.10 and the rest we have said about enthalpy that ΔH is not all that different from ΔU. Appearances can be deceiving! Admittedly, though, in practice the difference between ΔH and ΔU is often small enough to be neglected. For instance, if a reaction occurs in solution and gas is neither produced nor consumed, $\Delta V \approx 0$. Under such circumstances $\Delta U \approx q_p$, as we can see from Eqn. 2.5, and so $\Delta U \approx \Delta H$. If H and U are so similar, can the difference between them always be neglected? An example will help to illustrate that as a general rule it is a mistake not to take account of differences when there are reasons to suspect they might be significant.

From Eqn. 2.7,

$$\Delta H = \Delta U + \Delta(pV) \tag{2.11}$$

Assuming that the gas involved in our experiment can be modeled as an ideal gas, Eqn. 2.11 can be written as

$$\Delta H = \Delta U + \Delta(nRT) \tag{2.12}$$

where n represents the number of moles of gas, R is the gas constant ($8.31451 \, \mathrm{JK^{-1}mol^{-1}}$ in SI units. $1.9872 \, \mathrm{cal \, K^{-1} \, mol^{-1}}$, however, is also still in use. T is the absolute temperature, and we have simply exchanged one side of the gas law equation for the other. If we now require constant temperature, $\Delta(nRT) = RT(\Delta n)$, where Δn represents the change in the number of moles of gas in the reaction.

Now, suppose that a lump of graphite is oxidized to carbon dioxide, as follows:

$$2C(s) + O_2(g) \rightarrow 2CO_2(g) \tag{2.13}$$

Bomb calorimetry shows that when this reaction is carried out at 298 K and constant volume, $395\,980 \, \mathrm{J \, mol^{-1}}$ of heat are released. So, $\Delta U(298 \, \mathrm{K}) = -395\,980 \, \mathrm{J \, mol^{-1}}$. We have $\Delta n = 2 - 1 = 1$. Therefore, by Eqn. 2.12, $\Delta H(298 \, \mathrm{K}) = -395\,980 \, \mathrm{J \, mol^{-1}} + 2480 \, \mathrm{J \, mol^{-1}} = -393\,500 \, \mathrm{J \, mol^{-1}}$. This is a small difference between ΔH and ΔU – less than 1% – but it is nevertheless an honest to goodness difference. This example tells us that, although the oxidation heats of Chapter 1 are changes in internal energy, they are very close to the corresponding changes in enthalpy. Well-presented combustion data will indicate clearly whether or not the pV term has been taken into account. A process for which the change in enthalpy is negative is called **exothermic**, as it lets heat out of the system into the surroundings; a process for which the change in enthalpy is positive is called **endothermic**, because it lets heat into the system from the surroundings.

It might be far from obvious, but combustion of food in a bomb calorimeter tells us more than just how much heat is produced when food is burnt completely to a crisp. Indeed, tables of oxidation would be of little use to nutritionists if the numbers did not say something about the energetics of metabolism. The tables are useful to the physicist, the nutritionist and the biochemist alike because the laws of physics are assumed to be independent of time and location. In other words, the enthalpy of oxidation of glucose is not one value in a bomb calorimeter and some other value in the striated muscle connecting your foot to your big toe. By Hess's law, this enthalpy equivalence holds despite the fact glucose oxidation in the body occurs by a large number of sequential steps involving a large number of chemical intermediates (Chapter 5). This discussion also tells us that we can use machines like calorimeters to investigate the thermodynamic properties of the molecules our bodies are made of, and that our bodies themselves resemble machines.

To conclude this section, suppose we have a system of pure water. It is known from careful measurements that when ice melts at $+0.1\,°C$, barely above the melting temperature, $\Delta H = 1437.2$ cal mol^{-1}. When melting occurs at $-0.1\,°C$, just below the freezing point, $\Delta H = 1435.4$ cal mol^{-1}. The difference in enthalpy differences, $\Delta\Delta H$, is 1.8 cal mol^{-1}. This enthalpy change is only about *half* the amount we would expect on changing the temperature of water by $0.2\,°C$ *in the absence of melting*. (See Section H below.) The difference arises from the **change of phase** that occurs between the initial state and final state. This shows that one needs to account for any changes in the state of matter (solid, liquid or gas) when calculating ΔH in addition to changes in the amount of gas present.

F. Standard state

Changes in enthalpy (and other state functions) are usually given for processes occurring under a standard set of conditions. In tables of thermodynamic data, the **standard state** is usually defined as one mole of a pure substance at 298.15 K (25.00 °C) and 1 *bar* (1 bar = 10^5 Pa = 0.986 932 atm). An example is the standard enthalpy change accompanying the conversion of pure solid water to pure liquid water at the melting temperature and a pressure of 1 *bar*:

$$H_2O(s) \rightarrow H_2O(l) \qquad \Delta H^{\ominus}(273\ K) = 6.01\ kJ\ mol^{-1} \qquad (2.14)$$

Note that this standard enthalpy change is *positive*: heat must be added to ice at $0\,°C$ in order to melt it. The **standard enthalpy change** used by the biochemist, $\Delta H°$, is the change in enthalpy for a process in which the initial and final states of *one mole* of a substance in pure form are in their standard state: 25 °C and 1 *atm* pressure. The difference in enthalpy from the pressure difference between 1 bar and 1 atm is almost always small enough to be neglected in biochemical reactions. But one should nevertheless be aware of the different ways in which thermodynamic data of use to the biochemist are presented in tables.

G. | Some examples from biochemistry

Equation 2.10 is very useful to the biochemist. As we have seen, it helps to make sense of the oxidation heats measured by bomb calorimetry. It can also be used in the study of protein stability. Protein stability is an important subject for several reasons. Two of them are that about half of the dry mass of the human body is protein, and knowing how a poly-peptide folds up into its native state would be of tremendous value in making good use of all the protein-encoding DNA sequence data that has been revealed by the Human Genome Project.

How can Eqn. 2.10 be used to study the thermodynamic properties of proteins (or nucleic acids, for that matter)? The folded state of a protein is like an organic crystal. It is fairly rigid, and held together by a large number of different kinds of 'weak' non-covalent interactions, including hydrogen bonds (largely electrostatic in character), van der Waals interactions (named after the Dutch physicist Johannes Diderik van der Waals (1837–1923), who was awarded the Nobel Prize in Physics in 1911) and 'salt bridges' (electrostatic attractions between ionized amino acid side chains) (Tables 2.2 and 2.3). In the core of a folded protein, apolar amino acid side chains are interdigitated and tightly packed, forming rigid and specific contacts. The rigidity of a folded protein is important to its biological function; in favorable circum-stances, it also permits determination of its structure at atomic resolu-tion. This is not to say that a folded protein exhibits no fluctuations of structure or rotations of bonds. Native protein structure certainly does fluctuate, as we know for example by nuclear magnetic resonance studies, and such fluctuations can be important in the binding of small compounds to macromolecules (Chapter 7) and enzyme function (Chapter 8). But the folded state of a typical protein is nevertheless quite rigid. In contrast, the unfolded state of a protein is extremely flexible and fluid-like. Bonds in amino acid side chains rotate relatively freely, and in the ideal case all amino acid side chains are completely exposed to solvent (Table 2.4).

The non-covalent interactions that stabilize folded protein struc-ture (or double-stranded DNA) can be broken in a number of ways. One is by heating. If all the non-covalent bonds break simultaneously, in an all-or-none fashion ('cooperative' unfolding), then there are in essence just two states of the protein: the folded state and the unfolded state. The **transition** from the folded state to the unfolded state is like melting. So when one induces the unfolding of protein by heat or some other means, it is something like carrying out a process in which one melts a solid. This is true even if one is working not with a mass of freeze-dried protein but with the folded protein dissolved in aqueous solution. The **cooperativity** of the transition results from the simulta-neous breaking of a large number of weak interactions. As we know from secondary school chemistry, the melting of pure ice or any other pure solid is a cooperative phenomenon. That is, melting occurs at a single or over a very narrow range temperatures, not over a broad range.

Table 2.2 | Energetics of non-covalent interactions between molecules

Type of interaction	Equation	Approximate magnitude (kcal mol^{-1})
Ion–ion[a]	$E = q_1 q_2 / Dr$	14
Ion–dipole[b]	$E = q\mu\theta / Dr^2$	-2 to $+2$
Dipole–dipole[c]	$E = \mu_1\mu_2\theta' / Dr^3$	-0.5 to $+0.5$
Ion-induced dipole[d]	$E = q^2\alpha / 2D^2 r^4$	0.06
Dispersion[e]	$E = 3h\nu\alpha^2 / 4r^6$	0 to 10

Notes:

[a]Charge q_1 interacts with charge q_2 at a distance r in medium of dielectric D. [b]Charge q interacts with dipole μ at a distance r from the dipole in medium of dielectric D. θ and θ' are functions of the orientation of the dipoles. [c]Dipole μ_1 interacts with dipole μ_2 at an angle q relative to the axis of dipole μ_2 and a distance r from the dipole in medium of dielectric D. [d]Charge q interacts with molecule of polarizability α at a distance r from the dipole in medium of dielectric D. [e]Charge fluctuations of frequency ν occur in mutually polarizable molecules of polarizability α separated by a distance r. The data are from Table 1.1 of van Holde.

Table 2.3 | Characteristics of hydrogen bonds of biological importance

Bond type	Mean bond distance (nm)	Bond energy (kJ mol^{-1})
O—H ... O	0.270	-22
O—H ... O$^-$	0.263	-15
O—H ... N	0.288	-15 to -20
N$^+$—H ... O	0.293	-25 to -30
N—H ... O	0.304	-15 to -25
N—H ... N	0.310	-17
HS—H ... SH$_2$	—	-7

Note:

The data are from Watson (1965).

Table 2.4 | Principal features of protein structure

Folded (native) state	Unfolded (denatured) state
Highly ordered polypeptide chain.	Highly disordered chain – 'random coil'
Intact elements of secondary structure, held together by hydrogen bonds	No secondary structure
Intact tertiary structure contacts, as in an organic crystal, held together by van der Waals interactions	No tertiary structure
Limited rotation of bonds in the protein core	Free rotation of bonds throughout polypeptide chain
Desolvated side chains in protein core	Solvated side chains
Compact volume	Greatly expanded volume

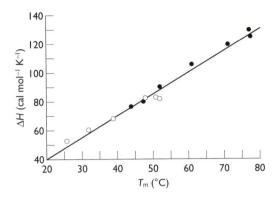

Fig. 2.7 Enthalpy of unfolding of hen egg white lysozyme as a function of transition temperature. Filled symbols: intact lysozyme. Open symbols: lysozyme in which one of the four native disulfide bonds has been removed. When folded, 3-SS lysozyme closely resembles the native state of intact lysozyme. The changes in transition temperature were induced by changes in pH. Note that ΔH is approximately linear in T_m. The data are from Cooper *et al.* (1991).

The same is true of cooperative protein unfolding or the melting of DNA.

A number of experimental studies have been carried out in order to measure the energy required to break a hydrogen bond at room temperature. This is pertinent not only to the unfolding of proteins but also to the 'melting' of double-stranded DNA, because DNA is held together by hydrogen bonds. Estimates of the bond energy vary, but a reasonable and generally agreed upon rough figure is 1 kcal mol^{-1}. Though weak on their own, hydrogen bonding 'networks' can be far stronger than just the sum of individual contributions.

The enthalpy of the folded state of a protein in terms of Eqn. 2.10 is $H_F{}^\circ$. The enthalpy of the unfolded state is $H_U{}^\circ$, and the difference, $H_U{}^\circ - H_F{}^\circ$, is the enthalpy of the unfolded state with respect to the enthalpy of the folded state or the **enthalpy of denaturation**, $\Delta H_d{}^\circ$. The subscript 'd' refers to denaturation. In this case the folded state of the protein is what is known as a **reference state**, because the enthalpy of the unfolded state is measured with respect to it. What is this enthalpy difference between conformations of a protein? An analogy will help us to form a clearer mental picture of it. As discussed above, the enthalpy change for a process is equal to the heat absorbed by the system at constant pressure. The rigid folded state of a protein is solid-like, and the flexible unfolded state is liquid-like. At a given pressure, a certain amount of heat must be added to a solid to get it to melt. For instance, one could put a block of ice in an insulated saucepan and measure the amount of heat needed to turn it into water. Similarly, the enthalpy difference between the unfolded and folded states of a protein is the amount of heat needed to unfold the protein. As we shall see below, the heat required depends on the temperature.

The temperature at which a protein unfolds (or double-stranded DNA melts) is called the melting temperature, T_m. This temperature depends not only on the number and type of non-covalent bonds in the folded state but also on the pH and other solution conditions. Changing the pH of solution changes the net charge on the protein surface. This can have a marked impact on T_m and $\Delta H_d{}^\circ$, as shown in Fig. 2.7 for the example of hen egg white lysozyme, a well-studied small globular protein that is found in abundance in egg white. The figure also illustrates that the relationship between $\Delta H°$ and T_m for this protein is very regular throughout the pH range shown.

We have already discussed how a bomb calorimeter can be used to obtain thermodynamic information. Here we introduce another type of calorimetry, isothermal titration calorimetry (ITC), which was first described in 1922 by the Belgian Théophile de Donder, founder of the Brussels School of thermodynamics. It can be used to measure the enthalpy of a biochemical process (Fig. 2.8) because, by Eqn. 2.10, the heat absorbed at constant pressure measures the enthalpy change. Suppose, for example, that we are interested in the energetics of the binding of the F_c portion of immunoglobulin G (IgG) to soluble protein A, a bacterial protein. We need not be terribly concerned at the moment what the F_c portion of IgG is; it is enough to know that antibodies can be dissected into components and that the F_c portion is one of them. The thermodynamic states of interest here are the unbound state, where protein A is free in solution, and the bound state, where protein A is physically associated with F_c. The heat exchanged at constant pressure upon injection of protein A into a calorimetric cell containing the antibody can thus be used to determine $\Delta H_b°$, **the enthalpy of binding**. The heat of injection will change as the number of vacant binding sites decreases.

What if we're interested in the energetics of an enzyme binding to its substrate? This can be measured if a suitable substrate analog can be found or the enzyme can be modified. For instance, ITC has been used to measure the enthalpy of binding of a small compound called 2′-cytidine monophoshate (2′CMP) to ribonuclease A, an enzyme that hydrolyzes RNA to its component nucleotides. 2′CMP binds to and inhibits the enzyme. If the enzyme of interest is, say, a protein phosphatase with a nucleophilic cysteine in the active site, mutation of the cysteine to serine or asparagine will abolish catalytic activity and the energetics of binding can be studied. The mathematics of binding will described in Chapters 5 and 7.

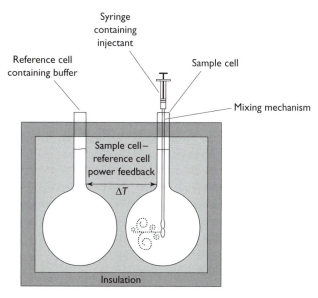

Fig. 2.8 Isothermal titration calorimeter. The temperature is constant. There are two identical chambers, the sample cell and the reference cell. In most cases, the sample cell will contain a macromolecule, and a syringe/stirrer is used to inject ligand into the cell. The syringe is usually coupled to an injector system under software control and rotated at a constant speed. The reference cell is filled with buffer; no reaction occurs in it. ΔT measures the temperature difference between the cells. The cells are surrounded by insulation, to minimize heat exchange with the surroundings. Electronic circuitry (power feedback) is used to minimize ΔT on a continuous basis. If injection of ligand results in binding, there will ordinarily be a change in the temperature of the sample. The sign of the change will depend on whether the reaction is exothermic or endothermic. An experiment consists of equal-volume injections from the syringe into the sample cell.

Table 2.5	Standard ion hydration enthalpies at 298K in kJ mol^{-1}		
H$^+$	−1090	Mg^{2+}	−1920
Li$^+$	−520	Ca^{2+}	−1650
Na$^+$	−405	Ba^{2+}	−1360
K$^+$	−321	Fe^{2+}	−1950
—	—	Zn^{2+}	−2050
NH^{4+}	−301	Fe^{3+}	−4430

Notes:
The data refer to X$^+$(g)→X$^+$(aq) at 1 bar and are from Table 2.6c in Atkins (1998). 1 bar = 10^5 Pa = 10^5 N m^{-2} = 0.987 atm. 1 Pa = 1 pascal. Blaise Pascal was a French scientist and religious philosopher who lived 1623−1662.

If you've spent any time in a biochemistry lab, you have probably experienced the large heat given off by a salt solution as the salt dissolves. There are several contributions to the effect, but the main one is the **enthalpy of hydration**. This is the enthalpy change that occurs when an ion in vacuum is dropped into a sea of pure water. Water molecules form what is called a hydration shell around the ion, the number of molecules depending on the radius of the ion and its net charge. Calorimetry can been used to measure the hydration enthalpy of biologically important ions. Values are given in Table 2.5. Why is this important? One reason is that some of the water molecules hydrating an ion must be stripped away before it can pass through a selective ion channel in the plasma membrane, and this requires an input of energy. Ion channels that are specific for the passage of certain types of ion are part of the molecular machinery underlying the transmission of nerve impulses. It is clear from Table 2.5 that there is a large energy change associated with ion dehydration.

H. Heat capacity

It is evident from Fig. 2.7 that the enthalpy of a protein rises as its temperature is increased. This is true of substances in general. The numerical relationship, however, between H and T depends on the conditions. Most *in vitro* biochemical reactions are carried out at constant pressure rather than constant volume, so we concern ourselves here with constant pressure only. The data shown in Fig. 2.7 were obtained at constant pressure. The slope of a graph of H versus T at constant pressure is called the **heat capacity at constant pressure**, C_p. The heat capacity per unit mass of material, or specific heat, was first described in detail by Joseph Black. This quantity is readily measured, and it can be used to calculate changes in the internal energy or, more importantly, changes in the enthalpy.

We are all familiar with heat capacity, even if we have not formally been introduced to the concept. Returning to our water-in-a-saucepan example, if water is heated at a high enough rate, it will eventually boil. The amount of heat that must be added to increase the temperature of

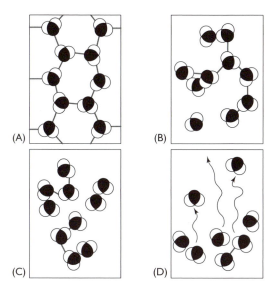

(A)　　　　(B)

(C)　　　　(D)

Fig. 2.9 Schematic diagram of the structure of water under different conditions: (A) solid state; (B) melting point; (C) liquid state; (D) boiling point. Oxygen is shown in black, hydrogen in white. The black bars represent hydrogen bonds. Hydrogen bonds are relatively persistent in the solid state. The number of bonds decreases at the melting point, as molecules move out of the lattice. In the liquid state, hydrogen bonds are present, but they are formed only transiently. Boiling water has such a high thermal energy that persistent hydrogen bonds are rare. As the temperature increases, there are increases in translational, vibrational and rotational energy. The change in translational energy can be seen, for example, when water is brought to the boil on the stove. Increases in vibrational and rotational motion of water can only be detected using sophisticated instruments. Based on Fig. 3-3. in Voet & Voet (1995).

one gram of water by one degree is called the **specific heat capacity** of water. The **heat capacity** of a substance is the amount of heat required to raise the temperature of one mole of that substance by 1 degree K. The heat capacity of liquid water at 1 atm pressure varies only slightly with temperature over the range 0–100 °C. The physical basis of this is the bonding structure of water (Fig. 2.9). Just as the structure of water changes substantially as it freezes or vaporizes, the heat capacity of water depends substantially on passing to the solid or gaseous state from the liquid state. This is true of substances in general. The number of hydrogen bonds formed by an individual water molecule is more or less constant throughout the temperature range 0–100 °C at 1 atm pressure.

Returning to proteins, just as the folded and unfolded states have different enthalpies, they have different heat capacities. The heat capacity of the folded state is $C_{p,F}$, whilst that of the unfolded state is $C_{p,U}$. $C_{p,U} - C_{p,F} = \Delta C_{p,d}$ is the heat capacity difference between the unfolded state and the folded state at constant pressure. In principle, $C_{p,F}$, $C_{p,U}$ and therefore $\Delta C_{p,d}$ are temperature dependent. In practice, however, the variation with temperature can be and often is sufficiently small to be ignored. An example of a case where $\Delta C_{p,d}$ is only slightly dependent on temperature is shown in Fig. 2.7; $\Delta C_{p,d}$ is approximately constant throughout the temperature range. The sign of $\Delta C_{p,d}$ is positive for proteins. This is related to the increase in hydrophobic surface – e.g. the aliphatic side chain of Leu (leucine; a full list of three-letter codes for amino acids is given in Appendix B) – that is in contact with the solvent.

Now we are in a position to write a general expression for the enthalpy of a substance as a function of temperature. It is

$$H(T_2) = H(T_1) + C_p(T_2 - T_1) \tag{2.15}$$

where T_1 is the temperature of the system when it is in state 1 and $H(T_2)$ is the enthalpy of the system when it is in state 2, not $H \times T_2$. Another way of writing this is

$$\Delta H = C_p \Delta T \tag{2.16}$$

Fig. 2.10 Differential scanning calorimetry. (A) Schematic diagram of the instrument. The reference cell contains buffer only, the sample cell contains the macromolecule dissolved in buffer. Both cells are heated very slowly (e.g. 1 °C min^{-1}) in order to maintain equilibrium, and feedback electronic circuit is used to add heat so that $\Delta T \approx 0$ throughout the experiment. (B) Data. This additional heat can be plotted as a function of temperature. The endothermic peak corresponds to the heat absorbed on denaturation. The peak maximum is very close to the transition temperature, or melting temperature. The area below the peak is $\Delta H_d(T_m)$. The heat capacity of the unfolded state minus the heat capacity of the folded state is $\Delta C_{p,d}$.

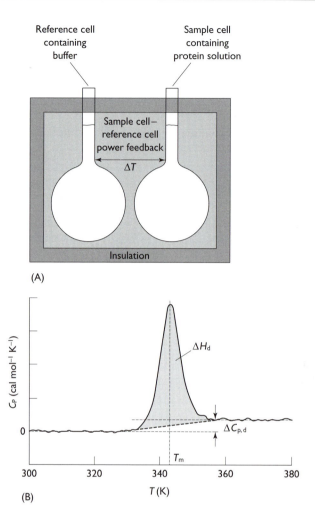

(A)

(B)

where $\Delta H = H(T_2) - H(T_1)$. Note that ΔH would have the same magnitude but the opposite sign if the labels attached to the states were reversed. This follows from the enthalpy's being a state function. From a mathematical point of view Eqn. 2.16, which can be written $C_p = \Delta H / \Delta T$, tells us that the constant pressure heat capacity is equal to the slope of a graph of H versus T. When C_p is constant throughout the temperature range, the slope of H versus T is constant. Eqn. 2.16 assumes that C_p is effectively constant over the temperature range $T_2 - T_1$, and as we have said this is approximately true for many materials in the absence of a change of phase. The unfolding of a protein, however, can be described as a phase change (melting of a solid), and we should therefore expect that there will be a difference in heat capacity between the folded and unfolded states. The corresponding expression to Eqn. 2.15 for the enthalpy difference between the unfolded and folded states of a protein is

$$\Delta H_d{}^\circ(T_2) = \Delta H_d{}^\circ(T_1) + \Delta C_{p,d}(T_2 - T_1) \tag{2.17}$$

where, as before, the heat capacity change is independent of temperature. Eqns. 2.15 and 2.17 apply to many different situations (not just protein folding/unfolding!) and are known as Kirchhoff's enthalpy law, after the German physicist Gustav Robert Kirchhoff (1824–1887).

One way of determining $\Delta C_{p,d}$ for protein unfolding is to denature the protein under different conditions. A popular method is to measure $\Delta H_d{}^\circ$ and T_m for different values of pH, as shown in Fig. 2.7. Under favorable conditions this can be done easily using a technique called differential scanning calorimetry (DSC), which measures the heat absorbed as a function of temperature (Fig. 2.10). The experiment is repeated at a variety of pH values to generate a curve like that shown in Fig. 2.7. As we shall see in Chapter 5, the relatively large $\Delta C_{p,d}$ of protein unfolding has a dramatic impact on how much work must be done to unfold a protein.

I. | Energy conservation in the living organism

We have seen that the First Law tells us that if a system does work, w makes a negative contribution to ΔU; the system loses energy. This implies that not even the most sophisticated 'machine' – the human body, as far as we know – can do work without an energy source. No matter how much the urge might conflict with ways we might rather spend our time, there is no getting around having to eat relatively often! But does this necessarily mean that the First Law applies to living organisms?

In Chapter 1 we noted that calorimetry experiments on whole organisms were carried out as early as 1781 by Lavoisier and Laplace. They measured the heat given off by animals (and other objects) as the amount of water produced by melting ice, relative to a control. The greater the volume of water at the end of the experiment, the greater the amount of heat given off during the experiment. Lavoisier and Laplace also collected and measured the amounts of gas produced by their experiments. The quantity of heat and carbon dioxide produced by a guinea pig was compared with what was found for the combustion of carbon. In this way, Lavoisier established that the combustion of food in animals leads to the production of heat, CO_2, and H_2O. About a century later, in 1904, a German physiologist named Max Rubner (1854–1932) reported on similar experiments with dogs. Rubner's work was effectively the final word on whether the principles of thermochemistry applied not only to inanimate substances but also to the physiology of living organisms. For he was able to show that the heat production of a dog can be accounted for by the carbon and hydrogen balance of its respiration and the heats of combustion of fat, protein, and excrement. And on that cheerful note we bring the text of this chapter to a close.

J. | References and further reading

Atkins, P. W. (1998). *Physical Chemistry*, 6th edn, ch. 2. Oxford: Oxford University Press.

Atkinson, D. E. (1977). *Cellular Energy Metabolism and Its Regulation*. New York: Academic Press.

Bergethon, P. R. (1998). *The Physical Basis of Biochemistry: the Foundations of Molecular Biophysics*, ch. 11. New York: Springer-Verlag.

Blandamer, M. J., Cullis, P. M. & Engberts, J. B. F. N. (1995). Differential scanning and titration calorimetric studies of macromolecules in aqueous solution. *Journal of Thermal Analysis*, **45**, 599–613.

Burton, R. F. (1998). *Biology by Numbers: an Encouragement to Quantitative Thinking*, ch. 3.1–3.3. Cambridge: Cambridge University Press.

Christensen, H. N. & Cellarius, R. A. (1972). *Introduction to Bioenergetics: Thermodynamics for the Biologist: A Learning Program for Students of the Biological and Medical Sciences*. Philadelphia: W.B. Saunders.

Cooper, A., Eyles, S. J., Radford, S. E. & Dobson, C. M. (1991). Thermodynamic consequences of the removal of a disulfide bridge from hen lysozyme. *Journal of Molecular Biology*, **225**, 939–943.

Creighton, T. E. (1993). *Proteins: Structure and Molecular Properties*, 2nd edn, ch. 4.4.3. New York: W. H. Freeman.

Encyclopædia Britannica CD 98, 'Enthalpy,' 'Heat Capacity,' 'Heat of Reaction,' 'Heat Transfer,' 'Internal Energy,' 'Kinetic Theory of Gases,' 'Latent Heat,' 'Maxwell–Boltzmann Distribution Law,' 'Principles of Thermodynamics,' 'Specific Heat,' 'Temperature,' 'Thermal Energy,' and 'Work.'

Feynman, R. P., Leighton, R. B. & Sands, M. (1963). *Lectures on Physics*, vol. I, cc. 14–1, 44–1, 45–1 & 45–2. Reading, Massachusetts: Addison-Wesley.

Fruton, J. S. (1999). *Proteins, Enzymes, Genes: the Interplay of Chemistry and Biology*. New Haven: Yale University Press.

Gillispie, Charles C. (ed.) (1970). *Dictionary of Scientific Biography*. New York: Charles Scribner.

Gislason, E. A. & Craig, N. C. (1987). General definitions of work and heat in thermodynamic processes, *Journal of Chemical Education*, **64**, 660–668.

Harold, F. M. (1986). *The Vital Force: a Study of Bioenergetics*, ch. 1. New York: W. H. Freeman.

Haynie, D. T. (1993). *The Structural Thermodynamics of Protein Folding*, ch. 2. Ph.D. thesis, The Johns Hopkins University.

Haynie, D. T. & Ponting, C. P. (1996). The N-terminal domains of tensin and auxilin are phosphatase homologues. *Protein Science*, **5**, 2643–2646.

Katchalsky, A. & Curran, P.F. (1967). *Nonequilibrium Thermodynamics in Biophysics*, ch. 1. Cambridge, Massachusetts: Harvard University Press.

Jones, C. W. (1976). *Biological Energy Conservation*. London: Chapman & Hall.

Klotz, I. M. (1986). *Introduction to Biomolecular Energetics*, ch. 1. Orlando: Academic Press.

Kondepudi, D. & Prigogine, I. (1998). *Modern Thermodynamics: from Heat Engines to Dissipative Structures*, ch. 2. Chichester: John Wiley.

Lawrence, C., Roger, A. & Compton, R. (1996). *Foundations of Physical Chemistry*. Oxford: Oxford University Press.

Lazarides, T., Archontis, G. & Karplus, M. (1995) Enthalpic contribution to protein stability: insights from atom-based calculations and statistical mechanics. *Advances in Protein Chemistry*, **47**, 231–306.

Morowitz, H. J. (1978). *Foundations of Bioenergetics*, ch. 3. New York: Academic.

Microsoft Encarta 96 Encyclopedia, 'Thermodynamics.'

Peusner, L. (1974). *Concepts in Bioenergetics*, ch. 2. Englewood Cliffs: Prentice-Hall.

Polanyi, M. (1946). *Science, Faith and Society*, ch. 1. Chicago: University of Chicago Press.

Price, G. (1998). *Thermodynamics of Chemical Processes*, cc. 1 & 2. Oxford: Oxford University Press.

Roberts, T. J., Marsh, R. L., Weyland, P. G., & Taylor, C. R. (1997). Muscular force in running turkeys: the economy of minimizing work, *Science*, **275**, 1113–1115.

Smith, C. A. & Wood, E. J. (1991). *Energy in Biological Systems*, cc. 1.2 & 1.3. London: Chapman & Hall.

Treptow, R. S. (1995). Bond energies and enthalpies: an often neglected difference. *Journal of Chemical Education*, **72**, 497–499.

van Holde, K. E. (1985). *Physical Biochemistry*, 2nd edn, ch. 1.1. Englewood Cliffs: Prentice-Hall.

Voet, D. & Voet, J. G. (1995). *Biochemistry*, 2nd edn, cc. 2–2, 3, 4, 11-2, 15-4–15-6, 16, 18-1, 19-1, 28-3 & 34-4B. New York: Wiley.

Watson, J. D. (1965). *The Molecular Biology of the Gene*. New York: Benjamin.

Williams, T. I. (ed.) (1969). *A Biographical Dictionary of Scientists*. London: Adam & Charles Black.

K. | Exercises

1. Invent three syllogisms.
2. Give the etymologies of *kinetic* and *potential*.
3. Give an example of a law of biology. What makes it a law?
4. Eqn. 2.1 involves a difference in internal energy. Differences in energy are much easier to measure than absolute magnitudes. Explain.
5. Fig. 2.3 shows a thermodynamic cycle. The state function is U, but in principle a cycle of this sort could be given for any state function. Suppose that each arrow represents an experimental process, and that each internal energy represents an experimentally determined quantity. Give representative values for each energy change so that the condition $\Sigma_{loop} X = 0$ is satisfied.
6. The 'Δ' in Eqn. 2.1 represents, effectively, a *measurable* change. What does this mean? Strictly speaking, the 'Δ' should be used with state functions only; it should not be used to represent changes in q or w, which are *path* functions. Given this, and referring to Fig. 2.5, suggest a definition of *path function*. Does it follow that q (or w) can *never* be considered a state function? Why or why not?
7. Show that the right hand sides of Eqns. 2.2 and 2.3 have the same dimensions.
8. We used Eqn. 2.2 to show that -1.96 kJ of work is done against gravity as a 50-kg woman climbs 4 m. Let the system be the woman. What is ΔU? Explain how energy is conserved.
9. How many joules are expended by a 70 kg man climbing up 6 m of stairway? Does this quantity represent a maximum or minimum energy expenditure? Why? How much work is done if the climbing takes place on the surface of the moon? (Assume that the acceleration due to gravity on the moon's surface is 1.6 m s^{-2}.)
10. How many meters of stairway could a 70 kg man climb if all the energy available in metabolizing an 11 g spoonful of sugar to carbon dioxide and water could be converted to work?

11. A cylinder of compressed gas has a cross-sectional area of 50 cm². How much work is done by the system as the gas expands, moving the piston 15 cm against an external pressure of 121 kPa?

12. Indicate whether the temperature increases, decreases or remains the same in the following four situations: an endothermic/exothermic process in an **adiabatic** (no heat exchange with surroundings)/non-adiabatic system.

13. A mathematical statement of the First Law of Thermodynamics is $\Delta U = q + w$. This holds for all processes. Assume that the only type of work done is pV-work. Show that $\Delta U = +w$ for an **adiabatic** process (no heat exchange with surroundings). Show that $\Delta U = 0$ for a process in an isolated system. Show that $\Delta U = q$ for a process that occurs at constant volume. Show that $\Delta H = 0$ for an adiabatic process at constant pressure.

14. When glucose is burned completely to carbon dioxide and water,

$$C_6H_{12}O_6 + 6O_2 \rightarrow 6CO_2 + 6H_2O$$

673 kcal are given off per mole of glucose oxidized at 25 °C. What is ΔU at this temperature? Why? What is ΔH at this temperature? Why? Suppose that glucose is fed to a culture of bacteria, and 400 kcal mol^{-1} of glucose is given off while the growing bacteria converted the glucose to CO_2 and H_2O. Why there is a discrepancy between the oxidation heats?

15. Conservation of energy is said to be implicit in the symmetrical relation of the laws of physics to time. Explain.

16. A person weighing 60 kg drinks 0.25 kg of water. The latter has a temperature of 62 °C. Assume that body tissues have a specific heat capacity of 0.8 kcal kg^{-1} K^{-1} (the specific heat of water is 1.0 kcal kg^{-1} K^{-1}). By how many degrees will the hot drink raise the person's body temperature from 37 °C? Explain how arriving at the answer involves the First Law of Thermodynamics.

17. Prove that Eqn. 2.17 follows from Eqn. 2.16.

18. Non-polar moieties in proteins make a positive contribution to $\Delta C_{p,d}$. This is known from measurements of the change in heat capacity of water on dissolution of non-polar compounds, e.g. cyclohexane. Is this true for polar moieties as well? What is the sign of their contribution to $\Delta C_{p,d}$? Explain your reasoning.

19. The earliest protein microcalorimetry studies were done by Peter Privalov, a Soviet biophysicist who emigrated to the United States in the early 1990s. One of the most thorough of all microcalorimetric studies of a protein is Privalov and Pfeil's work on hen egg white lysozyme, published in 1976. According to this work and later studies, $\Delta C_{p,d} = 1.5$ kcal mol^{-1}K^{-1}, and at pH 4.75, $\Delta H_d(25 °C) = 52$ kcal mol^{-1}. Calculate the enthalpy difference between the unfolded and folded states of lysozyme at (a) 78 °C, the transition temperature and (b) -10 °C. What is the physical meaning of ΔH in part (b)?

20. You have been asked to investigate the thermodynamic properties of a newly identified small globular protein by differential scanning calorimetry. The following results were obtained:

pH	T_m (°C)	$\Delta H_d(T_m)$ (kJ mol^{-1})
2.0	68.9	238
2.5	76.1	259
3.0	83.2	279
3.5	89.4	297
4.0	92.0	305
4.5	92.9	307
5.0	93.2	308
5.5	91.3	303
6.0	88.9	296
6.5	85.9	287
7.0	82.0	276
7.5	79.4	268
8.0	77.8	264

Plot $\Delta H_d(T_m)$ v. T_m. Describe the curve and rationalize its shape. Now plot $\Delta H_d(T_m)$ v. pH. What is happening?

21. ITC can be used to measure the enthalpy of protonation of amino acid side chains. Suppose three peptides were separately dissolved in weak phosphate buffer at pH 8 and injected into weak phosphate buffer at pH 2.5. There is a change in side chain ionization in going from one pH to the other. The peptides and the measured heats of reaction were Gly–Asp–Gly (-7.2 ± 0.8 μcal), Gly–Glu–Gly (-5.4 ± 0.8 μcal) and Gly–His–Gly (-5.5 ± 1.0 μcal). The data represent an average of 10 experimental data points, heat of injection minus background signal (injection of the pH 8 buffer into the pH 2 buffer in the absence of peptide). The peptide concentrations for the experiments were 0.64 mM, 0.57 mM and 0.080 mM, respectively. At pH 8, the side chains are approximately completely deprotonated, while at pH 2 they are approximately completely protonated. The solutions were injected into a sample cell in 10 μl aliquots. What is the physical basis of the background signal? What are the approximate protonation enthalpies of the Asp, Glu and His side chains? Suggest why tripeptides were used for these experiments rather than free amino acids. Would pentapeptides be any better? What could be done to account for the possible contribution to the measured effect of the terminal amino or carboxyl group?

22. Table C in Appendix C gives enthalpies of protonation for a number of popular biochemical buffers. Which five of these are likely to best for thermodynamic measurements? Why?

23. The conditions of the standard state are chosen arbitrarily. What additional condition(s) might a biochemist add to those given in the text? Why?

24. Explain in structural and thermodynamic terms how the unfolding of a protein is like the melting of an organic crystal.

25. A protein called α-lactalbumin is a close homolog of hen egg white lysozyme. Unlike lysozyme, α-lactalbumin binds Ca^{2+} with high

affinity. The measured enthalpy of binding, however, is much smaller in magnitude than the enthalpy of hydration. Explain.

26. Design a series of experiments to test whether the First Law of Thermodynamics applies to all living organisms.

27. Fig. 2.7 shows that the enthalpy change on protein unfolding is large and positive. Suggest what give rise to this.

28. Matter can neither be created nor destroyed, merely interconverted between forms. Discuss this statement in terms of the First Law of Thermodynamics.

29. Living organisms excrete excess nitrogen from the metabolism of amino acids in one of the following ways: ammonia, urea, or uric acid. Urea is synthesized in the liver by enzymes of the urea cycle, excreted into the bloodstream, and accumulated by the kidneys for excretion in urine. The urea cycle – the first known metabolic cycle – was elucidated in outline by Hans Krebs and Kurt Henseleit in 1932. As we shall see in Chapter 5, urea is a strong chemical denaturant that is used to study the structural stability or proteins. Solid urea combusts to liquid water and gaseous carbon dioxide and nitrogen according to the following reaction scheme:

$$CO(NH_2)_2(s) + 1.5O_2(g) \rightarrow CO_2(g) + N_2(g) + 2H_2O(l)$$

According to bomb calorimetry measurements, at 25 °C this reaction results in the release of 152.3 kcal mol^{-1}. Calculate ΔH for this reaction.

30. Giant sequoias, an indigenous species of California, are among the tallest trees known. Some individuals live to be 3500 years old. Water entering at the roots must be transported up some 300 m of xylem in order to nourish cells at the top of the tree. Calculate the work done against gravity in transporting a single water molecule this distance.

31. Suggest three proofs that although it has fluid-like properties, heat is not a fluid is the sense that liquid water is a fluid.

32. Describe energy expenditure in humans from a thermodynamic perspective.

For solutions, see http://chem.nich.edu/homework

Chapter 3

The Second Law of Thermodynamics

A. | Introduction

We have seen that a given amount of energy can be distributed in many different ways, just as a certain volume of fluid can adopt many different shapes and adapt itself to its container. In this chapter we shall learn about a thermodynamic function that enables us to *measure* how 'widely' a quantity of energy is distributed. The First Law of Thermodynamics relates heat, work and internal energy. It tells us that energy can neither be created nor destroyed, despite being able to change form; the total energy of a reaction, and indeed of the universe, is *constant*. The First Law tells us with breathtaking generality that a boundary on the possible is a basic characteristic of our universe.

It is not hard to see, though, that the First Law does not tell us other things we would like to know. For instance, if we put a hot system into contact with a cold one and allow them to come spontaneously to thermal equilibrium, we find that the final temperature of the two objects, which persists indefinitely if the objects are insulated, is at some intermediate value. ΔU, however, is 0. Similarly, if we mix a concentrated solution of substance A with a dilute solution of substance A, we find that the final concentration, which persists indefinitely, is between the initial concentrations (Fig. 3.1). Again, $\Delta U = 0$. We conclude that unless $\Delta U = 0$ implies spontaneity of reaction, the magnitude of ΔU probably does not indicate the direction of spontaneous change. But $\Delta U = 0$ for a system that undergoes no change at all! So we draw the much stronger conclusion that **in general ΔU does not indicate the direction of spontaneous change**! What about our two objects in thermal equilibrium? How could we get them to return from two warm objects to one that's hot and one that's cold? Could we get the solution at intermediate concentration to spontaneously unmix and return to the concentrated solution and the dilute one? No! For in both cases something has been lost, something has changed, and the change is **irreversible**. The First Law, useful as it is, does not provide even the slightest clue about what that something is. Nor does it answer any of the following important questions. In which direction will a reaction go spontaneously? Is

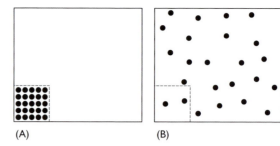

(A)   (B)

Fig. 3.1 Distribution of substance A before and after mixing. Panel (A) shows the situation before mixing, panel (B) when mixing is complete. Let substance A be perfume molecules on a heavily scented person who enters a small room, say an elevator (lift). The molecules are initially very close together; the concentration of perfume is high. In such high doses, it might be hard to distinguish perfume from insect repellent; visual inspection of the wearer might not help too much! After a while, maybe by the time you've reached the fifty-seventh storey, the perfume molecules are spread approximately randomly throughout the accessible space; the perfume is much more diffuse than at first; the concentration is uniform. This process is irreversible! The driving force for change is a movement of the system toward the most probable distribution of perfume molecules. Entropy, the key thermodynamic function of this chapter, measures the change in distribution. We can squeeze a few more drops of understanding from this figure. Imagine that the dots correspond not to molecules but to heat energy. Let the region where all the heat is collected in panel (A) be one object, and let the rest of the area be another object. The small object is hot, the big one is cold. Panel (B) shows the situation some time later, when the two objects have reached thermal equilibrium. (500 ml cans of Guinness draught say that the beer should be chilled at $4\,^{\circ}C$ for 3 h, giving the gas plenty of time to dissolve in the liquid. The head is not quite so creamy if the beer has been chilled for less time.) The heat energy has been redistributed throughout the matter, and both objects are at the same temperature. The driving force for change is a movement of the system toward the most probable distribution of heat energy. We should not fail to notice from a comparison of perfume and heat that there is something exceptionally similar about matter and energy. Indeed, this relationship lies at the heart of $E = mc^2$. (See Exercises 1.20 and 1.21.)

there a limit to how much work we can obtain from a reaction? If so, what is it? Why do highly purified enzymes degrade even when stored in the cold?

To be able to do more than just wave our hands in response, we must turn from the First to the Second Law of Thermodynamics. Like the First Law, the Second is an empirical result and a window on the relationship of heat to work. In less qualitative terms, the Second Law provides a way of describing the conversion of heat into work. It gives a precise definition of a thermodynamic state function called the entropy, and the sign of this function (plus or minus, not Leo or Libra!) tells us whether a process will occur spontaneously or not. This is something neither ΔU nor w nor q alone can do. After all, ice changes spontaneously into liquid water at $1\,^{\circ}C$ and 1 atm, despite the increase in translational motion of the water molecules (kinetic energy).

Our approach in this chapter is mainly that of 'classical' (pre-twentieth century) thermodynamics. But we should be careful not to prejudge the discussion and think that the age of the development makes it somehow less useful or illuminating. People who think that way are common, and few of them are good at thermodynamics. Besides, in learning a foreign language, one starts with relatively simple prose and not with poetry of the highest art! This chapter, though, is not an end in itself: it is the foundation of what comes next. In Chapters 4 and 5 we shall see how the entropy plays a role in the Gibbs free energy, the biochemist's favorite thermodynamic function. And in Chapter 6 we'll face up to the limitations of the classical approach and turn the spotlight on statistical thermodynamics. Then we shall see how the statistical behavior of a collection of identical particles underlies the classical concept of entropy and the other thermodynamic functions we'll have met by then.

We are all familiar with becoming aware of the cologne or aftershave someone near us is wearing (if not of the malodorous output of their underarm bacteria!). Movement of the nose-titillating particles through the air plays a key role in this. In some cases, for instance when we walk by someone who is wearing cologne, our motion relative to the other person is important in explaining the effect (not to mention the amazing powers of our olfactory apparatus, and how its membrane-bound receptor proteins connect up to brain-bound neurons). But the relative motion of one's neck and another's nose is not an essential ingredient here. For when you enter a relatively small and enclosed room in which a heavily scented person has been present for even just a few minutes, you can smell the perfume immediately.

How does that work from the point of view of thermodynamics? The sweet-smelling molecules in perfume are highly volatile aromatic amines. When heated by the body to 37 °C and exposed to the atmosphere, these molecules quickly become airborne. Then convection currents resulting from differences of temperature in patches of air help to spread the cologne about. Finally, and most important in the present discussion, constant bombardment of the aromatic amines by randomly moving gaseous nitrogen and oxygen also moves them around a good deal – even in the absence of convection currents – by a process called **diffusion**. After enough time, the concentration of perfume molecules will be *uniform* throughout the room. Amazing! And no machine is required to bring about this state of affairs! Experience tells us that this phenomenon is the *only* thing that will happen (as long as the situation is not basically altered by including an open window or something like that). That is, leaving a perfume-filled room and returning to find that all the aromatic amines had somehow gone back into the bottle like an Arabian genie seems extremely highly improbable. It *is* extremely highly improbable! As we shall see, diffusion is 'governed' by the Second Law of Thermodynamics.

In a nutshell, the Second Law is about the tendency of *particles* to go spontaneously from being concentrated to being spread out in space. It is also about the tendency of *energy* to go spontaneously from being 'concentrated' to being 'spread out.' Consider a mass in a gravitational field, for example a football that has been kicked high above the pitch. The driving force for change of the airborne ball is motion towards the most probable state, the state of lowest potential energy. In this analogy, that's just the ball at rest on the football field. Similarly, the tendency of concentrated particles to become more uniformly dispersed is a reflection of the tendency of (chemical) energy to disperse itself into its most probable distribution. We see this tendency in action whenever we add cold milk to a cup of coffee or tea. Initially, the milk is seen in distinct swirls, but before long the color and temperature of the liquid become uniform. The Second Law, which describes this process, is marvelously general: it applies not just to the mixing of milk and coffee but also (and equally well) to the movement of cologne molecules, the release of pent-up energy from a thundercloud during a lightning storm, the cooling of a hot sauce pan when removed from the stove, and the expansion of a gas into vacuum.

Some students find the Second Law of Thermodynamics difficult to grasp. One reason is that there are numerous equivalent formulations of it, each of which sheds new light on the topic and enables one to understand it in a different way. Let's look at one of the earliest ways of stating the Second Law from the historical point of view, one that is of particular interest here because it helps to reveal the practical nature of the human activity out of which the science of thermodynamics developed: **it is *impossible* for a system to turn a given amount of heat into an equivalent amount of work**. In other words, if we put q into the system, whatever work comes out will be $w < q$. This comes to us from the work of the visionary French military engineer Nicolas Léonard Sadi Carnot (1796–1832). Publication of Carnot's *Réflexions sur la puissance motrice du feu et les machines propre à développer cette puissance*[1] (at the age of 28) outlined a theory of the steam engine and inaugurated the science of thermodynamics. We shall encounter M. Carnot again shortly.

B. | Entropy

The foregoing discussion brings us to the thermodynamic state function S, **entropy** (Greek, *en*, in + *trope*, transforming; also coined by Clausius). Being a state function, the change in entropy for a process is independent of path, regardless of whether the change takes place reversibly or irreversibly. The entropy is an index of the tendency of a system toward spontaneous change; it is a measure of the state of differentiation or distribution of the energy of system; it is the key to understanding energy transformation. As such, the entropy function enables us to rationalize why solutes diffuse from a concentrated solution to a dilute one without exception, why smoke leaves a burning log and never returns, why wind-up clocks always run down, why heat always flows from a hot body to a cold one. All this might suggest to us that entropy is something physical, as indeed many people have believed. It is important to realize, however, that the **entropy is not so much a thing as a highly useful mathematical object that provides insight to the nature of change in the material world**.

---

[1] *Report on the driving force of heat and the proper machines to develop this power.*

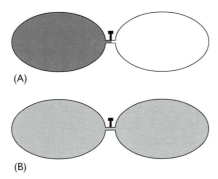

(A)

(B)

Fig. 3.2 The irreversible expansion of a gas. The bulbs are of equal volume. In panel (A), when the stopcock is closed, the gas is concentrated and confined to the left bulb. The right bulb is completely evacuated. After the opening of the stopcock, as shown in panel (B), the gas flows rapidly from the bulb on the left to the bulb on the right. After a sufficiently long time the condition known as equilibrium is reached; the concentration of gas is the same in both bulbs and the net flow of molecules between the bulbs is zero. This is just another way of thinking about the perfume on the heavily scented person we met in the elevator.

As we have said, entropy is a measure of the order of a system. Throughout this chapter *order* and *disorder* are somewhat arbitrarily defined, but the meaning will usually be clear enough. When we reach Chapter 6, more precise definitions will be given. Entropy is less a 'thing' than a way of describing how things are *arranged* in space (e.g. perfume in the bottle or distributed throughout the room) and how particle arrangement changes as a system is subjected to changes of temperature, pressure, number of particles, volume, etc. Entropy measures how close a system is to the state corresponding to no further change, or equilibrium. The tendency toward a condition of *no further change* that we have seen in the examples above is so basic to all of physics that **most scientists consider the Second Law of Thermodynamics the most universal 'governor' of natural activity that has ever been revealed by scientific study.**

To avoid drifting too far into the ethereal world of abstractions, let's come back to earth by way of another illustration, one that is closely related to the phenomenon of diffusion. Suppose we have two glass bulbs connected by a stopcock, as shown in Fig. 3.2. Initially, the stopcock is closed. A gas occupies one bulb only; the other bulb is evacuated. When the stopcock is opened, the gas molecules zoom into the evacuated bulb. There is a net flow of molecules into the formerly empty bulb until the concentration of molecules is the identical (on the average) throughout the accessible volume. Such expansion of a gas is accompanied by an *irreversible* increase in entropy. The process is irreversible because a substantial amount of *work* would have to be done to corral all the gas molecules back into one bulb. The state in which all the molecules are distributed at random throughout the volume of the two bulbs – state 2 at equilibrium – is less ordered than the state in which all the molecules were randomly distributed throughout the volume of one bulb – state 1 at equilibrium. In this context, *equilibrium* just means a state of no further change. Similarly, perfume is more ordered when it's still in the bottle than when it's been applied and scattered to the right corners of an elevator.

In Fig. 2.4C we saw how heat can be used to make a rubber band contract and lift a weight; heat can be used to do work. Now let's see what insights can be gained by shrinking several orders of magnitude in size and examining the goings on inside our rubber band machine. Rubber

Fig. 3.3 Schematic diagram of a rubber band in the unstretched (equilibrium) state (left) and stretched state (right). In the unstretched state, the molecules are highly disordered. When the band is stretched, the molecules form a relatively orderly array. The entropy of the molecules is lower in the stretched state than in the unstretched state.

consists of long, chainlike molecules (Fig. 3.3). Stretching makes them align and become more orderly, decreasing the number of different spatial arrangements. Like forcing all the gas molecules back into one bulb, stretching a rubber band requires *work* to be done *on* the system, and $w > 0$. When you use your fingers, hands and arms to do the stretching, the energy comes from the cleavage of chemical bonds in your muscles. The contracting muscle is then used to stretch the rubber beyond its relaxed, 'equilibrium' position, and there is a change in mechanical energy.

Now, how do we explain work being done by the *contraction* of rubber? Just as heat decreases the order of perfume molecules by spreading them throughout a volume, the addition of heat decreases the order of the rubber molecules and the stretched rubber band contracts to its equilibrium position. A more subtle point here is that similar to muscle, contraction of a rubber band permits work to be done. In this case, however, work is not done *on* the system, it is done *by* the system. We can *calculate* the magnitude of this work knowing gravitational acceleration, the mass of the object lifted by the rubber band, and the distance the mass is moved against the force of gravity (Δx in Fig. 2.4C). By Newton's Second Law $\Delta \mathbf{p}/\Delta t = F = ma = mg = -w/\Delta x$, where m is mass, a is acceleration and g is the gravitational acceleration.

Stretching increases the order of the long molecules. In the stretched state, the molecules make fewer random collisions with each other; the entropy of the system is reduced relative to the equilibrium, unstretched state. When the rubber band returns to being unstretched, the orderliness of the molecules is lost and the entropy of the system increases. The closer the molecules are to being randomly ordered, the greater the entropy of the system. So, in view of all that we have said about rubber bands, we can see that changes in entropy are related to transfers of heat.

Here's another illustration of the relationship of entropy to heat. When a scented candle burns, there is an increase in the thermal energy of the aromatic molecules embedded in the wax. The greater the thermal energy of a light aromatic molecule, the easier it is can overcome the attractive forces of other molecules in the melted wax, break forth from the surface, escape into the air, and collide with a receptor molecule in the cavernous recesses of your nasal passage. (What happens after that is very complicated – not to mention far from being fully understood – and we shall not able to go into it here.) The point is that the addition of heat increases the entropy of the system, the 'breadth' of the distribution of the energy of the system. Heat is a form of energy related to random molecular motion, as discussed in Chapter

2. This example might not appear to have all that much to do with biology, but in fact it does. For the plant *Arum maculatum* (both Lords and Ladies) attracts pollinating insects to its flowers by heating and vaporizing aromatic molecules. At certain times of the day, the temperature of a part of the flower called the appendix increases rapidly, resulting in the evaporation of volatile compounds produced within the flower. Insects detect the aromatic compounds, find them stimulating, and move in the direction of increasing concentration until they reach the flower. Insect movement within the flower leads to pollination.

Let's have another quick look at Fig. 2.4. Panel (A) shows heat being transferred from a hot system to a cold one. This is similar to panel (C), where heat is being transferred from the candle to the rubber band. Back in panel (A), kinetic energy is being shared as heat by means of multiple collisions between the molecules of the heat source and the molecules of the heat sink. A relatively warm object is in contact with a relatively cool one, and the warm molecules are banging on the cool ones harder than the cool ones are banging on the warm ones (or on each other!). Heat transfer from the warm object to the cool one enables work to be done, as in the rubber band heat engine, and this situation continues *as long as there is a temperature difference between the two systems*.

When a thing is cold it does not move very much. Kinetic energy is proportional to thermal energy, and this is proportional to absolute temperature. So addition of heat to something cold can disorder that thing more than the addition of the same amount of heat to the same thing at a higher temperature. In view of this, we can guess that if a given amount of heat q is transferred to a system, it will increase the randomness of the system by an amount that is inversely proportional to the absolute temperature. Because entropy is measure of the breadth of the distribution of energy, we should expect $\Delta S \propto q/T$, though at this point we don't know if it's the first power of the temperature that's needed or something else.

Now we are ready to become quantitative about heat and entropy. The entropy change, ΔS, on heat transfer at *absolute* temperature T, is *defined* as:

$$\Delta S \geq q/T \qquad\qquad (3.1)$$

where, by the convention adopted here, $q > 0$ if heat is added *to* the system. This comes to us from work on quantifying the maximum amount of work that can be obtained from an ideal reversible engine. The equality holds only under a rather extraordinary but none the less important constraint: when heat transfer is carried out slowly and any change in the system is **reversible**, i.e. when both the system *and* its surroundings can be returned to their *original* states. **A reversible process is one that occurs through a succession of equilibrium or near-equilibrium states**. The inequality corresponds to any irreversible process. No wonder the expansion of a gas into vacuum is irreversible! Eqn. 3.1 tells us that although heat is a path function, the *heat* exchanged in a reversible *and* isothermal process is independent of path.

Eqn. 3.1 also tells us that when a quantity of heat is transferred from a hot system to a cold one, $\Delta S_{hot} < 0$, $\Delta S_{cold} > 0$, $|\Delta S_{hot}| < |\Delta S_{cold}|$ and $\Delta S_{total} = \Delta S_{hot} + \Delta S_{cold} > 0$. Regardless of the magnitudes of q, T_{hot}, and T_{cold}, the *total* entropy change *must* be greater than zero. To make this more concrete, let's look at an example. Suppose we wish to calculate the entropy change in the surroundings when 1.00 mol of $H_2O(l)$ is formed from H_2 and O_2 at 1 bar and 298 K. This is of interest to us here because liquid water is the only known 'matrix' in which life occurs. We require the reaction to occur slowly. A table of standard thermodynamic quantities tells us that $\Delta H = -286$ kJ; the reaction is exothermic. Heat is transferred from the system to the surroundings, and $q_{sur} = +286$ kJ. Substituting this into Eqn. 3.1 and solving for ΔS_{sur}, we obtain 286 kJ/298 K $= +959$ J K^{-1}. The entropy of the surroundings increases as heat is transferred *to* it. To put the numbers into perspective, one mole of water has a mass of 18 g and a volume of 18 ml, about the same as a large sip. Formation of a mole of water from hydrogen and oxygen at room temperature and ambient pressure increases the entropy of the universe by ~ 1000 J K^{-1}.

C. | Heat engines

In this section we encounter some reasoning that is important because it shows how living organisms do *not* behave. You might guess therefore that this will be the usual sort of useless, academic exercise that one is often required to do to earn a degree. But in fact an important general principle is that it can be very helpful in trying to understand what something is, to seek to know *why* it is not what it is not. That's the spirit in which we discuss heat engines.

Let's suppose, as Carnot did, that heat q is transferred from a heat source to a heat sink. How much of this heat is available to do work? (See Fig. 2.4A.) No more than can be done without violating the Second Law, and usually a lot less than that! To find the limit, we require that

$$\Delta S_{hot} + \Delta S_{cold} = 0 \tag{3.2}$$

Plugging in Eqn. 3.1, and calling the cold sink the place where the waste heat goes, we have

$$-q_{transferred}/T_{hot} + q_{waste}/T_{cold} = 0 \tag{3.3}$$

Rearranging gives

$$q_{waste} = q_{transferred} T_{cold}/T_{hot} \tag{3.4}$$

In this equation q_{waste} is the *minimum* amount of heat transferred to the cold system; this heat *cannot* be used to do work. If we were designing a heat engine, we would want to make q_{waste} as small as possible, probably by making T_{cold} as small as possible and T_{hot} as large as possible. The *maximum* work one can do is to use whatever heat is left over, and that is $q_{transferred}$ less q_{waste}:

$$w_{max} = q_{transferred} - q_{transferred}T_{cold}/T_{hot} = q_{transferred}(1 - T_{cold}/T_{hot}) \qquad (3.5)$$

A simple numerical example is the following. If 30 J is transferred from a heat source at 300 K to a heat sink at 200 K, the maximum work that can be done is 30 J×[1−(200 K/300 K)= 10 J. The **efficiency** of this process = $w_{max}/q_{transferred}$ = 10/30 = 33%. One concludes from this that an engine in which *all* the heat was converted to mechanical work *cannot* exist, a view that seems suspiciously like another limit on the possible in our universe. Indeed, it is yet another way of stating the Second Law, one due to Carnot himself: **heat of itself cannot pass from a colder body to a hotter one; work is required**. But what does this have to do with biology? It helps us to realize that **cells cannot do work by heat transfer because they are isothermal systems** (Fig. 2.4B). This applies not only to terrestrial mammals like armadillos, but also to the hyperthermophilic bacteria that live on the ocean floor in thermal vents, and probably to any living thing anywhere in the universe.

We have seen that the transfer of heat can be used to do work but that this process generates waste heat, q_{waste}. By Eqn. 3.1, $S_{irreversible}$, the *minimum* irreversible entropy produced by heat transfer, is

$$S_{irreversible} = q_{waste}/T_{cold} \qquad (3.6)$$

Does this mean that irreversible entropy increases are always a write-off? No! To see how, let's pay another visit to our friend the rubber band. In the stretched state, it has a relatively low entropy, as the long rubber molecules are ordered. Release of tension results in decreased ordering of the molecules, for which process $\Delta S > 0$. We should therefore expect $q > 0$ on release of tension, and this is easily verified by experiment (test it yourself!). The heat is lost to the environment (surroundings). This heat is not completely useless because the contraction of the rubber could be used to do something constructive like lift a weight, as in Fig. 2.4C. So we see that an irreversible increase in entropy can be used to do work.

Having covered the necessary background, let's look at a biological example of an irreversible increase in entropy being used to work. Grasshoppers (and crickets and fleas and other such organisms) store elastic energy in the compressed form of a protein called resilin, from *resilient*. This is something like a compressed spring, for instance the spring in a loaded jack-in-the-box. When the insect jumps (when the lid of the jack-in-the-box is opened), elastic energy is released and the resilin becomes much less ordered. ΔS for this process is large and positive. This form of energy release is just about as fast as the transmission of a nerve impulse, and much faster than a typical metabolic reaction, enabling the grasshopper to make tracks if it senses danger from a predator. You will certainly know something about this if you have ever tried to catch a grasshopper or cricket in the prime of its life!

Now, before getting too far out of equation mode, we ask: which is greater in magnitude, $q_{transferred}$ or q_{waste}? Or suppose we have a process that can be carried out either reversibly or irreversibly. For which process will q be larger? Combining Eqn. 3.1 with the First Law, we obtain

$$\Delta U \leq T\Delta S + w \tag{3.7}$$

or, upon rearrangement,

$$w \geq \Delta U - T\Delta S \tag{3.8}$$

The most negative value of w that this expression can yield, and therefore the greatest amount of work that can be done *by* the system, is

$$w_{max} = \Delta U - T\Delta S \tag{3.9}$$

That is, the maximum work is obtained when the process is carried out reversibly. (Note that if $w_{max} = 0$ and $\Delta U = 0$, then $\Delta S = 0$ at any T.) By the First Law,

$$\Delta U_{rev} = q_{rev} + w_{rev} \tag{3.10}$$

for a reversible process, and

$$\Delta U_{irrev} = q_{irrev} + w_{irrev} \tag{3.11}$$

for an irreversible one. But if the starting and ending points of the processes are the same, then $\Delta U_{rev} = \Delta U_{irrev} = \Delta U$. If work is done by the system on the surroundings, then the sign of w is negative, and Eqns. 3.10 and 3.11 are, respectively,

$$\Delta U = q_{rev} - w_{rev} \tag{3.12}$$

and

$$\Delta U = q_{irrev} - w_{irrev} \tag{3.13}$$

Combining these equations, which we can do because the change of state is identical in the two cases, gives

$$\Delta U = q_{rev} - w_{rev} = q_{irrev} - w_{irrev} \tag{3.14}$$

or, upon rearrangement,

$$q_{rev} - q_{irrev} = w_{rev} - w_{irrev} \tag{3.15}$$

Above we found that $w_{rev} \geq w_{irrev}$, so both sides of Eqn. 3.15 must be positive. This implies that

$$q_{rev} \geq q_{irrev} \tag{3.16}$$

and we have answered the question we set ourselves. What does this result mean? In an endothermic process, the heat extracted from the surroundings will always be greatest when the process is reversible. In an exothermic process, the heat released to the surroundings will always be smallest when the process is reversible. So, living organisms would release the least amount of energy as heat if the processes going on inside them were reversible. But those processes are on the whole not reversible, so we release a lot of heat to the surroundings! To keep from cooling down to the temperature of the surroundings, we need energy, and that comes from food. Biological thermodynamics.

| Table 3.1 | Comparison of the 'orderliness' of different types of energy. Note how the entropy of a given amount of energy increases as it is transformed from a nuclear reaction to the heat given off by biological organisms on the surface of Earth |

Form of energy	Entropy per unit energy
Nuclear reactions	10^{-6}
Internal heat of stars	10^{-3}
Sunlight	1
Chemical reactions	$1-10$
Terrestrial waste heat	$10-100$

D. Entropy of the universe

As we have seen, **the total entropy of an isolated system increases in the course of a spontaneous change.** Put another way, the Second Law says that **no natural process can occur unless it is accompanied by an increase in the entropy of the universe** (Fig. 3.4 and Table 3.1). The meaning of this is that ultimately every process that occurs in nature is irreversible and unidirectional, with the direction being dictated by the requirement of an overall increase in entropy. This can be symbolized in a compact mathematical form as $\Delta S_{total} = \Delta S_{hot} + \Delta S_{cold} > 0$. Rewriting this in a more general way, we have:

$$\Delta S_{system} + \Delta S_{surroundings} = \Delta S_{universe} > 0 \qquad (3.17)$$

In order for a physical change to occur spontaneously, the entropy of the universe must increase. Looking back at the earlier pages of this chapter, ΔS in Eqn. 3.1 is seen to be $\Delta S_{universe}$, and in the previous two sections the entire universe consisted of just a heat source and a heat sink, one being the system and the other the surroundings.

It is important to realize that Eqn. 3.17 does *not* say that entropically 'unfavorable' reactions (those for which the entropy change is negative) cannot occur; they can and do occur, albeit *not spontaneously*. When an entropically unfavorable process is made to occur, the *overall* change in the entropy of the universe *must* be greater than zero, according to the Second Law of Thermodynamics. What if we are working with a complicated thing like an amoeba or a nematode or an ant and want to *measure* its entropy production? Can we do that and decide whether the change occurred spontaneously? No! Because the sign of ΔS for a system indicates whether a reaction will proceed spontaneously *only if* the system is isolated from its surroundings or both the entropy change of the system *and* the surroundings have been measured. These criteria are *not* met very easily. In fact, they cannot be met at all. That's because $q_p = T\Delta S$ only if a reaction is reversible, and the sum total of biochemical reactions within an organism is definitely not reversible. How would one go about measuring the entropy change of the surroundings, which is the entire rest of the universe? As we shall see in the next chapter, we can

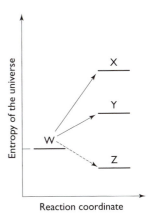

Entropy of the universe / Reaction coordinate

Fig. 3.4 The Second Law of Thermodynamics. No process will occur spontaneously unless it is accompanied by an increase in the entropy of the universe. This applies to an isolated system, a closed system, and an open system.

circumvent difficulties posed by measuring entropy and determine whether the change in a system occurred spontaneously. We'll do that by employing a less general but in many respects more useful index of reaction spontaneity, the Gibbs free energy.

E. Isothermal systems

Now we wish to look more closely at the specialized situation of constant temperature. Isothermal conditions are of great importance to the biochemist, because living organisms maintain body temperature within a fairly narrow temperature range and it obviously makes good sense for bench-top experiments to be done at constant temperature. In measuring enzyme activity, say, one would not want to report that the temperature at which the reaction took place fluctuated over a range of 7 degrees during the experiment!

Human beings and many other organisms cannot tolerate changes in body temperature of more than a few degrees. We have terribly sophisticated negative feedback systems that, when functioning normally, keep everything at about the same temperature. Suppose we have a system, a cell deep within your body. And suppose this cell undergoes a reversible change at constant pressure that results in the transfer of heat q to the surroundings (the other cells of the tissue). Then

$$\Delta S_{surroundings} = -q/T \tag{3.18}$$

where $T_{surroundings}$ is so close to T_{system} that we write both temperatures as T (the temperature difference is too small to be measured). Because $\Delta T \approx 0$, the pV-work that can be done from the heat transfer is practically negligible and $q = \Delta H$ (Chapter 2). Substituting into Eqns. 3.17 and 3.18 gives

$$\Delta S_{system} - \Delta H/T > 0 \tag{3.19}$$

which, after rearrangement, can be written

$$\Delta H - T\Delta S_{system} < 0 \tag{3.20}$$

$H - TS$ is a thermodynamic state function called the Gibbs free energy, and as we can see here it enables us to predict the direction of spontaneous change of a system under the constraints of constant temperature and constant pressure. ΔH measures the heat exchanged at constant pressure. The quantity $T\Delta S$ can be thought of as 'isothermally unavailable' energy. Transfer of this energy out of the system would result in a decrease in the temperature of the system, but we have stipulated that the temperature of the system is constant.

We can approach the idea of a portion of the heat released by a reaction not being available for work from another direction. First, let's convince ourselves again that some of the energy cannot be used to do work. Suppose we have a system which undergoes a process that results in a decrease in the entropy of the system. One such system is liquid water that comes to equilibrium in the solid state at $-1\,^{\circ}\text{C}$. If all the heat released by this process, about 6 kcal mol^{-1}, were exported to the surroundings as *work*, there would be *no* increase in the entropy of the surroundings, and the overall entropy for the process would be negative. This, however, contradicts Eqn. 3.17, the requirement of the Second Law that any real process should result in an increase in the entropy of the universe. So, we conclude that at least some of the heat generated by the process must be unavailable to do work. But how much?

We can answer the question by continuing with our simple qualitative discussion. Let's suppose that the reaction reduces the entropy of the system by some amount ΔS. In order for the reaction to be spontaneous, the entropy of the surroundings must increase by at least ΔS. According to Eqn. 3.6, this entropy is just q_{waste}/T. Solving for the heat supplied, $q_{waste} = T\Delta S$, where the magnitude of ΔS is the amount of decrease in the entropy of the system. This is the energy that *is not* available to do work, as we saw above. **The energy that *is* available to do work must be the difference between the total energy and the energy that is not available to do work.** This energy is called the Gibbs free energy. Note the resemblance of this qualitative example to Eqn. 3.20. The Gibbs free energy is so important to the biological sciences that the next two chapters will be devoted to it, and Chapters 6 and 7 will build directly on it.

Above we combined the First and Second Laws to arrive at the Gibbs free energy. Now we wish to combine these laws again, but in a slightly different way. The result will be different and provide new insight into the expansion of a gas into vacuum that was discussed above and prepare the way for a discussion of osmosis in Chapter 5. Moreover, it will provide a good starting point for the chapter on statistical thermodynamics, as we shall see when we get to Chapter 6. Let's require the system to be adiabatic, so that no energy can enter or leave. $\Delta U = 0$. By the First Law, $q = -w$. If we require that any work done be pV-work, then $q = p\Delta V$; and if the system is an ideal gas, then $q = nRT\Delta V/V$. Suppose now that a process we'd like to carry out is reversible. Then, by the Second Law, $q = T\Delta S$. Combining the First Law and the Second Law gives

$$T\Delta S = nRT\Delta V/V \tag{3.21}$$

The Ts cancel out, and we are left with

$$\Delta S = nR\Delta V/V \tag{3.22}$$

This is the increment in entropy when an ideal gas is allowed to expand by a small but measurable volume ΔV. If the expansion is carried out reversibly over a large change in volume, the total change in entropy is the sum of all the small changes, and the result, which can be found using a little calculus, is

$$\Delta S = nR\ln(V_f/V_i) \tag{3.23}$$

where V_f is the final volume and V_i is the initial volume. Because the entropy is a state function, as long as the initial and final states are the same, the entropy difference between states is independent of how the change occurred. Therefore, the entropy change on opening the stopcock in the example shown in Fig. 3.2 can be found using Eqn. 3.22b. It is $\Delta S = nR\ln(2V_i/V_i) = nR\ln 2$.

F. | Protein denaturation

Let's see how Eqn. 3.18 (or Eqn. 3.1) can be used to describe the reversible isothermal entropy change of any system we like. Here we apply it to protein denaturation, but it describes equally well the 'melting' of DNA, i.e. the dissociation of the double-stranded helix into two single strands. In Chapter 4 we shall see how it is yet even more general.

Rewriting Eqn. 3.18 using symbols introduced in the previous chapter, we have

$$\Delta S_d = \Delta H_d/T_m \tag{3.24}$$

The minus sign has vanished because heat is being transferred *to* the system and we are describing the entropy change of the protein system not of the surroundings. Suppose T_m is 342 K and we wish to know ΔS_d at 25 °C. What can we do? If the heat transfer is carried out reversibly, according to Eqns. 3.1 and 3.18, $\Delta S = -q/T$. The heat transferred, q, will increase the enthalpy of the system according to Eqn. 2.10 and Eqn. 2.16, if the system is at constant pressure. Combining these equations, we obtain

$$\Delta S = C_p \Delta T/T \tag{3.25}$$

If we now sum up all the small contributions to find the entropy change over a measurable range of temperatures $(T_1 - T_2)$ and use a bit of mathematical wizardry (the same sort used in the previous section), the result is

$$\Delta S(T_2) = \Delta S(T_1) + C_p\ln(T_2/T_1) \tag{3.26}$$

where '$\Delta S(T_i)$' means 'the change in entropy evaluated at temperature T_i,' not $\Delta S \times T_i$, and we have assumed that C_p is constant throughout the temperature range. It can be shown that Eqn. 3.26 becomes

$$\Delta S(T_2) = \Delta S(T_1) + \Delta C_p\ln(T_2/T_1) \tag{3.27}$$

if states 1 and state 2 differ in heat capacity and ΔC_p is constant throughout the relevant temperature range. As an example, suppose that $\Delta S = 354$ cal mol^{-1} K^{-1} at 80 °C and $\Delta C_p = 1500$ cal mol^{-1} K^{-1}. Then $\Delta S(25 °C) = 354$ cal mol^{-1} K$^{-1} + (1500$ cal mol^{-1} K$^{-1}) \times \ln(298.16$ K/353.16 K$) = 100$ cal mol^{-1} K^{-1}.

There is another way of thinking about proteins and entropy, one that does not involve a large change in heat capacity. As the pH decreases, acidic amino acid side chains become protonated. From an exercise in Chapter 2, the protonation enthalpy change is about

1 kcal mol^{-1}. This is so small as to be negligible in comparison with the enthalpy change of unfolding the protein in the absence of protonation effects. Changes in pH can nevertheless have a dramatic effect on protein stability; we have already seen how lowering the pH reduces the transition temperature of hen egg white lysozyme. We surmise that protonation of Glu and Asp is mainly an *entropic* effect, not only with regard to the binding of the proton to the amino acid side chain but also with regard to the effect on protein unfolding. The T_m of the protein decreases upon reduction of pH because the entropy difference between the folded and unfolded states decreases at a faster rate than the enthalpy difference, making $\Delta H_d / \Delta S_d$ progressively smaller. The ionization of food molecules in the low-pH environment of the gut denatures proteins, facilitating their breakdown into short peptides by digestive proteases. Short peptides and amino acids are absorbed in the GI tract and utilized as a source of energy to power biochemical reactions within the body.

How do we *interpret* pH denaturation of proteins, explain it on a more detailed level? As the pH goes down there is a change in the ionization state of the acidic side chains. This results in a net increase in the charge on the surface of the protein. So at low pH, the positive charges repel each other by electrostatic interactions more than at high pH, and this destabilizes the folded conformation. The situation can be represented as follows:

$$P_{folded} + nH^+ \rightarrow P^* \rightarrow P_{unfolded} \tag{3.28}$$

where P* is an unstable folded conformation. The effect of charge on protein stability was first described mathematically in 1924 by the Danish physical biochemist Kaj Ulrik Linderstrøm-Lang (1896–1959). The earliest known experiments on the use of a strong acid to denature proteins were done about a century earlier by a Dutch chemist named Gerardus Johannes Mulder (1802–1880), who is also said to have stimulated the eminent Swedish chemist Jöns Jacob Berzelius (1779–1848) to coin the word *protein* (Greek, of the highest importance). Linderstrøm-Lang's mentor at the Carlsberg Laboratory, Copenhagen, Denmark, was the Danish biochemist Søren Peter Lauritz Sørensen (1868–1939). The latter developed the now universally adopted pH scale for measuring the acidity of an aqueous solution and ushered in the modern era of protein chemistry. We shall meet Linderstrøm-Lang again in Chapters 6 and 8.

G. | The Third Law and biology

Yet another way of stating the Second Law brings us to the Third Law. **Any system not at absolute zero has some minimum amount of energy that is a necessary property of that system at that temperature.** This energy, which has magnitude TS, is the 'isothermally unavailable' energy from above (Section E). Now, the Third Law of Thermodynamics states that **the entropy of a *perfect* crystal is zero when the absolute temperature is zero.** A perfect crystal is one with no imperfections, like the ideal diamond in which each atom is at its proper place in an orderly array. The reason why we care about this is that the Third Law implies that the rapid and complex changes exhibited by living organisms, for instance in the migration of a eukaryotic cell, can only occur far from thermodynamic equilibrium.

There is an extensive literature on studies of living organisms that were frozen, taken to an extremely low temperature (e.g. 4 K), and allowed to thaw again. Remarkably, relatively 'simple' organisms like bacteria, plant seeds and some types of microscopic animals can return to room temperature from the deadly cold and function normally. Even nematodes, which are complex eukaryotes and have on the order of 10^3 cells at adulthood, can survive this process.

As discussed in Chapter 2, temperature measures the average kinetic energy of a collection of molecules. So, when the temperature is made to approach 0 K, all molecular motion ceases, except that required by the Heisenberg uncertainty principle; named after the German mathematical physicist and philosopher Werner Karl Heisenberg (1901–1976), son of a professor of ancient history and recipient of the Nobel Prize in Physics in 1932). Thus, near absolute zero, the only 'memory' a biological system has of its life-before-deep-freeze is the *information* contained in the structure and arrangement of the macromolecules of which it is composed. When the organism is allowed to thaw, no new information is added; in fact, information is removed, because heating is a disordering process, as we have seen throughout this chapter.

This would suggest that 'all' one would have to do to 'create' a living organism would be to construct an appropriate configuration of atoms. If the configuration (structure) were 'the right type,' the cell would function 'on its own.' From this point of view, it seems that a cell of an organism or indeed an entire organism is not qualitatively different from any other collection of organic molecules. On another view, a cell is simultaneously both more highly organized and more complex than any other collection of organic molecules of the same mass that we can imagine!

H. | Irreversibility and life

In the first chapters of this book we have looked at ways in which living organisms can be thought of as machines. (Note: this does not mean that living organisms are machines!) Back in Chapter 1 we discussed energy 'consumption' as though such machines really do 'consume' energy. Now that we've covered the First and Second Laws of Thermodynamics in reasonable depth, we wish to take a more critical look at energy consumption. If food is potential energy, and we consume food, then we consume energy; a logically sound argument, as long as we agree on what is meant by *consume*. There is, however, another way of looking at energy consumption, and that is what we want to do now. While we do of course consume food, that should not be understood to mean that we consume energy. For all a living organism or any type of system whatsoever can do is transform energy from one form to another; the total energy of the universe remains the same throughout such transformations, by the First Law of Thermodynamics. The amount of energy returned to environment, for instance as excretory products or as heat, is equivalent in magnitude to the energy taken in, assuming no change in the weight of the organism. In this sense, living things do not consume energy at all; energy simply flows through them.

More importantly, the energy an organism returns to the environment *must* be in a less *useful* form than the form in which it was 'consumed.' Sure, excretory products make great fertilizer, but there are many good reasons why higher eukaryotes would not want to feed on them! As we have seen, any real process *must* increase the entropy of the universe; any change in the universe must result in an overall decrease in order. As we shall see in the next chapter, biologically useful energy, or free energy, is the energy that can be used to do work under isothermal conditions, and that is exactly what humans, goats and sea slugs need to satisfy the energy requirements of life. As the example of the heat engine shows, heat transfer cannot be used to perform a substantial amount of work in biological systems, because all parts of a cell are effectively at the same temperature (and pressure). We have therefore eliminated a major class of ways in which cells could conceivably do work, at least within the constraints of the physical properties of our universe, and narrowed the path towards understanding. Whew! In playing the game of 20 questions, the ideal strategy is to pose a query whose yes-or-no answer will eliminate the greatest number of possible answers and enable you to home in on the correct one. The direction of spontaneous change in an isothermal system from a non-equilibrium state to an equilibrium one is determined by the requirement that the extent of change be a maximum at every point on the reaction pathway.

The total energy of the universe is a constant and the universe always goes from 'order' to 'disorder.' The Second Law of Thermodynamics tells us that the disorder of the universe increases in any real process, and there is no known exception to the Second Law. We have seen an example of the increase of entropy in the spontaneous increase in the breadth and uniformity of the distribution of perfume. The prediction of the scientific theory is in excellent agreement with our ordinary experience of the world. Indeed, the suggestion that, say, uniformly dispersed smoke particles could somehow move spontaneously from all corners of a room back into a burning cigar would seem absurd (except in a videotape run backwards).

There are various causes of the irreversibility of real-world processes. These include friction between two objects during relative motion, unrestrained expansion of a gas or liquid without production of work, the mixing of different substances that would require the input of work to separate them – all very common phenomena. Because all atoms interact with each other, even noble gases, it would appear that there must be at least a small amount of irreversibility in any actual process. What we're driving at here in a rather roundabout way is that the inexorable increase in the entropy of the universe that attends any real process is very similar in character to the unidirectional flow of time. For as far as anyone knows, time moves in one direction only: forward!

Why should this be so remarkable? Time moves on and the past is, well, the past. This apparent conjunction of an interpretation of a scientific theory (the Second Law of thermodynamics) our ordinary (psychological?) perception of time is none the less intriguing, because *all* organisms come into being and pass out of being *in time* and *all* the fundamental laws of physics are *time-reversible*. (The decay of kaons and other sub-nuclear particles violate time symmetry; these particles appear to possess an intrinsic 'sense' of past-future. See Christenson *et al.* (1964) *Phys. Rev. Lett.* **13**:138.) Newton's laws of motion work equally well in either direction. Maxwell's equations of electromagnetism work equally well forwards and backwards. The time-dependent Schrödinger equation of quantum theory works equally well whether time is positive or negative. (The Austrian physicist Erwin Schrödinger (1887–1961) was awarded the Nobel Prize in Physics in 1933.) Einstein's theory of relativity works equally whether time moves forward or backward. The time-reversibility or time-symmetry of the laws of physics is related to the conservation of energy (the First Law).

Because the widely accepted mathematical formulations of physical law help us to rationalize many aspects of the nature of the universe and, moreover, provide tools for the creation of technology, we cannot but be convinced that physics gives us at least an approximately right sense of the nature of reality. Nevertheless, and regardless of one's familiarity with physics, time marches on. The only law of physics that jibes with this aspect of our everyday experience of the world is the Second Law of Thermodynamics. This is all the more noteworthy in this book because life on Earth has grown increasingly complex since the advent of the first cell: humans, which are composed of billions of cells, are good deal more complex than one-celled beasts like bacteria. We'll come back to this point in Chapter 9.

One can think about the irreversibility of chemical processes and life on a number of different levels. Just as the increase in complexity of life forms on Earth is irreversible, in that it cannot be undone (though we could, of course, blast ourselves to oblivion by means of a well-placed and sufficiently large nuclear bomb!): so at certain points during the development of an individual organism 'commitment' occurs. For instance, in higher eukaryotes, once tissue has differentiated into mesoderm or ectoderm, it does not become endoderm. If you have had the extreme misfortune of losing a limb, you will be acutely aware of the fact that a new one won't grow in to take its place. And some researchers think that biological ageing can be described in terms of the Second Law of Thermodynamics. On this view, what we call ageing is the process whereby a biological system moves from a point far from equilibrium toward equilibrium, a state of no further change. Another way of stating this is that order is a basic property of a living organism, disorder of a dead one.

We can guess from all this that the concepts of entropy and irreversibility (and energy conservation) have had a huge impact on our view of the universe. Indeed, the concept of entropy throws into high relief philosophies of progress and development. 'How is it possible to understand life when the entire world is ordered by a law such as the second principle of thermodynamics, which points to death and annihilation?'[2]

Finally, we wish to comment briefly on the origin of irreversibility in many-body systems (large collections of small interacting particles). The thermodynamic description of such systems is so useful precisely because, in the usual case, there is no detailed knowledge or *control* over the (microscopic) variables of position and momentum for each individual particle. If such control were possible, the dynamics of many-body systems would presumably be reversible. When the number of microscopic variables is large, the state of maximum entropy is *overwhelmingly* probable, and the only lack of *certainty* that the entropy is maximal is the requirement that statistical fluctuations be allowed to occur. As we shall see in Chapter 6, under given constraints, the equilibrium (maximum entropy) state is identical to the macroscopic state that can be formed in the greatest number of microscopic ways.

I. References and further reading

Atkins, P. W. (1994). *The Second Law: Energy, Chaos, and Form*. New York: Scientific American.

Atkinson, D. E. (1977). *Cellular Energy Metabolism and Its Regulation*. New York: Academic Press.

Barón, M. (1989). With Clausius from energy to entropy. *Journal of Chemical Education*, **66**, 1001–1004.

Bennet, C. H. (1987). Demons, engines and the Second Law, *Scientific American*, **257**, no. 5, 108–112.

Bent, H. A. (1965). *The Second Law*. New York: Oxford University Press.

Bergethon, P. R. (1998). *The Physical Basis of Biochemistry: the Foundations of Molecular Biophysics*, cc. 12.1–12.2. New York: Springer-Verlag.

Christensen, H. N. & Cellarius, R. A. (1972). *Introduction to Bioenergetics: Thermodynamics for the Biologist: A Learning Program for Students of the Biological and Medical Sciences*. Philadelphia: W.B. Saunders.

Craig, N.C. (1988). Entropy analyses of four familiar processes. *Journal of Chemical Education*, **65**, 760–764.

Clugston, M.J. (1990). A mathematical verification of the second law of thermodynamics from the entropy of mixing. *Journal of Chemical Education*, **67**, 203–205.

Cropper, W.H. (1988). Walther Nernst and the last law. *Journal of Chemical Education*, **64**, 3–8.

Davies, P.W.C. (1995). *About Time: Einstein's Unfinished Revolution*. London: Penguin.

Djurdjevic, P. & Gutman, I. (1988). A simple method for showing that entropy is a function of state. *Journal of Chemical Education*, **65**, 399.

Einstein, A. (1956). *Investigations on the Theory of Brownian Movement*. New York: Dover.

---

[2] Léon Brillouin, *Life, Thermodynamics, and Cybernetics*.

Encyclopædia Britannica CD 98, 'Absolute Zero,' 'Entropy,' 'Metabolism,' 'Onsager, Lars,' 'pH,' 'Reversibility,' and 'Principles of Thermodynamics.'

Entropy: An International and Interdisciplinary Journal of Entropy and Information Studies. See http://www.mdpi.org/entropy/

Fenn, J.B. (1982). *Engines, Energy and Entropy*. New York: W.H. Freeman.

Feynman, Richard P., Robert B. Leighton and Matthew Sands, *Lectures on Physics* (Reading, Massachusetts: Addison-Wesley, 1963) Vol. I, cc. 44-2–44-6.

Fruton, J. S. (1999). *Proteins, Enzymes, Genes: the Interplay of Chemistry and Biology*. New Haven: Yale University Press.

Gardner, Martin (1967) Can Time Go Backward? *Scientific American*, **216**, no. 1, 98–108.

Gillispie, Charles C. (ed.) (1970). *Dictionary of Scientific Biography*. New York: Charles Scribner.

Gutfreund, H. (1951). The nature of entropy and its role in biochemical processes. *Advances in Enzymology*, **11**, 1–33.

Hale, F. J. (1993). Heat engines and refrigerators. In *Encyclopedia of Applied Physics*, vol. 7, ed. G. L. Trigg, pp. 383–403. New York: VCH.

Harold, F. M. (1986). *The Vital Force: a Study of Bioenergetics*, ch. 1. New York: W. H. Freeman.

Haynie, D. T. (1993). *The Structural Thermodynamics of Protein Folding*, ch. 3. Ph.D. thesis, The Johns Hopkins University.

Hollinger, H.B. & Zenzen, M.J. (1991). Thermodynamic irreversibility: 1. What is it? *Journal of Chemical Education*, **68**, 31–34.

Holter, H. & Møller, K. M. (ed.) (1976). *The Carlsberg Laboratory 1876–1976*. Copenhagen: Rhodos.

Katchalsky, A. & Curran, P.F. (1967). *Nonequilibrium Thermodynamics in Biophysics*, ch. 2. Cambridge, Massachusetts: Harvard University Press.

Klotz, I. M. (1986). *Introduction to Biomolecular Energetics*, ch. 2. Orlando: Academic Press.

Kondepudi, D. & Prigogine, I. (1998). *Modern Thermodynamics: from Heat Engines to Dissipative Structures*, cc. 3 & 4. Chichester: John Wiley.

Microsoft Encarta 96 Encyclopedia, 'Thermodynamics.'

Morowitz, H. J. (1970). *Entropy for Biologists*. New York: Academic Press.

Morowitz, H. J. (1978). *Foundations of Bioenergetics*, cc. 4 & 5. New York: Academic Press.

Ochs, R. S. (1996). Thermodynamics and spontaneity. *Journal of Chemical Education*, **73**, 952–954.

Peusner, L. (1974). *Concepts in Bioenergetics*, cc. 3 & 10. Englewood Cliffs: Prentice-Hall.

Planck, M. (1991). The Second Law of Thermodynamics. In *The World Treasury of Physics, Astronomy, and Mathematics*, ed. T. Ferris, pp. 163–169. Boston: Little Brown.

Price, G. (1998). *Thermodynamics of Chemical Processes*, ch. 3. Oxford: Oxford University Press.

Prigogine, I. (1967). *Introduction to Thermodynamics of Irreversible Processes*. New York: John Wiley.

Schrödinger, E. (1945). *What is Life? The Physical Aspect of the Living Cell*. Cambridge: Cambridge University Press.

Seidman, K. & Michalik, T. R. (1991). The efficiency of reversible heat engines: the possible misinterpretation of a corollary to Carnot's theorem. *Journal of Chemical Education*, **68**, 208–210.

Skoultchi, A.I. & Morowitz, H.J. (1964). Information storage and survival of biological systems at temperatures near absolute zero. *Yale Journal of Biology and Medicine*, **37**, 158.

Smith, C. A. & Wood, E. J. (1991). *Energy in Biological Systems*, cc. 1.2, 1.3 & 2.3. London: Chapman & Hall.

Williams, T. I. (ed.) (1969). *A Biographical Dictionary of Scientists*. London: Adam & Charles Black.

Wrigglesworth, J. (1997). *Energy and Life*, cc. 1.3–1.4.1. London: Taylor & Francis.

J. | Exercises

1. Is the word *entropy* a misnomer? Why or why not?
2. State whether the following phrases pertain to (A) the First Law of Thermodynamics, (B) the Second Law, (C) both the First and Second Law, or (D) neither of the Laws.
   (1) Is concerned with the transfer of heat and the performance of work.
   (2) Is sufficient to describe energy transfer in purely mechanical terms in the absence of heat transfer.
   (3) Indicates whether a process will proceed quickly or slowly.
   (4) Predicts the direction of a reaction.
   (5) Is a statement of the conservation of energy.
   (6) Says that the capacity to do work decreases as the organization of a system becomes more uniform.
   (7) Is a statement of the conservation of matter.
   (8) Says that a quantity of heat cannot be converted into an equivalent amount of work.
   (9) Says that the capacity to do work decreases as objects come to the same temperature.
3. Examine Eqn. 3.1. What happens to ΔS as $T \to 0$? In order to ensure that this equation remains physically meaningful as $T \to 0$, what must happen to ΔS? The answer to this question is a statement of the Third Law of thermodynamics.
4. Heat engine. Suppose 45 J is transferred from a heat source at 375 K to a heat sink at 25 °C. Calculate the maximum work that can be done and the efficiency of the process.
5. We have said that heat engines do not tell us very much about living organisms work. Show that if the human body depended on *thermal* energy to do work, it would cook before it could demonstrate its efficiency as a heat engine. Assume that the 'engine' has an efficiency of 20%.
6. 1 calorie is produced for every 4.1840 J of work done. If 1 cal of heat is available, can 4.1840 J of work be accomplished with it? Why or why not?
7. In Chapter 2 we learned about thermal equilibrium. In the approach to thermal equilibrium when two objects of differing initial temperature are brought into contact, although no energy is lost (by the First Law of Thermodynamics), *something* certainly is lost. What is it?

8. Entropy change of protein unfolding. Suppose that $\Delta H_d(25\,°C) = 10$ kcal mol^{-1}, $T_m = 68\,°C$ and $\Delta C_p = 1650$ cal mol^{-1} K^{-1}. Calculate $\Delta S_d(T_m)$, $\Delta S_d(37\,°C)$ and $\Delta S_d(15\,°C)$. At what temperature does $\Delta S_d = 0$? Speculate on the thermodynamic significance of ΔS_d in molecular terms at each temperature.

9. Recall Exercise 20 from Chapter 2. Use the same data to evaluate $\Delta S_d(T_m)$ at each pH value. Rationalize the entropy values.

10. For irreversible pathways, q/T is generally dependent on the path. How can one discover the entropy change between two states? Knowing that $q_{reversible} > q_{irreversible}$, use the First Law to write down a similar inequality for $w_{reversible}$ and $w_{irreversible}$.

11. Explain why water freezes in thermodynamic terms.

12. Suppose you have a cyclic process, as in Fig. 2.3. The entropy change for the system must be 0. Is there any inconsistency with The Second Law of Thermodynamics? Explain.

13. In his book *What is Life?* Erwin Schrödinger says 'an organism feeds with negative entropy.' What does he mean? (Hint: consider an organism that is able to maintain its body temperature and weight in an isolated system.)

14. Consider a gas, a liquid and a crystal at the same temperature. Which system has the lowest entropy? Why?

15. Can a machine exist in which energy is continually drawn from a cold environment to do work in a hot environment at no cost? Explain.

16. There are many different causes of undernutrition. Some of these are: failure of the food supply; loss of appetite; fasting and anorexia nervosa; persistent vomiting or inability to swallow; incomplete absorption, comprising a group of diseases in which digestion and intestinal absorption are impaired and there is excess loss of nutrients in the feces; increased basal metabolic rate, as in prolonged fever, overactivity of the thyroid gland, or some cancers; and loss of calories from the body; e.g., glucose in the urine in diabetes. Rationalize each type in terms of the First and Second Laws of Thermodynamics.

17. The macroscopic process of diffusion can be identified with microscopic Brownian motion, which subjects molecules to repeated collisions with the molecules of their environment and results in random rotation and translation. Some people say that the time-asymmetry in the inevitable increase of randomness of the universe is not strictly true as Brownian motion may contravene it. What is your view? Support it with well-reasoned arguments. (Brownian motion is named after the Scottish botanist Robert Brown, who was the first to observe it, in 1827.)

18. Consensus is a weak but nonetheless important criterion of truth, particularly in the scientific community. Doig and Williams have claimed that disulfide bonds make a substantial contribution to the enthalpy change of protein unfolding. Their view is rejected by most researchers who study protein thermodynamics. In the light of the results of the study by Cooper *et al.*, and considering the structure of a disulfide bond, rationalize the long-standing view of the of the sci-

entific community to the thermodynamic role of disulfide bonds in proteins. (*References*: Doig, A.J. and D.H. Williams (1991) Is the hydrophobic effect stabilizing or destabilizing in proteins – the contribution of disulfide bonds to protein stability. *J. Mol. Biol.*, **217**, 389–398; Cooper, A., S.J. Eyles, S.E. Radford and C.M. Dobson (1992) Thermodynamic consequences of the removal of a disulfide bridge from hen lysozyme. *J. Mol. Biol.*, **225**, 939–943.)

19. The Gibbs paradox. Consider two gas bulbs separated by a stopcock. The stopcock is closed. Both bulbs are filled with the same inert gas at the same concentration. What is the change in entropy when the stopcock is opened?

20. Is it possible for heat to be taken in to a system and converted into work with no other change in the system or surroundings? Explain.

21. Organisms are highly ordered, and they continually create highly ordered structures in cells from less-ordered nutrient molecules. Does this mean that organisms violate the Second Law of Thermodynamics? Explain.

22. The process whereby the Earth was formed and living organisms grew increasingly complex with time is 'essentially irreversible,' says Thomas Huxley. It 'gives rise to an increase of variety and an increasingly high level of organization.' Thus, this process appears not to square with the Second Law of Thermodynamics. Explain.

23. It would appear that all living organisms on Earth are, essentially, isothermal systems. Relatively few organisms live where the surroundings are at a substantially higher temperature than they are. Rationalize this observation in thermodynamic terms.

24. Tube worms are animals that thrive at black smokers at the bottom of the ocean. These invertebrates live as long as 250 years, longer than any other known spineless animal. Tubeworms have no mouth, stomach, intestine, or way to eliminate waste. The part of the worm that produces new tube material and helps to anchor the worm in its protective tube, a chitin proteoglycan/protein complex, is often planted deep within the crevices of the black smoker. The soft, bright-red structure (made so by hemoglobin) at the other end of the worm serves the same purpose as a mouth and can be extended or retracted into the surrounding water. Giant tubeworms are over 1 m long, and they have to cope with a dramatic temperature gradient across their length. The temperatures at a worm's plume is about 2 °C, just above the freezing point of pure water at 1 atm, while that at its base is about 30 °C! Can tube worms be modeled as isothermal systems? Why or why not?

25. Individual model hydrogen bond donors and acceptors do not often form hydrogen bonds in aqueous solution. Why not?

26. You may have noted that Carnot's formulation of the Second Law of Thermodynamics involves a very bold and unusually strong word: 'impossible.' Is the Second Law *always* true when stated this way? Why or why not?

27. The contraction of rubber is largely an entropic phenomenon. What are the sources of the enthalpic component?

28. Recall the example used to illustrate the entropy change in the surroundings when a mole of liquid water is formed from molecular hydrogen and molecular oxygen at 298 K. Use the data given in the text to calculate the entropy change per water molecule formed.
29. Protonation of the side chains of Glu and Asp is mainly an entropic effect. Why is this not true of protonation of His as well?
30. Show that when a system gains heat reversibly from surroundings held at constant temperature, there is no change in entropy.
31. 'The entropy change during an irreversible process is higher than the entropy change during a reversible process.' Is the statement true? Under what conditions?
32. What bearing does the Second Law have on pollution?
33. Discuss Fig. 1.5 in terms of the concepts of Chapter 3.

For solutions, see http://chem.nich.edu/homework

Chapter 4

Gibbs free energy – theory

A. | Introduction

This chapter discusses a thermodynamic relationship that is a basis for explaining spontaneous chemical reactivity, chemical equilibrium, and the phase behavior of chemical compounds. This relationship involves a thermodynamic state function that allows us to predict the direction of a chemical reaction at *constant temperature and pressure*. These constraints might seem annoyingly restrictive, because any method of prediction of spontaneity based on them must be less general than the Second Law of Thermodynamics, but as far as we are concerned the gains will outweigh the losses. Why is that? One reason is at any given time an individual organism is practically at uniform pressure and temperature (an 'exception' is discussed in one of the exercises). Another is that constant T and p are the very conditions under which nearly all bench-top biochemistry experiments are done. Yet another is that, although the total entropy of the universe must increase in order for a process to be spontaneous, evaluation of the *total* entropy change *requires* measurement of both the entropy change of the system *and* the entropy change of the surroundings. Whereas ΔS_{system} can often be found without too much difficulty, albeit only indirectly, $\Delta S_{surroundings}$ *cannot* really be measured. How could one measure the entropy change of the rest of the universe? The subject of the present chapter provides a way around this difficulty.

A particularly clear example of the inadequacy of ΔS_{system} to predict the direction of spontaneous change is given by the behavior of water at its freezing point. Table 4.1 shows the thermodynamic properties of water for the liquid→solid phase transition. The decrease in internal energy (which is practically identical to the enthalpy as long as the number of moles of gas doesn't change; see Chapter 2) would suggest that water freezes spontaneously in the range 263–283 K. Going on internal energy alone, this would seem even more probable at $+10\,°C$ than at $-10\,°C$, because ΔU for this system becomes increasingly negative with increasing temperature. The entropy change too is negative at all three temperatures, consistent with the solid state being more ordered than the liquid one. So the sign and magnitude of the entropy

Table 4.1 The thermodynamics of the liquid→solid transition of water at 1 atm pressure

Temperature (°C)	ΔU (J mol^{-1})	ΔH (J mol^{-1})	ΔS (J mol^{-1} K^{-1})	$-T\Delta S$ (J mol^{-1})	ΔG (J mol^{-1})
-10	-5619	-5619	-21	5406	-213
0	-6008	-6008	-22	6008	0
$+10$	-6397	-6397	-23	6623	$+226$

Table 4.2 The sign of ΔG and the direction of change

Sign of ΔG	Direction of change
$\Delta G > 0$	the forward reaction is energetically unfavorable, the reverse reaction proceeds spontaneously
$\Delta G = 0$	the system is at equilibrium, there is no further change
$\Delta G < 0$	the forward reaction is energetically favorable, the forward reaction proceeds spontaneously

of the *system alone* does not predict the direction of spontaneous change, unless the system is in solitary confinement (i.e. isolated).

In contrast to $\Delta U (\Delta H)$ and ΔS, the last column, ΔG, matches what we know about the physical chemistry of water: below 0 °C, it freezes spontaneously ($\Delta G < 0$), at 0 °C solid water and liquid water coexist ($\Delta G = 0$), and above 0 °C, ice is unstable ($\Delta G > 0$). This example illustrates something of very general importance: **ΔG is negative for a spontaneous process** (Table 4.2). As we know from experience, a stretched rubber band will contract when released. What is the sign of ΔG for this process? Negative! ΔS is large and positive, making $-T\Delta S$ negative, and ΔH is negative. So, as we shall see below, ΔG can only be negative for a contracting rubber band. **$\Delta G < 0$ is a basis for the thermodynamic explanation of chemical reactivity, equilibrium, and phase behavior.** Providing a good understanding of the Gibbs free energy and how it can be used by the biochemist is one of the most important aims of this book.

The thermodynamic state function of chief interest in this chapter is G, the **Gibbs free energy**. The eponym of this quantity is Josiah Willard Gibbs (1839–1903), the American theoretical physicist who was the first to describe it. Like its cousins U and H, which we met in Chapter 2, G is measured in joules. If all these state functions are measured in the same units, what distinguishes G from U and H? What sort of energy is the Gibbs free energy?

Free energy is energy that is available in a form that can be used to do work. We should not find this statement terribly surprising, since we know from the last chapter that *some* energy is not free to do work; heat transfer *always* generates waste heat, and that heat *cannot* be used to do work. The Gibbs free energy measures the *maximum* amount of work that can be done by a process going from a non-equilibrium state to an equilibrium state (at constant temperature and pressure). In other words, ΔG sets an upper limit on the work that can be done by any real process (at constant temperature and pressure). **The free energy of a system represents its capacity to do work**. Despite its importance, though, there is not a law of free energy in the way that there is a First Law and Second Law of Thermodynamics. The First and Second Laws place boundaries on the possible in extremely general terms. The Gibbs free energy tells us how much work can be done by a system under the constraints of the First Law, the Second Law, constant temperature, and constant pressure.

Like H and U, G is defined for **macroscopic systems**, ones that involve a very large number of particles. This implies that, although measurement of the Gibbs free energy difference between two states, ΔG, does tell us how much work must be done to convert one state into the other, or how much work could be done by a process, it does not explain *why* that much work should done. Despite this inherent roadblock to *understanding*, which can only be cleared by constructing a theory of molecular interactions (Chapter 6), thermodynamics is often all the more useful for its very power in dealing with systems described in the most qualitative of terms. For instance, a microcalorimeter enables the enthalpy change, say of protein unfolding, to be measured directly and with extraordinary accuracy, regardless of how little or much one knows about the structure of the protein. And a bomb calorimeter can be used to measure the heat of combustion of a Krispy Kreme doughnut or a *crème brûlé* without knowing the slightest thing about ingredients, shape, or structure of the molecules involved, or the nature of the interactions in either sweet. Indeed, one need not have heard of Krispy Kreme doughnuts or *crème brûlé* to record good combustion data! This character of things can be exceedingly useful to the biochemist doing experiments with very pure protein or DNA molecules whose three-dimensional structures are not known in any detail, but whose thermodynamic properties and effects on other molecules can be measured with relative ease. The 'low-level' description of classical thermodynamics is also of use to the plant biologist, who might be very good measuring temperature changes in the appendix of *Arum maculatum* but would be very much at a loss if asked to give a detailed description of the mechanisms involved! When a molecular interpretation of thermodynamic quantities is needed, one turns to a branch of physical chemistry called statistical mechanics (Chapter 6).

In the language of the physical biochemist (usually someone with a degree in physics or chemistry who is interested in biology), G is a **thermodynamic potential function**. As such, it is analogous to the gravitational potential function of classical mechanics, which describes how the gravitational energy of an object varies with position in a gravitational field. For instance, if you take a coin out of your pocket and let go, it will change position spontaneously and rapidly! What's so profound about that? The coin falls quickly because its gravitational potential energy is much greater in the air than on the ground, and air gives relatively little resistance to a change of position. A potential function permits one to predict whether a system will change or stay the same under given conditions.

When changes in a system do occur, they are said to take place at some rate – a measure of the amount of change per unit time. As we shall see in this chapter, we do not need to know the rate of a reaction in order to predict whether it will occur spontaneously; and, at the same, rate of reaction is not a predictor of the free energy difference between reactants and products. So to keep things simple, in-depth discussion of reaction rates will be deferred to Chapter 8.

Finally, mastering the mathematics of this chapter will be a somewhat rougher ride than earlier on, so fasten your seatbelt. Don't worry, though, because we'll find that we can make the complete journey without anything more sophisticated than algebraic equations and the odd logarithm. Moreover, the ideas themselves will always be considered more important than the specific mathematical tricks by which the results are obtained. This is not to imply that the mathematics is unimportant; it's to put our priorities in proper order. One should always remember that. Those who are prepared for greater mental punishment, however, and wish to explore an off-road vehicle-level treatment of the subject, should consult the more advanced references at the end of the chapter.

B. Equilibrium

We turn now to the very important concept of **equilibrium**. It was first proposed in the context of physical chemistry by the Norwegian chemists Cato Maximilian Guldberg (1836–1902) and Peter Waage (1833–1900) in the 1860s, in the form of the law of **mass action**: when a system is at equilibrium, an increase (decrease) in the amount of reactants (products) will result in an increase (decrease) in the amount of products (reactants); an equilibrium system responds to change by minimizing changes in the relative amounts of reactants and products. For example, suppose we have a solution of our favorite protein molecule. At equilibrium some molecules will be in the folded state (reactant), some in the unfolded state (product). Now add a dash of protease. In the usual case, unfolded proteins are much more susceptible to proteolytic attack than folded proteins. If a proteolyzed protein is unable to refold, proteolysis will change the balance of folded proteins and unfolded ones. By the law of mass action, the response of the system to the decrease in the number unfolded proteins will be for folded proteins to unfold, in order to minimize the change in the relative amounts of reactants and products. Equilibrium is so important that the Swede Svante August Arrhenius (1859–1927) has called it 'the central problem of physical chemistry.' That was in 1911, eight years after he had won the Nobel Prize in Chemistry. We'll learn about his work on the temperature-dependence of kinetics in Chapter 8.

We have mentioned equilibrium before but avoided a detailed discussion until now because neither the First Law nor the Second Law depends on it. Another reason we haven't said much about equilibrium is that no living organism functions and practically no real process occurs under such conditions! We would, however, be very wrong

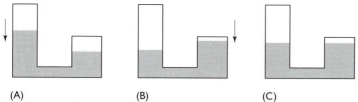

(A) (B) (C)

Fig. 4.1 The movement of a system toward equilibrium. Water is in a U-shaped container. In panels (A) and (B) the system is not at equilibrium; we know from experience that change will occur. The driving force for change is a difference in hydrostatic pressure. The pressure difference is proportional to the difference in the height of the fluid in the two arms of the vessel. Water flows downhill! The rate of flow in one direction is greater than the rate of flow in the opposite direction. The system will continue to change until the fluid level is the same on both sides of the vessel. In panel (C), there is no difference in fluid height and consequently no difference in pressure. The flow rate is the same in both directions; the system is at equilibrium; no further change occurs.

indeed to think that equilibrium is therefore unimportant. Abstractions and idealizations play an extremely important role in scientific study, serving as models of reality or simple generalizations of otherwise exceeding complex phenomena. As we shall see, a central consideration of thermodynamics is that *any* physical system will *inevitably* and *spontaneously* approach a stable condition called equilibrium. This concept is bound up in the Second Law, and it closely resembles the proposed relationship between entropy and ageing that we touched at the close of the previous chapter.

A system will exhibit net change if it is *not* at equilibrium, even if the rate of change is imperceptible. An example of a system that is not at equilibrium is a canary. It's fine to keep it caged to prevent it becoming the pet cat's supper, but unless the bird is fed its protected perch won't lengthen its lifespan. This is because the metabolic reactions of its body require a continual input of chemical energy, and when the energy requirements are not met the body winds down and dies. All living organisms are highly non-equilibrium systems. Panels (A) and (B) of Fig. 4.1 are clearly not at equilibrium, because the fluid height is not level. There will be a net flow of water from one side to the other. The flow rate in one direction will be greater than the flow rate in the opposite direction. In contrast, the system in panel (C) is at equilibrium. There is no net flow of water. The flow rates in opposite directions are equal. **The equilibrium state is the one in which no further macroscopic change takes place because all the forces acting on the system are balanced.**

A system will not show net change if it is *at equilibrium* and left unperturbed. For instance, a plugged test-tube filled with a biochemical buffer and kept at constant temperature will not change (barring bacterial contamination and chemical degradation). This system is in a very stable equilibrium. Also at equilibrium, but a less stable one than the buffer in a test tube, might be a heavy snow lying peacefully on a mountainside. Demonstrating the tenuous nature of this equilibrium could be a regretful affair for anyone or anything nearby, for the slightest disturbance could turn the resting snow into a raging avalanche! Another example of an unstable equilibrium is a perfectly balanced seesaw. The weights on either side need not be equal, but if they are not, adjustments must be made in the distances of the weights from the fulcrum in order to achieve a balance, in accordance with Newton's laws of mechanics (Fig. 4.2). What might be called a semi-stable equilibrium is

Fig. 4.2 Equilibrium. We know from experience that if the weights are the same, as in panel (A), the distance of each weight from the fulcrum must be the same for the weights to balance. If the weights are different, however, as in panel (B), the distance from the fulcrum will not be the same. The lighter weight must be father from the fulcrum than the heavier weight. By Newton's Second Law, at equilibrium the clockwise torque equals the counterclockwise torque, where torque = mass $\times$ gravitational acceleration $\times$ horizontal distance from fulcrum.

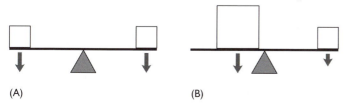

(A) (B)

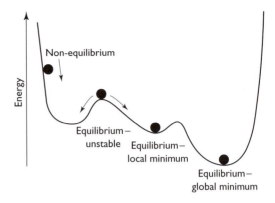

Fig. 4.3 Different types of equilibrium. The non-equilibrium situation will change immediately. The unstable equilibrium, like the snow at rest on a mountainside, will not change without the addition of energy. The amount of energy needed, however, is very small. There are two main types of energy minimum: local minimum and global minimum. A local minimum has the lowest energy in the vicinity. The global minimum has the lowest free energy of all. An ongoing debate in protein folding research is whether the folded state of a protein is a local minimum or a global minimum of the energy surface.

in which the energy of the system is at a minimum but not at the lowest possible minimum. For example, suppose a rock is resting on the floor of a ravine between two mountains. The rock will not leap up and move elsewhere spontaneously! If given a hard enough kick, though, it might roll up the side of the ravine, reach the crest, and tumble down the mountain into the valley below. The various types of equilibrium are summarized in Fig. 4.3.

To find out more about equilibrium let us return to the subject of diffusion. This type of transport process is extremely important to many different biochemical reactions, including for example the chemical transmission of nerve impulses across synaptic junctions. Fig. 4.4 shows one of the many ways in which diffusion can be conceived. We have two miscible liquids and, for the sake of argument, we suppose that one can be beside the other, giving a sharp boundary, until we say 'go.' The right solution is colorless, like water, the left one opaque, like ink. The densities of the liquids are about the same. Immediately after mixing has begun, which occurs spontaneously, the concentration changes abruptly going from one side of the boundary to the other. We know from experience that this situation is unstable and will not persist. The result after a very long time is that the dark liquid will be distributed uniformly throughout the volume, as will the clear liquid.

As we shall see in the discussion that follows, the top panel of Fig. 4.4 corresponds to the minimum entropy and maximum free energy of the system, while the bottom panel, representing equilibrium, corresponds to the maximum entropy and minimum free energy of the system. We might guess that equilibrium will only be achieved if the temperature of the system is uniform throughout (discounting statistical fluctuations). For when the temperature is not uniform, convection currents and differences in particle concentration will be present, as for example in the swirls one sees shortly after pouring milk into a cup of coffee or tea. The convection currents and inhomogeneities will eventually go away, and when they have vanished, the system is at equilibrium. To take a deeper view, if we have two systems **A** and **B**, and these are in thermal equilibrium, the distribution of the kinetic energy of the molecules of one system is identical to that of the other system. Some of the molecules will be moving very rapidly, others not so fast, but the distributions will be the same in **A** and **B**. At equilibrium, the kinetic

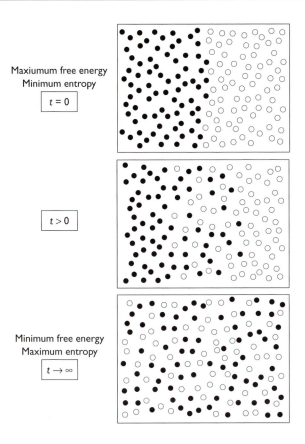

Maxiumum free energy
Minimum entropy

$t = 0$

$t > 0$

Minimum free energy
Maximum entropy

$t \rightarrow \infty$

Fig. 4.4 Mixing. At time $t = 0$, when the virtual partition is removed, the dark and light liquids are completely separate. The liquids are miscible, so mixing occurs. After a sufficiently long time, the liquids are completely mixed. The unmixed state has a maximum of free energy for the system and a minimum of entropy. The completely mixed state has a minimum of free energy and a maximum of entropy. Based on Fig. 4.5 of van Holde (1985).

energy distribution of the molecules is one of maximum probability; the entropy, both of the individual systems and in combination, is a maximum; and the free energy, both of the individual systems and in combination, is a minimum.

One of the exercises in the previous chapter asked you to show that under the three constraints of pV-work only, constant internal energy, and reversibility, $\Delta S = 0$ at any temperature. We will now build on this foundation to develop the concept of equilibrium. Recall that when we required all changes to a system to be reversible, we were requiring that the system be at (or extremely close to) equilibrium throughout the process. In this context we interpret $\Delta S = 0$ to mean that the entropy must be at an extreme value when equilibrium is reached. By the Second Law, we know that the extremum is a maximum and not a minimum (the entropy of the universe always increases). In any non-equilibrium state of a system, $S < S_{max}$, and the system will change *spontaneously* until $S = S_{max}$ (in the absence of energy input to the system). Once equilibrium has been reached there are no further increases in the entropy, so $\Delta S = 0$. Somehow or other, an isolated system 'knows' when to stop generating entropy. This provides yet another way of stating the Second Law: **an *isolated* system will change spontaneously until a maximum state of disorder is obtained**.

To get an even better sense of what equilibrium is, let's consider the reaction A⇔B. How far will the reaction go? Until equilibrium is reached; until there is no longer a tendency for a (macroscopic) change to occur spontaneously. At equilibrium, the average concentrations of A and B are constant in time; there is no further *macroscopic* change to the system. This, however, does not mean that the particles are no longer moving! For as long as the temperature is above 0 K, all the particles will be in motion, the amount depending on the temperature and whether the substance is in the solid, liquid or gas phase. If follows that at equilibrium A and B can and will interconvert, even if this happens only very slowly. Chemical equilibrium is a **dynamic equilibrium**. But the *concentrations* of A and B do not change.

If a process occurs through a succession of near-equilibrium states, the process must be slow enough to allow the system to come to equilibrium after each small change. Real processes necessarily occur at finite rates, so the best one can achieve in practice is a near-reversible process. Many small proteins exhibit very highly (>95%) reversible order–disorder transitions on equilibrium thermal or chemical denaturation. It would appear in such cases that all the information required for the protein to interconvert between its folded and unfolded states is present in the matter it's made of, i.e. in the amino acid sequence. We'll return to this subject in Chapters 5 and 9.

There are a number of other important features of the equilibrium state. One is that **for a system to be truly at equilibrium, it *must* be closed**. For example, if an unopened Coke can has reached thermal equilibrium, its contents will be at equilibrium. The amount of carbonic acid present and the amount of gaseous CO_2 will be constant, even if individual CO_2 molecules are constantly escaping from the liquid and returning thither. On opening the can, however, there is a very rapid change in pressure and a jump to a non-equilibrium state. This leads to a net loss of CO_2 for three principal reasons: because the system is no longer closed, gaseous CO_2 can escape immediately; the decrease in the abundance of CO_2 gas near the liquid–gas interface will promote a loss of CO_2 from the liquid; and if the can is held in your hand and not kept on ice, its contents will begin warming up, and this will drive off CO_2 because solubility of the gas varies inversely with temperature. We infer from all this that for a system to remain at equilibrium, variables such as T, p, V and pH must be constant. If any of them should change, or if the concentration of any component of the system should change, a non-equilibrium state would result, and the system would continue to change until equilibrium was reached.

C. | Reversible processes

Now it's time for a mathematical statement of the Gibbs free energy, G:

$$G = H - TS \tag{4.1}$$

We can see from this that the thermodynamic potential G is a sort of combination of the First and Second Laws, as it involves both enthalpy and entropy. We must bear in mind, though, that the temperature and pressure are required to be constant for G to predict the direction of spontaneous change in a system.

Let's see what we can do with Eqn. 4.1. For an incremental measurable change in G, Eqn. 4.1 becomes

$$\Delta G = \Delta H - T\Delta S - S\Delta T \qquad (4.2)$$

The last term on the right-hand side vanishes at constant temperature, leaving us with $\Delta G = \Delta H - T\Delta S$. ΔG tells us that the gain in *useful* work that one can obtain from a system as a result of change must be *less* than the gain in energy or enthalpy (ΔH). The difference is measured by the gain in entropy (ΔS) and the temperature at which the reaction occurs. $T\Delta S$ is what is called 'isothermally unavailable energy.' At the same time, ΔG is the *minimum* useful work required for a system to undergo a specified change from an equilibrium state to a non-equilibrium state.

If we require pV-work only, then, because $\Delta U = T\Delta S - p\Delta V$, Eqn. 4.2 simplifies to

$$\Delta G = V\Delta p - S\Delta T \qquad (4.3)$$

If we further require p and T to be constant, then $\Delta G = 0$. Just as with ΔS, we interpret this to mean that a reversible system has an extreme value of G when T and p are constant and the system is at equilibrium. In this case, and opposite to the entropy, the extremum is a minimum, just as with gravitational potential. In other words, the magnitude of ΔG measures the extent of displacement of the system from equilibrium, and when $\Delta G = 0$ the system is at equilibrium (Fig. 4.4).

Now we have come far enough to see how the Gibbs free energy is of great utility in predicting the direction of a process. Have another look at Table 4.1, particularly the ΔG column. If ΔG is *positive* (the energy change is **endergonic**), the process will *not* occur *spontaneously*. This is because the final state of the process has a higher free energy than the initial state, and this can only be achieved at the expense of the energy of the surroundings. If ΔG is *negative* for a process (the energy change is **exergonic**), the reaction proceeds *spontaneously* in the direction of equilibrium, and when this is reached no further change occurs. **For any real process to occur spontaneously at constant temperature and pressure, the Gibbs free energy change must be negative.** It should be clear from Eqn. 4.1 that the lower the enthalpy (energy), the lower G, and the higher the entropy, the lower G. Hence, spontaneity of reaction is favored by reductions of the enthalpy (exothermic reactions) and by increases of the entropy (reactions leading to increases in disorder). It must be emphasized that, **while the magnitude of ΔG tells us the size of the driving force in spontaneous reactions, it says *nothing at all* about the time required for the reaction to occur.** The rate of a reaction can be treated separately and will be developed in Chapter 8.

Though the general principles hold, the situation with real chemical reactions is more complicated than the ideal world we've been discussing. Real physical, chemical, and biological processes occur at a finite rate; the slope of the Gibbs function between states is finite (the

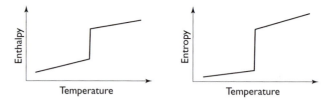

Fig. 4.5 First-order phase transition. The graphs show the behavior of the enthalpy and entropy functions in a first-order phase transition. Both of these thermodynamic quantities are *discontinuous* at the transition temperature, as is C_p.

step between G_1 and G_2 of a two-state system is not infinitely steep); *all real chemical reactions are, to some extent, irreversible.* Nevertheless, as mentioned above, reversibility can be approached in the real world of biochemistry, by having the process take place in a controlled manner, and the biochemist can aim to do this by studying reactions *in vitro*.

D. | Phase transitions

We have already encountered phase transitions in two different contexts in this book: in likening the thermal denaturation of a protein molecule to the melting of an organic crystal, and in describing the physical properties of water. A **phase** is a system or part of a system that is homogeneous and has definite boundaries. A phase need not be a chemically pure substance. Phase transitions are ordinarily caused by heat, and when they do they occur at a definite temperature and involve a definite amount of heat. Phase changes are associated with many fascinating and general aspects of biological thermodynamics, so we had better take a few minutes to know more about them.

Despite possible appearances to the contrary, phase transitions are *very* complicated phenomena. No one really knows how to describe them on the level of individual particles, though of course everyone knows that quantum mechanics must be involved somehow. Nevertheless, it is possible to give a description of a phase change on the macroscopic level in terms of classical thermodynamics. Some energetic quantities, for instance enthalpy and entropy, exhibit a *discontinuous* change at a phase boundary. What this means is that the shape of the enthalpy function and entropy function with temperature is like a step; the state functions change abruptly, not smoothly, as the solid becomes a liquid or the liquid a solid (Fig. 4.5). The definite amount of heat absorbed on melting or boiling is a heat capacity change, the latent heat of melting or vaporization, respectively. There is a relatively large change in heat capacity over an extremely small temperature range upon a change of phase. Transitions of this type are 'all-or-none' transitions; the material is completely in one phase on one side of the phase boundary and completely in another phase on the other side of the phase boundary, though the two phases can coexist at the transition temperature. Such transitions are known as first-order phase transitions, and they are so abrupt that they have been likened to the catastrophes of an area of mathematics imaginatively known as catastrophe theory. In many cases protein folding/unfolding transitions closely resembles first-order phase transitions (e.g. the case of hen egg white lysozyme discussed in Chapter 2); in more complicated situations, however, such a description is clearly inadequate (see Chapter 6). The phase transition in water from the solid state to the liquid state is a first-order phase transition.

The liquid–solid phase boundary of water plays an especially important role in life on Earth in more ways than one. For instance, when the temperature drops, it is rather significant that water begins to freeze on the surface of a pond, not at the bottom, and that it remains up top. Water on the surface loses its heat to the surroundings, the temperature of which can fluctuate more wildly than the temperature of the floor of a pond and dip well below 0 °C. The density of ice is lower than that of liquid water. Water is a very peculiar substance indeed! The physical properties of water also play a critical role in determining the level of the oceans and in shaping the world's weather, by determining the fraction of the oceans' water that is liquid and the amount in the polar icecaps.

Less well-known and obvious, perhaps, than the phases changes of water and proteins, are those of lipids. These mainly water-insoluble molecules can undergo changes in state just as other compounds do. We say 'mainly water-insoluble' because lipids are made of two parts, a small water-soluble 'head' and a long water-insoluble 'tail' (Fig. 4.6). In bilayers, which are surfaces of two layers of lipids in which the tails face each other, lipids are distinguished by their ability to exhibit what is called the liquid crystalline or gel state. This is an intermediate level of organization between the solid state, which is rigid, and the liquid state, which is fluid. The gel state is the one in which lipids are found in the membranes of the cells of living organisms. In pure lipid bilayers, there is a definite melting temperature between the solid phase and the gel

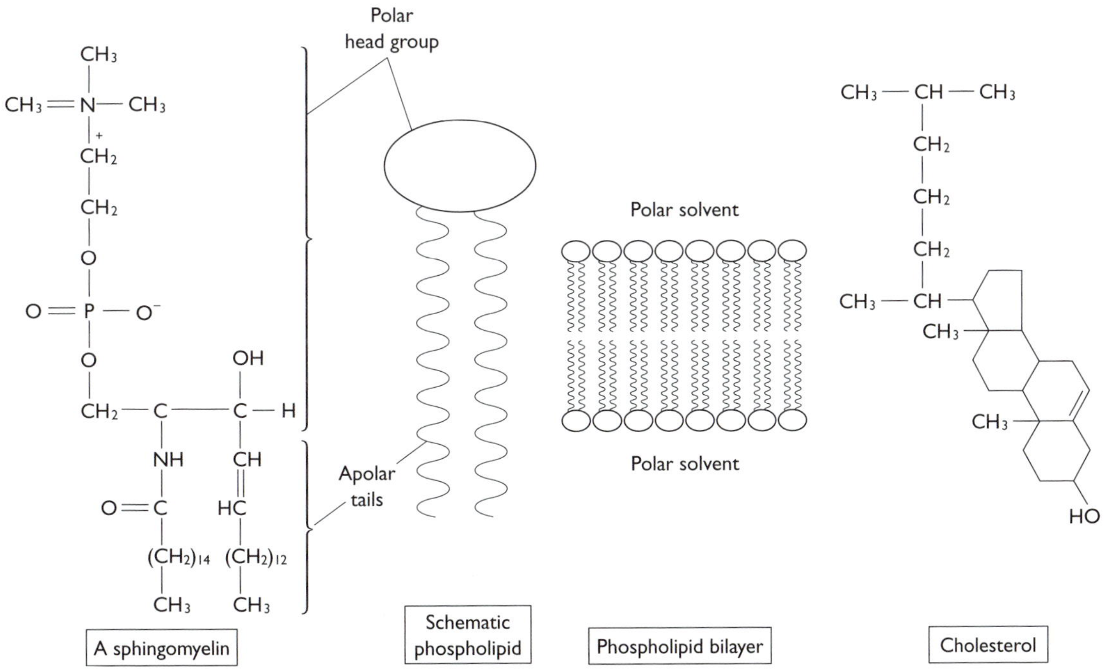

Fig. 4.6 Lipids bilayers. Lipids, or phospholipids, have two main regions, a polar 'head' and aliphatic 'tails.' The head group is in contact with the solvent in a lipid bilayer, as shown. Sphingomyelins are the most common of the sphingolipids, a major class of membrane component and just one type of phospholipid. The myelin sheath that surrounds and electrically insulates many nerve cell axons is rich in sphingomyelin. Cholesterol is the most abundant steroid in animals and the metabolic precursor of steroid hormones and a major component of animal plasma membranes. Biological membranes are highly heterogeneous. They include not just several kinds of lipids and cholesterol, but membrane-spanning and membrane-associated proteins as well. At physiological temperatures, membranes are gel-like and allow lateral diffusion of their components.

A sphingomyelin | Schematic phospholipid | Phospholipid bilayer | Cholesterol

Fig. 4.7 Membrane melting temperature. In general, the melting temperature of a phospholipid bilayer decreases with increasing heterogeneity. One exception to this rule is cholesterol, which increases the melting temperature by increasing the rigidity of the bilayer. Increasing the number of double bonds in the aliphatic tail decreases the melting temperature by decreasing the ability of molecules to pack against each other. A double bond introduces a kink in a tail. In contrast, increasing the length of the tail increases the melting temperature, because the aliphatic portions of lipids can interact favorably with each other by means of van der Waals forces. Redrawn from Fig. 21.7 of Bergethon (1998).

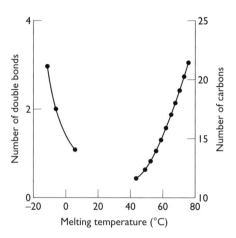

phase, just as there is for water at the liquid–gas phase boundary and for a protein like lysozyme between its folded and unfolded states. Moreover, the solid–gel transition of pure lipid bilayers is highly cooperative, again resembling the behavior of water and some proteins. A **highly cooperative phase transition occurs over a very narrow temperature range.** When one decreases the purity of a bilayer, for instance by introducing a second type of lipid that is miscible with the first, the cooperativity of the solid–gel transition decreases proportionately. Biological membranes are highly heterogeneous: they are made of lipids, proteins and small amounts of carbohydrate. Such membranes therefore do not exhibit very cooperative transitions, and melting occurs over a range of temperatures, usually 10–40 °C. Thus, biological membranes are in a gel-like fluid lamellar phase in the living organism.

Other physical properties of lipids influence bilayer fluidity. One is length of the non-polar hydrocarbon tails (Fig. 4.7). The longer the chain, the higher the transition temperature. This is because the hydrophobic stuff of one lipid can interact fairly strongly with the hydrophobic stuff of another lipid. Another is the degree of saturation of the carbon–carbon bonds in the tails. Unsaturated bonds (double bonds) introduce kinks into the chain, making it more difficult for lipids in a bilayer to form an orderly array. Variation in the number and location of double bonds ensures that biological membranes do not become too rigid. Yet another contributor to membrane rigidity is lipid-soluble cholesterol, which decreases membrane fluidity by disrupting orderly interactions between fatty acid tails. Cholesterol itself is a rigid molecule (Fig. 4.6). Finally, the melting temperature also depends on added solute and counter ion concentration! The physical properties of lipids, which are extremely complex, are of vital importance to the cell. Biological membranes permit membrane-bound proteins some degree of lateral movement, enable the cell to change shape and migrate, and make tissue formed by millions of cells relatively soft to the touch. What baby would enjoy nourishing itself at its mother's breast if her nipples were no more flexible than fingernails?

E. | Chemical potential

Life as we know it could not exist without water. No wonder Thales of Miletus (fl. *c.* 2500 yr B.P.) considered water the ultimate substance or *Urstoff*, the stuff of which all things are made! All physiological bio-chemical reactions take place in a largely aqueous environment, from enzyme catalysis to the folding of proteins to the binding of proteins to DNA to the assembly of large macromolecular complexes. It is probable that life on Earth began in a primordial sea of salt water. We had there-fore better devote some energy to learning the basics of the thermo-dynamics of solutions. The path we shall take towards greater understanding is one that involves scaling a few jagged mathematical formulas, but the view we shall have at the end of the journey will make the effort worthwhile.

Now, if you have doubted the usefulness of mathematics in your study of biochemistry up to this point, it might help to remember that mathematics is to biochemistry as a protocol is to producing a biochem-ical result. That is, mathematics is the handmaiden of biochemistry, not its master. However, protocols themselves can be extremely useful, particularly when they are very general in scope. One of the most highly cited scientific journal articles of all time is one of the most boring biology articles ever published. This is because it has to do with a tech-nique for separating proteins on the basis of size and says nothing specific about biology. Yet, the paper in question is very important because the method it outlines (polyacrylamide gel electrophoresis) can be used in an extremely broad range of situations – it doesn't matter at all which tissue the protein came from or which organism; in fact, the technique can also be used to separate nucleic acids. In this sense, the protocol is even more important than any particular result it might be used to produce. In the same way, getting to know something of the mathematical background to a formula is worth the time and effort, because the sort of thinking involved is of very general utility.

Being quantitative about free energy changes is a matter of both being careful in making measurements and being clear about the con-ventions one has adopted. Teaching the first of these is beyond the scope of this book! To do the latter, we need to return to a topic introduced in Chapter 2: the standard state. Previously, we *defined* the standard enthalpy change, $\Delta H°$, as the change in enthalpy for a process in which the initial and final states of one mole of a pure substance are at 298 K and 1 atm. Now we wish to define the **standard free energy change**, $\Delta G°$. The superscript naught indicates *unit activity* at standard temperature (298.16 K, though three significant digits are usually enough) and pres-sure (1 atm). The **activity** of a substance (introduced by Gilbert Newton Lewis (1875–1946), an American) is its concentration after correcting for non-ideal behavior. There are many sources of non-ideality, an impor-tant one being the ability of the solute to interact with itself. In the sim-plest case, the activity of substance A, a_A, is defined as

$$a_A = \gamma_A[A] \tag{4.4}$$

where γ_A is the **activity coefficient** of A on the *molarity* scale. When a different concentration scale is used, say the molality scale, a different activity coefficient is needed. The concept of activity, however, is always the same. Ideal solute behavior is approached only in the limit of infinitely dilute solute: as $[A] \rightarrow 0$, $\gamma_A \rightarrow 1$. Activity is the *effective* concentration; it measures the tendency of a substance to function as a reactant in a given chemical environment. According to Eqn. 4.4, $0 < a_A < [A]$, because $0 < \gamma_A < 1$. Activity is a *dimensionless* quantity; the units of the *molar* activity coefficient are l mol^{-1}.

Defining $\Delta G°$ at unit activity, while useful, is nevertheless somewhat problematic for the biochemist. This is because free energy changes depend on the concentrations of reactants and products, and the products and reactants are practically never maintained at molar concentrations throughout a reaction! Moreover, most reactions of interest do not occur at standard temperature. So, we'll need a way to take all these considerations into account when discussing free energy changes.

The relationship between the concentration of a substance A and its free energy is *defined* as

$$\mu_A - \mu_A{}° = RT \ln a_A \tag{4.5}$$

μ_A is the **partial molar free energy**, or **chemical potential**, of A, and $\mu_A{}°$ is the chemical potential of A in the standard state. The partial molar free energy of A is, in essence, just $\Delta G_A / \Delta n_A$, or how the free energy of A changes on when the number of molecules of A in the system changes by one (Fig. 4.8). Eqn. 4.5 does not involve a volume term or an electrical term, because we are assuming for the moment that the system does not expand against a constant pressure and that no charged particles are moving in an electric field; we must keep in mind that there are numerous kinds of work (Chapter 2) and that they must be taken into account when they are relevant to the system under consideration. It is particularly appropriate to call μ the chemical potential, since at constant T and p, G is a function of chemical composition alone. It is relatively easy to switch between μ and G, and in cases where we are dealing with one mole of the substance we'll often use G instead of μ since either would be correct.

Eqn. 4.5 tells us that when $a_A = 1$, $\mu_A - \mu_A{}° = 0$. That is, $\mu_A - \mu_A{}°$ measures the chemical potential of A relative to the standard state conditions; **the activity of a substance is unity in the standard state**. The chemical potential also depends on temperature as shown, and the units of the gas constant puts things on a per mole basis. (If inclusion of the gas constant here seems arbitrary, compare Eqn. 4.5 to Eqn. 3.23 after the latter has been multiplied by T.) Eqn. 4.5 also says that the chemical potential of the solvent decreases as solute is added. This is because the activity of a substance is always highest when that substance is pure. The reduction in chemical potential, however, occurs even if the solution is ideal (enthalpy of mixing of zero), as in the case where the solute is 'inert' and does not interact at all with the solvent. This tells us that that the lowering of the chemical potential is fundamentally an entropic effect, though for any real solvent and solute it will of course contain an enthalpic component stemming from inter-

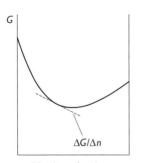

Fig. 4.8 Thermodynamic potential and solute concentration. The Gibbs free energy of a solute varies with concentration. The chemical potential measures the rate of change of G with n, or the slope of the curve at a given value of n ($\Delta G/\Delta n$). Note that G can decrease or increase on increases in concentration.

particle interactions. The similarity of Eqns. 4.5 and 3.23 should no longer seem coincidental.

The given definition of an ideal solution is less general than it could be. We could say instead that an ideal solution is one for which the enthalpy of mixing is zero *at all temperatures*, implying that the solute always interacts with the solvent as it interacts with itself. An even more technical definition is that an ideal solution is a homogeneous mixture of substances whose physical properties are linearly proportional to the properties of the pure components, which holds for many dilute solutions. This is known as Raoult's law, after François-Marie Raoult (1830–1901), a French chemist.

Before we get too far down the road, let's look at an example that is aimed at helping to clarify the difference between $\Delta G°$ and ΔG (and $\Delta\mu°$ and $\Delta\mu$). Suppose we're involved in a study of the binding of a peptide hormone to a receptor situated in the plasma membrane. The cheapest way to obtain large quantities of pure hormone is just to synthesize it chemically and purify it using a type of chromatography. Let's suppose that the sequence of the peptide we're studying is X–X–X–X–X–X–X–X–Ala–Gly, and because chemical synthesis of peptides is done from the C-terminus to the N-terminus (the reverse of how it occurs during translation of mRNA on a ribosome), we first join the Ala to the Gly. We'd like to know something about the energetics of formation of the peptide bond between these two amino acids. When two free amino acids join to form a peptide bond a water molecule is produced; the reaction is a type of dehydration synthesis. When the reaction occurs *in aqueous solution* under standard state conditions – all products and reactants are at a concentration of 1 M *except water* (we'll say why later on) – the free energy change, $\Delta G°$, is 4130 cal mol^{-1}. We are far from equilibrium, and the driving force for change is in the direction of the reactants. When Ala and Gly are at 0.1 M, and Ala–Gly is at 12.5 μM, the reactants and products are no longer in their standard states, and the free energy difference is not $\Delta G°$ but ΔG. On doing the experiment, however, one finds that the reaction is at equilibrium and that no change in the concentrations of reactants takes place unless the system is perturbed. In other words, $\Delta G = 0$. If the concentrations are changed again, so that Ala and Gly are at 1M, as in the standard state reaction, but Ala–Gly is at 0.1 mM, one finds that the reaction proceeds in the direction of the products and $\Delta G = -1350$ cal mol^{-1}. ΔG on its own measures how far away a reaction is from equilibrium, and $\Delta G - \Delta G°$ measures how much the conditions of a reaction differ from those in the standard state.

Here the discussion is going to become a little more difficult. We have a two-component solution made of solvent and solute. The solute is some general substance A that is soluble in the solvent. The solvent could be water and the solute a metabolite. Assuming ideal behavior, $a_A = [A]$ and Eqn. 4.5 becomes

$$\mu_A - \mu_A° = RT \ln[A] \tag{4.6}$$

Now we construct a *notional* partition between two regions of the solution and require that the system *not* be at equilibrium. In more general terms, our boundary could just as well be a liquid–gas interface or a membrane permeable to A. Substance A in any case can move across the boundary and back again by *random* motion. Calling the two regions of the solution α and β, we have

$$\Delta\mu_A = \mu_{A,\beta} - \mu_{A,\alpha} = RT\ln([A]_\beta/[A]_\alpha) \tag{4.7}$$

The standard state terms have vanished because the standard state free energy is the same in both regions. Intuitively, we would expect $\Delta\mu_A < 0$ when $[A]_\alpha > [A]_\beta$. When $[A]_\alpha > [A]_\beta$, the argument of the logarithm ($[A]_\beta/[A]_\alpha$) is less than one, and because $\ln x < 0$ for $x < 1$, our expectation is met (Fig. 4.9). When $\Delta\mu_A$ is negative, the solute particle moves down its concentration gradient spontaneously from α to β. If the concentration is greater in region β than in region α, $\Delta\mu_A > 0$, and A moves spontaneously from β to α.

So far, so good. Backing up a couple of steps and combining Eqns. 4.4 and 4.5, we have

$$\mu_1 = \mu_1^\circ + RT\ln\gamma_1[1] \tag{4.8}$$

which is written for component 1, the solvent. We can rewrite Eqn. 4.8 as

$$\mu_1 = \mu_1^\circ + RT\ln f_1 X_1 \tag{4.9}$$

where f_1 is the activity coefficient of component 1 on the **mole fraction** scale. The mole fraction X_i is the *number of molecules* of i (i.e. n_i) expressed as a fraction of the total *number of molecules* in the system, n. In other words, $X_i = n_i/n$. We are still dealing with a two component system, so $X_1 = n_1/(n_1 + n_2) = 1 - X_2$, where X_2 is the mole fraction of the solute; the mole fractions of the individual components of a solution must sum to 1, which is how the mole fraction gets its name. Writing $RT\ln f_1 X_1$ as $RT\ln f_1 + RT\ln X_1$ (which we can do because $\ln ab = \ln a + \ln b$), using $\ln(1 + x) = x - x^2/2 + x^3/3\ldots$ (a relationship from mathematics which is valid for $-1 < x < 1$), and rearranging terms, Eqn. 4.7 becomes

$$\mu_1 - \mu_1^\circ = RT(-X_2 + \ldots) + RT\ln f_1 \tag{4.10}$$

We can simplify things in two ways if the solution is dilute ($n_2 \ll n_1$). One is that the 'higher order' terms in X_2 (namely, the square of X_2, the cube of X_2, etc.) are small because X_2 is small, and we shall assume we can neglect them. The other assumption we shall make is that

$$X_2 \approx C_2 V_1^\circ/M_2 \tag{4.11}$$

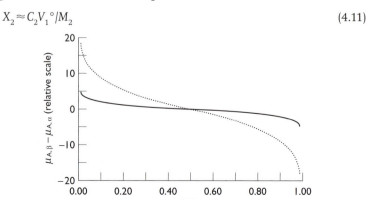

Fig. 4.9 Chemical potential difference as a function of concentration. This figure illustrates the behavior of the left-hand side of Eqn. 4.7 when the total concentration of A is fixed; $[A]_\alpha + [A]_\beta = $ constant. The dashed line represents a change in temperature.

where C_2 is the concentration of the solute in *molal* units $(g\,l^{-1})$, M_2 is the molecular weight of the solute, and $V_1{}^\circ$ is the molar volume of pure solvent. Eqn. 4.11 comes from $X_2 = n_2/(n_1 + n_2) \approx n_2/n_1$. But because $n_1 V_1{}^\circ = V$, the volume of *solvent*, $X_2 = n_2 V_1{}^\circ/V = M_2 n_2 V_1{}^\circ/M_2 V = (M_2 n_2/V)(V_1{}^\circ/M_2) = C_2 V_1{}^\circ/M_2$. Looking at Eqn. 4.11, we have $(g\,l^{-1}) \times (l\,mol^{-1})\,/\,(g\,mol^{-1}) = 1$ (i.e., no units). That is, X_2 is dimensionless, as it must be! Substitution of Eqn. 4.11 into Eqn. 4.10 gives

$$\Delta\mu_1 = \mu_1 - \mu_1{}^\circ \approx -RTC_2 V_1{}^\circ/M_2 + RT\ln f_1 \tag{4.12}$$

for a dilute solution of component 2, the solute. We have now developed a relatively simple and useful mathematical expression. Assuming that the second term on the right-hand side is small enough to be neglected (because $f_1 \approx 1$ and $\ln 1 \approx 0$), as is often the case in biochemical experiments, we see that the chemical potential of a single-component solution relative to the pure solvent $(\Delta\mu_1)$ is directly proportional to the concentration of the solute and inversely proportional to solute mass. The higher the concentration of a solute, the higher its chemical potential. A **substance always diffuses from a region of higher concentration to a region of lower concentration.** The greater the mass of a particle, the lower its chemical potential, as long as its volume is constant. If we start to think about real biological macromolecules like proteins or DNA, things become even more complex, because these molecules are almost always charged. In Chapter 5 we'll see how to take charge into account.

F. | Effect of solutes on boiling points and freezing points

Our study of phase transitions has suggested that biochemists should be aware of them. And the previous section showed that the chemical potential of a substance depends on its concentration. Now we want to combine these bits of knowledge to obtain an expression for the changes in boiling and freezing points of a substance resulting from the addition of a solute. We know that salt is used to keep roads clear of ice in the winter, and that salt is added to the water when cooking rice, but why either of these things is done has perhaps been a mystery up to now. Once we've slogged through a few mathematical formulas, we'll look at examples of how to apply this knowledge in biochemistry.

From Eqn. 4.9 we have

$$\mu_1 - \mu_1{}^\circ = RT\ln f_1 X_1 \tag{4.13}$$

This can also be expressed as

$$(\mu_1 - \mu_1{}^\circ)/RT = \Delta G_{1,m}/RT = (\Delta H_{1,m} - T\Delta S_{1,m})/RT = \ln f_1 X_1 \tag{4.14}$$

where $\Delta G_{1,m} = \Delta G_1/\Delta n_1$, and the '$\Delta$' represents the difference in the state function between the one phase and the other. Because ΔH_m and ΔS_m are relatively insensitive to temperature over short temperature ranges, $\Delta H_m = \Delta H_{tr}$ and $\Delta S_m = \Delta H_{tr}/T_{tr}$, where 'tr' stands for 'transition.' Substitution of these relations into Eqn. 4.14 gives

$$\Delta H_{tr}/R \times (1/T - 1/T_{tr}) = \ln f_1 X_1 \tag{4.15}$$

The difference $T - T_{tr} = \Delta T$ is very small, so $1/T - 1/T_{tr} \approx -\Delta T/T_{tr}^2$. When this and Eqn. 4.10 are substituted in, we obtain

$$-\Delta H_{tr}/R \times \Delta T/T_{tr}^2 = -(X_2 + \ldots) + \ln f_1 \tag{4.16}$$

If the concentration of component 2, the solute, is small, the solvent is almost pure, $\ln f_1 \approx 0$, and Eqn. 4.16 this simplifies to

$$\Delta T \approx R X_2 T_{tr}^2/\Delta H_{tr} \tag{4.17}$$

That's it! It is easy to see that ΔT varies proportionately with X_2; the greater the concentration of solute, the more the temperature of the phase transition will differ from that of the pure solvent. Because this effect depends on the mole fraction of the solute but is independent of the solute's identity, it is called a colligative property, one that depends 'on the collection' of solute molecules.

Are there practical applications of this hard-won knowledge? Yes! In both cases, the addition of the 'solute' changes the temperature of the phase change, in the one case lowering the freezing point and in the other raising the boiling point. There are also applications in the biochemistry lab. For instance, glycerol is often used to reduce the freezing point of aqueous protein solutions. Commercially produced enzymes so necessary for molecular biology are shipped in c. 50% glycerol, as this lowers the freezing point of water to below $-20\,°C$. The enzymes are much more thermostable at this temperature than at $4\,°C$, and this preserves them for much longer than if stored in the fridge. However, maintaining enzymes at $-20\,°C$ entails a cost – that of keeping a freezer running – so this is not a case of getting something for nothing. Perhaps more interesting than the other examples we've seen here are proteins that either raise or lower the freezing point of water with biological effect. The ones that lower it are called, colorfully enough, anti-freeze proteins, and they are found in polar icecap fish that live in sea water at $-1.9\,°C$. Anti-freeze proteins bind to and arrest the growth of ice crystals that enter the fish and thereby prevent the fish from freezing to death. Anti-freeze proteins are composed of repeats of glycotripeptide monomer (Thr–Ala/Pro–Ala) with a disaccharide attached to the hydroxyl group of each threonine residue. In contrast, some bacterial proteins *increase* the probability that supercooled water (liquid water cooled below its normal freezing point) will freeze. Biological thermodynamics.

G. | Ionic solutions

Ions in solution are called electrolytes. Charged particles get this name from their ability to conduct an electric current. Discussion about them takes up space in this tome because water is all over the place in the biological world, life as we know it cannot exist without it, and water is usually full of ions. Moreover, three of the four major classes of biological *macromolecule* – proteins, nucleic acids, and lipids – are charged at neutral pH, even if their net charge might be zero (usually not). Charge properties help to give biomacromolecules important physical properties that are closely bound to their physiological function.

Ionic solutions are dominated by electrical forces that can be very strong. Referring back to Table 2.2, we see that the electrostatic energy of two electronic charges can be as great as 14 kcal mol^{-1}, an energy as large as or larger than the free energy difference between the folded and unfolded states of a protein at room temperature! Often, though, electrostatic interactions are much weaker than this, as they depend not only distance between the interacting charges but also on the dielectric constant of the medium, D. In the middle of a protein, which is a bit like an oily solid, $D \approx 4$, and this reduces the electrostatic interaction by just 25% in comparison to its vacuum value. In bulk aqueous solution, however, which is very polar, $D \approx 80$! Water thus greatly reduces the distance over which the strength of the electric field created by a charge is significant. The strength of charge–charge interactions in water is reduced even further by the alignment of a few water molecules in a 'solvation shell' around an ion.

EDTA is a cation chelator. (Positive ions, e.g. Na$^+$, are called cations. Cl$^-$ and other negative ions are anions.) It is a useful tool in the biochemist's kit because it can be used in a variety of practical situations. For instance, when preparing dialysis tubing, EDTA is used to 'trap' divalent metal ions, 'removing' them from solution. ΔH for the binding of Mg^{2+} is positive, but because $-T\Delta S$ is very negative, $\Delta G < 0$ and binding occurs. One wants to rid the solution of such metal ions because certain proteases require them for enzymatic activity; the activity of these proteases is greatly reduced in the presence of EDTA. Anyone who has ever tried to make an aqueous EDTA solution will know very well that the sodium salt of EDTA does not dissolve very quickly at room temperature. If enough EDTA is present, when the solution comes to equilibrium only some of the salt will have dissolved.

What we want to do now is find some general ways of thinking about thermodynamic properties of electrolytes. We'll do this by way of the example of EDTA. But we must always bear in mind that the example is intended to show how the thinking involved is much broader in scope. The solubility equilibrium of EDTA can be written as

$$Na_4EDTA(s) \Leftrightarrow 4Na^+ + EDTA^{4-} \tag{4.18}$$

Based on this equation, at equilibrium,

$$\mu_{NaEDTA} = 4\mu_{Na^+} + \mu_{EDTA^{4-}} \tag{4.19}$$

The positive and negative ions appear as a pair because it is not possible to make separate measurements of the chemical potentials on the right hand side. To take this doubling effect into account, we define the **mean chemical potential**, $\mu_\pm$, which in this case is

$$\mu_\pm = \frac{4}{5}\mu_{Na^+} + \frac{1}{5}\mu_{EDTA^{4-}} \tag{4.20}$$

The coefficients account for the stoichiometry of dissociation of EDTA; for each EDTA ion there are four sodium ions. Eqn. 4.19 can now be rewritten as

$$\mu_{\text{NaEDTA}} = 5\mu_{\pm} \tag{4.21}$$

In the more general case, which admittedly is very abstract,

$$W \Leftrightarrow v_+ A^{z+} + v_- B^{z-} \tag{4.22}$$

In this equation W is a neutral compound, A and B are positive and negative ions with ion numbers z^+ and z^-, and v_+ and v_- are stoichiometric coefficients. The mean chemical potential is

$$\mu_{\pm} = \frac{(v_+ A^{z+} + v_- B^{z-})}{v_+ + v_-} = \frac{\mu_{\text{salt}}}{v_+ + v_-} \tag{4.23}$$

This is just W divided by or 'weighted by' the sum of the stoichiometric coefficients. If the path to Eqn. 4.23 has left you perplexed, try working through the equations using EDTA as the example.

Substituting Eqn. 4.9 into Eqn. 4.20, we have

$$\mu_{\pm} = \frac{4}{5}\left(\mu_{\text{Na}^+}^{\circ} + RT \ln(f_{\text{Na}^+} X_{\text{Na}^+})\right) + \frac{1}{5}\left(\mu_{\text{EDTA}^{4-}}^{\circ} + RT \ln(f_{\text{EDTA}^{4-}} X_{\text{EDTA}^{4-}})\right) \tag{4.24}$$

Making use of $x \ln a = \ln a^x$, a handy formula from mathematics, gives

$$= \mu_{\pm}^{\circ} + RT \ln \sqrt[5]{(f_{\text{Na}^+} X_{\text{Na}^+})^4 f_{\text{EDTA}^{4-}} X_{\text{EDTA}^{4-}}} \tag{4.25}$$

where the standard state chemical potentials of Na^+ and pure $EDTA^{4-}$ have been combined in the first term on the right-hand side of Eqn. 4.25.

Just as the chemical potentials of the ions cannot be measured separately, neither can one measure the activity coefficients separately. We therefore define a mean ionic activity coefficient, which for the example is

$$f_{\pm} = f_{\text{Na}^+}^{4/5} f_{\text{EDTA}^{4-}}^{1/5} \tag{4.26}$$

In the more general case, Eqn. 4.26 this looks like

$$f_{\pm} = (f_+^{v_+} f_-^{v_-})^{1/(v_+ + v_-)} \tag{4.27}$$

where f_+ and f_- are the activity coefficients of the positive and negative ions on the mole fraction scale.

Knowing the mean activity coefficient of a salt can be important in the interpretation of results in experimental biochemistry. This will be especially true whenever the solution conditions are far from ideal, particularly whenever the salt concentration is high. The bacteria that live in the Dead Sea, known as halophiles, thrive in a high salt environment. Somehow or other the molecular machinery of these bugs must be able to cope with high salt conditions. A high salt solution is far from ideal, so the activity coefficients of the ions in halobacteria must deviate substantially from unity. The salt guanidinium chloride (GuHCl) is a strong protein denaturant. Most proteins are unfolded at a concentration of about 5 M GuHCl (a large concentration indeed, but one still many times smaller than the concentration of pure water, which is about ten-fold greater). Like HCl, GuHCl dissociates completely in aqueous solution to guanidinium ion and chloride ion if the solubility limit is not exceeded (it is well above 5 M at 25 °C). To try to explain in molecular terms what GuHCl does to protein structure, one needs to know its activity coefficient. We'll learn more about guanidinium chloride-induced unfolding of proteins in Chapter 5.

Now we want to think about electrolytes in a slightly different way. What follows is a simplified version of the theory of strong electrolytes developed by the Dutchman Petrus Josephus Wilhelmus Debye (1884–1966; awarded the Nobel Prize in Chemistry in 1936) and the German Erich Hückel (1896–1980). The work was published in 1923.

The activity of an ion depends on a quantity known as the ionic strength, I, which is defined as

$$I = \frac{1}{2} \sum_i z_i^2 C_i \qquad (4.28)$$

where C_i, the *molality*, is the ratio of the mole fraction of solute i to the molecular mass of the solvent in *kilograms*. Note that if the salt we're working with is relatively simple, like NaCl, there is no problem in computing I: NaCl dissociates completely below its solubility limit, each particle carries just one charge, so a one molal solution of a this salt has an ionic strength of $[(1^2 \times 1) + (1^2 \times 1)]/2 = 1$. $CaCl_2$ is somewhat more complicated, because the ions involved no longer carry the same charge. Things are much more complicated when dealing with polyvalent ions like proteins, because the degree of ionization is sensitive to pH. Polyvalent ions and ionization will be treated in greater depth below.

It can be shown (see any good physical chemistry text) that the activity coefficient of ion i on the molality scale is

$$\log \gamma_i = - H z_i^2 \sqrt{I}$$

where γ_i is the activity coefficient on the *molality* scale. H is a very complicated expression that depends the density of the solvent, the absolute temperature, the charge on an electron, the dielectric constant of the solvent .... For ions in water at 25 °C, $H = 0.509$. This provides a way of calculating γ_i. Enough about ions for now!

H. | Equilibrium constant

We have already done a fair amount with the concept of equilibrium, and by now we should have a good sense of how important a concept it is. But in fact, we have only scratched the surface of what could be known. Many things in life work that way. To dig deeper, in this section we approach thinking about equilibrium in a new way, one that is extremely useful to biological scientists, particularly biochemists.

Table 4.3 | Relationship between $\Delta G°$ and K_{eq}

Free energy change	Equilibrium constant
$\Delta G° < 0$	$K_{eq} > 1$
$\Delta G° = 0$	$K_{eq} = 1$
$\Delta G° > 0$	$K_{eq} < 1$

Given a general reaction

$$a\text{A} + b\text{B} \Leftrightarrow c\text{C} + d\text{D} \qquad (4.30)$$

the overall free energy change is

$$\Delta G = c\mu_C + d\mu_D - a\mu_A - b\mu_B \qquad (4.31)$$

Substituting Eqn. 4.31 into Eqn. 4.5, we have

$$\Delta G = \Delta G° + RT \ln\left(\frac{[\text{C}]^c[\text{D}]^d}{[\text{A}]^a[\text{B}]^b}\right) \qquad (4.32)$$

where $\Delta G° = c\mu_C° + d\mu_D° - a\mu_A° - b\mu_B°$ and we have assumed that the reaction occurs at infinite dilution (i.e. $a_X \approx [\text{X}]$ for each X). The argument of the logarithm ($[\text{C}]^c[\text{D}]^d/[\text{A}]^a[\text{B}]^b$) is called the **mass action ratio**. The right-hand side of this equation shows that the free energy change of a reaction has two parts: a constant term that depends only on the particular reaction taking place, and a variable term that depends on temperature, concentrations of reactants and products, and stoichiometric relationships.

Now, at equilibrium, the forward reaction exactly balances the reverse reaction, and $\Delta G = 0$. It follows that

$$\Delta G° = - RT \ln\left(\frac{[\text{C}]_{eq}^c[\text{D}]_{eq}^d}{[\text{A}]_{eq}^a[\text{B}]_{eq}^b}\right) = - RT \ln K_{eq} \qquad (4.33)$$

where the subscript 'eq' signifies 'equilibrium.' The concentrations shown are the concentrations at equilibrium. K_{eq}, the **equilibrium constant** of the reaction, is *defined* as

$$K_{eq} = \left(\frac{[\text{C}]_{eq}^c[\text{D}]_{eq}^d}{[\text{A}]_{eq}^a[\text{B}]_{eq}^b}\right) = \left(\frac{\text{rate constant for forward reaction}}{\text{rate constant for reverse reaction}}\right) \qquad (4.34)$$

The relationship between K_{eq} and reaction rates, which was put forward by van't Hoff, has been included here as a foretaste of Chapter 8. Eqn. 4.33 indicates that K_{eq} can be calculated from standard state free energies. The form of relationship is illustrated in Fig. 4.10. We can get some feel for magnitudes by substituting in values (see Table 4.3). For instance, when $\Delta G° = 0$, $K_{eq} = 1$ (no units). A ten-fold change in K_{eq} at 25 °C corresponds to a change in $\Delta G°$ of 5.7 kJ mol^{-1}, an energy difference between two and three times greater than thermal energy at room temperature. As a rule of thumb, if $K_{eq} > 10^4$, the reaction will go to completion. According to Eqns. 4.33 and 4.34, deviations from equilibrium will stimulate a change in the system towards the equilibrium concentrations of reactants and products. This is known as **Le Châtelier's princi-**

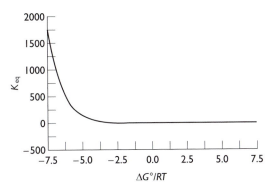

Fig. 4.10 Variation of K_{eq} with $\Delta G°$. K_{eq} is a function of the negative logarithm of $\Delta G°$. When $\Delta G°$ is large and negative, K_{eq} is very large. When $\Delta G°$ is large and positive, K_{eq} is very small. In most biochemical reactions, $\Delta G°$ will fall within the range shown here.

ple (named after the French chemist Henri Louis Le Châtelier (1850–1936)). Comparison of Eqn. 4.32 with Eqn. 4.34 tells us that alteration of the direction of a reaction with a very large or very small value of K_{eq} by changing the mass action ratio is difficult, though not impossible. In the next chapter we shall see how this is important in glycolysis.

Here's a worked example of the relationship between Gibbs free energy change and equilibrium constant. Consider the reaction in which a phosphoryl group is moved from carbon 1 of glucose to carbon 6:

$$\text{Glucose-1-phosphate} \Leftrightarrow \text{glucose-6-phosphate} \tag{4.35}$$

The equilibrium constant for this reaction, which is catalyzed by phosphoglucomutase, is 19. What is the value of $\Delta G°$ at room temperature? By Eqn. 4.33, $\Delta G° = -(8.314\,\text{J mol}^{-1}\text{K}^{-1}) \times 298\,\text{K} \times \ln(19) = -7.3\,\text{kJ mol}^{-1}$. Thus, the reaction proceeds spontaneously to the right *under standard conditions*. We need not have calculated $\Delta G°$ to know this. For as Table 4.3 shows, when $K_{eq} > 1$, $\Delta G° < 0$, and this meets the requirement for spontaneity. Let's see what happens when we alter the concentrations of reactants and products in Eqn. 4.35 and require the concentrations of glucose-1-phosphate and glucose-6-phosphate to be held at 10 mM and 1 mM, respectively. By Eqn. 4.32, $\Delta G = -7.3\,\text{kJ mol}^{-1} + (8.314\,\text{J mol}^{-1}\text{K}^{-1}) \times 298\,\text{K} \times \ln(1/10) = -13\,\text{kJ mol}^{-1}$. The reaction is still exergonic, but much more so than under standard state conditions. This tells not only that we are far from equilibrium, where $\Delta G = 0$, but also the direction in which the reaction will proceed in order to reach equilibrium. We see that the magnitude of ΔG can depend significantly on the concentrations of reactants and products. In fact, the concentrations can be different enough in some cases (when $\Delta G \approx 0$) to reverse the direction of the reaction, a phenomenon that is relatively common in metabolism and occurs in glycolysis. Now, what effect does the enzyme have on all this? Will it speed up the reaction? Yes! Will it change the free energy difference between products and reactants? No! Why not? The Gibbs free energy is a thermodynamic state function, so what it measures depends only on the state of the system and not at all on how that system was prepared.

I. | Standard state in biochemistry

Most *in vitro* biochemical experiments are done at constant temperature and pressure in dilute aqueous solution near neutral pH. To be maximally useful, the **biochemist's definition of the standard state** should take all these conditions into account. This will seem to make things more complicated at first, but in fact it simplifies life.

We *define* the standard state of water to be that of the pure liquid. This means that the activity of water is *set* to 1, even though its concentration is 55.5 M. One justification for this is that if a reaction is carried out in dilute aqueous solution, the percentage change in the concentration of water is negligible. Another is that it is easy to multiply and divide by 1! (While many *in vitro* experiments are carried out under relatively close-to-ideal conditions, one could justifiably object that such conditions might differ greatly from those in a cell. We'll come back to this point.) The standard temperature is *defined* as 25 °C, a convenient value for bench top experiments and one close to the physiological temperature of many organisms. The hydrogen ion activity is *defined* as unity at pH 7 (neutral solution), not pH 0 (highly acidic solution where the activity of H_3O^+ is 1; H_3O^+ symbolizes the hydronium ion, the proton in aqueous solution). (A more thorough treatment of acids and bases is given below.) Finally, no account is taken of the various ionization states that might be present at pH 7. This is particularly relevant to biological macromolecules like proteins, which have a multitude of ionization states. When all these conditions are accounted for, the standard state free energy change is symbolized as $\Delta G^{\circ\prime}$. See Table 4.4 for free energies of important biochemical reactions.

To see how all this works, consider the following chemical reaction:

$$A + B \Leftrightarrow C + D + nH_2O \tag{4.36}$$

From Eqn. 4.33,

$$\Delta G^{\circ} = -RT \ln K_{eq}$$

$$= -RT \ln([C]_{eq}[D]_{eq}[H_2O]^n/[A]_{eq}[B]_{eq})$$

$$= -RT \ln([C]_{eq}[D]_{eq}/[A]_{eq}[B]_{eq}) - nRT \ln[H_2O] \tag{4.37}$$

which, on invoking the biochemistry convention described above, is

$$\Delta G^{\circ\prime} = -RT \ln K_{eq}{}' = -RT \ln([C]_{eq}[D]_{eq}/[A]_{eq}[B]_{eq}) \tag{4.38}$$

Thus, the relationship between ΔG° and $\Delta G^{\circ\prime}$ is

$$\Delta G^{\circ\prime} = \Delta G^{\circ} + nRT \ln[H_2O] \tag{4.39}$$

What if protons are 'consumed' in the reaction? This can be modeled as

$$A + \nu H^+(aq) \rightarrow P \tag{4.40}$$

where ν is the number of protons transferred to reactant A in forming product P. By Eqn. 4.31,

$$\Delta G = \mu_P - \mu_A - \nu \mu_{H^+} \tag{4.41}$$

Table 4.4 | Values of $\Delta G^{\circ\prime}$ for some important biochemical reactions

Reaction	$\Delta G^{\circ\prime}$ (kcal mol^{-1})
Hydrolysis	
Acid anhydrides:	
Acetic anhydride $+ H_2O \rightarrow 2$ acetate	-21.8
$PP_i + H_2O \rightarrow 2P_i$	-8.0
$ATP + H_2O \rightarrow ADP + 2P_i$	-7.3
Esters:	
Ethylacetate $+ H_2O \rightarrow$ ethanol $+$ acetate	-4.7
Glucose-6-phosphate $+ H_2O \rightarrow$ glucose $+ P_i$	-3.3
Amides:	
Glutamine $+ H_2O \rightarrow$ glutamate $+ NH_4^+$	-3.4
Glycylglycine $+ H_2O \rightarrow 2$ glycine (a peptide bond)	-2.2
Glycosides:	
Sucrose $+ H_2O \rightarrow$ glucose $+$ fructose	-7.0
Maltose $+ H_2O \rightarrow 2$ glucose	-4.0
Esterification	
Glucose $+ P_i \rightarrow$ glucose-6-phosphate $+ H_2O$	$+3.3$
Rearrangement	
Glucose-1-phosphate $\rightarrow$ glucose-6-phosphate	-1.7
Fructose-6-phosphate $\rightarrow$ glucose-6-phosphate	-0.4
Glyceraldehyde 3-phosphate $\rightarrow$ dihydroxyacetone phosphate	-1.8
Elimination	
Malate $\rightarrow$ fumarate $+ H_2O$	$+0.75$
Oxidation	
Glucose $+ 6O_2 \rightarrow 6CO_2 + 6H_2O$	-686
Palmitic acid $+ 23O_2 \rightarrow 16CO_2 + 16H_2O$	-2338
Photosynthesis	
$6CO_2 + 6H_2O \rightarrow$ six-carbon sugars $+ 6O_2$	$+686$

Note:
The data are from Lehninger, A.L. *Biochemistry*, 2nd ed. (New York: Worth, 1975), p. 397.

If A and P are in their standard states, then

$$\Delta G = \mu_P^\circ - \mu_A^\circ - \nu\mu_{H^+} \qquad (4.42)$$

As to the chemical potential of H^+, according to Eqn. 4.5,

$$\mu_{H^+} = \mu_{H^+}^\circ + RT \ln a_{H^+} = \mu_{H^+}^\circ - 2.303RT(\text{pH}) \qquad (4.43)$$

where we have assumed ideal conditions, used the definition of pH ($\log[H^+] \approx -\text{pH}$). Combining Eqns. 4.42 and 4.43 gives, and accounted for the difference between $\ln x$ and $\log x$

$$\Delta G = \mu_P^\circ - \mu_A^\circ - \nu(\mu_{H^+}^\circ - 2.303RT(\text{pH})) = \Delta G^\circ + \nu 2.303RT(\text{pH}) \qquad (4.44)$$

The biochemist's standard state, however, is defined for pH $= 7$. So,

$$\Delta G^{\circ\prime} = \Delta G^\circ - \nu 16.121RT \qquad (4.45)$$

The free energy difference varies linearly with T and v, the number of protons transferred in the reaction. If neither H_2O nor H^+ is involved in the reaction, then $n = 0$ and $v = 0$, and $\Delta G^{\circ\prime} = \Delta G^\circ$.

J. | Effect of temperature on K_{eq}

Just as the concentrations of reactants and products depend on the physical conditions, so too does the equilibrium constant. K_{eq} varies with temperature as follows:

$$\ln K_{eq} = -\Delta G^\circ/RT = -(\Delta H^\circ/R)(1/T) + \Delta S^\circ/R \qquad (4.46)$$

(See Fig. 4.11.) As before, the superscript symbol indicates standard state, but because we have not specified pH 7, we do not need the prime. In general, ΔH° and ΔS° are temperature dependent. Often, though, the enthalpy and entropy changes will be such weak functions of temperature that one can plot $\ln K_{eq}$ and get something that is very nearly a linear function of $(1/T)$. A plot of $\ln K_{eq}$ versus $(1/T)$ is called a **van't Hoff graph**, and the slope and intercept of the line are $-\Delta H^\circ/R$ and $\Delta S^\circ/R$, respectively. The upshot of this is that if K_{eq} can be measured at several temperatures within a suitable range, ΔH° and ΔS° can be *measured*, albeit indirectly. ΔH° determined from the behavior of K_{eq} is called the **van't Hoff enthalpy**, or ΔH_{vH}. Jacobus Henricus van't Hoff, a Dutch physical chemist, lived 1852–1911. He was awarded the Nobel Prize in Chemistry in 1901, the first year the prestigious prizes were given.

If ΔH° and ΔS° do depend significantly on temperature, as with proteins because of the large ΔC_p on unfolding, as long as folding/unfolding is cooperative, the unfolding phase transition will occur over a narrow range of temperatures, as in the melting of a pure solid, and ΔH° and ΔS° will be *approximately* constant within that range. A van't Hoff plot can therefore be used to estimate these thermodynamic functions *at the transition temperature*. Measurements of protein energetics made in this way have been corroborated by differential scanning calorimetry (Fig. 2.10B). This is important because calorimetry provides the only means of making a direct measurement of the heat absorbed upon protein unfolding, q. The measurement occurs at constant pressure, so $q = \Delta H = \Delta H_{cal}$, the **calorimetric enthalpy**, the area below the heat absorption peak. If the folding/unfolding reaction is highly cooperative and involves effec-

Fig. 4.11 A van't Hoff plot. This approach to data analysis will be useful when two states only are involved in the transition *and* if either of two conditions is met: ΔC_p is of negligible magnitude; or, if ΔC_p is large, the temperature range is small enough that ΔH° and ΔS° can be considered approximately independent of temperature. A van't Hoff analysis is often used when data are collected using a technique that does not make a direct measurement of heat (e.g. any type of optical spectroscopy, NMR spectroscopy, viscometry, X-ray scattering, electrophoresis. . .). Calorimetry measures heat exchange directly. For a discussion of multi-state equilibria, see Chapter 6.

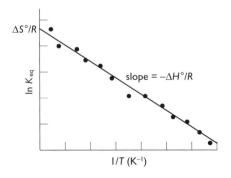

tively only two states, then the ratio $\Delta H_{vH}/\Delta H_{cal} = 1$. Deviations from such behavior, whether due to oligomerization or stabilization of partly folded species, have enthalpy ratios greater than or less than one, respectively. We shall return to this subject in later chapters.

How does a fluctuation in T translate into a fluctuation in $\ln K_{eq}$? In other words, how does $T \to (T + \Delta T)$ affect $\ln K_{eq}$? If the change brought about in $\ln K_{eq}$ by the temperature fluctuation is $\Delta \ln K_{eq}$, by Eqn. (4.44) we have

$$\ln K_{eq} + \Delta \ln K_{eq} = -(\Delta H^\circ/R)\big(1/(T+\Delta T)\big) + \Delta S^\circ/R \qquad (4.47)$$

To evaluate $1/(T+\Delta T) = (T+\Delta T)^{-1}$, we make use of a famous result in mathematics known as the binomial theorem. According to the theorem, $(x+y)^n = x^n + nx^{n-1}y + n(n-1)x^{n-2}y^2/2 + \ldots$, as long as $y < x$. For us the requirement is $\Delta T < T$, which is definitely the case! Rewriting Eqn. 4.47, we have

$$\ln K_{eq} + \Delta \ln K_{eq} = -(\Delta H^\circ/R)\big(T^{-1} + (-1)T^{-2}\Delta T + (-1)(-2)T^{-3}(\Delta T)^2/2 + \ldots\big)$$

$$+ \Delta S^\circ/R = -(\Delta H^\circ/R)(1/T) + (\Delta H^\circ/R)(T^{-2})(\Delta T)$$

$$- (\Delta H^\circ/R)(T^{-3})(\Delta T)^2/2 + \ldots + \Delta S^\circ/R \qquad (4.48)$$

The first term and last term on the right-hand side of Eqn. (4.48) sum to $\ln K_{eq}$, so we can cancel these from both sides. This leaves us with

$$\Delta \ln K_{eq} = (\Delta H^\circ/R)(T^{-2})(\Delta T) - (\Delta H^\circ/R)(T^{-3})(\Delta T)^2/2 + \ldots \qquad (4.49)$$

If ΔT is a small quantity, as a slight deviation from T is supposed it to be, then the second term on the right hand side and all subsequent terms will be very small, negligible in comparison to the first term. This leaves us with

$$\Delta \ln K_{eq} \approx (\Delta H^\circ/R)(T^{-2})(\Delta T) \qquad (4.50)$$

an important thermodynamic relationship. To convince ourselves that the approximation works, suppose that $\Delta H^\circ = 50\,000$ cal mol^{-1} at 300 K, a realistic value for a small protein. If $\Delta T \sim 1$ K, then by Eqn. (4.49) $\Delta \ln K_{eq} = (50\,000$ cal mol$^{-1}) \times (300$ K$)^{-2} \times (1$ K$)$ / $(1.9872$ cal mol^{-1} K$^{-1})$ $- (50\,000$ cal mol$^{-1}) \times (300$ K$)^{-3} \times (1$ K$)^2$ / $(1.9872$ cal mol^{-1} K$^{-1})$ $/2 = 0.2796 - 0.0009$. That is, in neglecting the second term, we incur an error of less than 1%, which is smaller than the usual uncertainty in measuring protein concentration! The error of the neglected terms would be greater if all of them were to be taken into account, but it can be shown that the total error would still be very small. If, however, we were working at temperatures much closer to absolute zero, the calculation would not involve division by such a large number and the terms we are neglecting would be a lot bigger. Is this why the matrix of life is in the liquid state so far from absolute zero?

Eqn. (4.50) does indeed provide us with a very handy way of calculating the effect of temperature on $\ln K_{eq}$. You may have noticed, though, that we made a very important simplifying assumption in our calculation. We ignored the well-known experimental certainty that ΔH° and ΔS° vary with temperature! As we have said, though, the dependence on temperature is often small, and when it is not small over a large temperature range, as for proteins, it is nevertheless small over a small range (c. 10 degrees centigrade).

K. | Acids and bases

One of the most important examples of equilibrium in biology is that between acids and bases in solution. According to the **Brønsted–Lowry definitions**, an **acid** is a proton donor, a **base** a proton acceptor. (Johannes Nicolaus Brønsted, a Danish physical chemist, lived 1879–1947. Thomas Martin Lowry was an Englishman. Brønsted and Lowry introduced their definitions simultaneously but independently in 1923.) Of central interest to biological scientists are aqueous solutions and their properties. One of the most important properties of an aqueous solution is its pH, defined as

$$pH = -\log a\,H_3O^+ \tag{4.51}$$

The pH of a solution plays a key role in determining the extent of proton dissociation from ionizable chemical groups in biological macromolecules, which can have a profound effect on enzyme activity, protein–protein association, protein–DNA binding, any other types of biochemical reaction. We had therefore better put some effort into this and eventually know it cold.

Suppose we have an acid HA. It participates in the following reaction in water:

$$HA(aq) + H_2O(l) \Leftrightarrow H_3O^+(aq) + A^-(aq) \tag{4.52}$$

The **acidity constant** for this reaction is *defined* as

$$K_a = \left(\frac{a_{H_3O^+} a_{A^-}}{a_{HA} a_{H_2O}} \right) \approx \left(\frac{a_{H_3O^+} a_{A^-}}{a_{HA}} \right) \tag{4.53}$$

The approximation is justified on the same ground as before: if we concern ourselves with dilute aqueous solutions, the activity of water is close to 1 and unchanging. At low concentrations, the activity of hydronium ions is roughly equal to their molar concentration, and the acidity constant is usually written as

$$K_a \approx [H_3O^+][A^-]/[HA] \tag{4.54}$$

where the concentrations are in $mol\,l^{-1}$. It is important to realize that this approximation is valid only when *all* the ions in solution are present in low concentrations. For instance, 1 mM imidazole buffer has an activity coefficient of about 0.95, and ignoring this can affect the equilibrium constant by as much as 10%! (Imidazole is a common buffer in biochemistry. Its structure is the same as that of the side chain of histidine.) Acidity constant values are often tabulated in terms of their pK_a:

$$pK_a = -\log K_a \tag{4.55}$$

Table 4.5 pK_a' values of acidic and basic groups in proteins

Group	Amino acid residue	pK_a' (25 °C)
α-Carboxyl		3.0–3.2
Carboxyl	Aspartic acid	3.0–4.7
	Glutamic acid	~4.5
Imidazolyl	Histidine	5.6–7.0
α-Amino		7.6–8.4
Sulfhydryl	Cysteine	9.1–10.8
Phenolic hydroxyl	Tyrosine	9.8–10.4
Guanidino	Arginine	11.6–12.6

Note:

That a range of values is given. This is because the acidity constant will depend in part on the specific electronic environment of the acidic or basic group.

The pK_a is related to the magnitude of the standard state free energy of ionization as

$$pK_a = \Delta G° / (2.303RT) \tag{4.56}$$

If we take the logarithm of both sides of Eqn. 4.54 and rearrange, we obtain the **Henderson–Hasselbalch equation**:

$$pH = pK_a - \log([HA]/[A^-]) \tag{4.57}$$

named after American biochemist Lawrence Joseph Henderson (1878–1942) and Danish biochemist Karl A. Hasselbalch (1874–1962). The form of this equation is exactly like that of Eqn. 4.7. Eqn. 4.57 tells us that the pK_a of an acid is the pH at which half of the protons are dissociated (when $[HA] = [A^-]$, $\log([HA]/[A^-]) = 0$). It is also the pH where a buffer exhibits is greatest **buffering capacity**; i.e., where a given amount of change in $[H_3O^+]$ or $[OH^-]$ has the smallest effect on the solution pH. The origin of this effect is the relative abundance of A^- ions present, which can react with most of the hydronium ions produced on addition of strong acid. Similarly, the relative abundance of HA present can react with any strong base that is added. The pK_a values of acidic and basic groups in proteins are given in Table 4.5. Proteins themselves are the main buffering systems in biological organisms. For example, about 80% of the buffering capacity of human blood is due to proteins, principally serum albumin and hemoglobin.

Let's use Eqn. 4.57 to calculate the pH of a common biological buffer solution containing 0.200 mol l^{-1} KH$_2$PO$_4$ (monobasic potassium phosphate) and 0.100 mol l^{-1} K$_2$HPO$_4$ (dibasic potassium phosphate). The equation describing the equilibrium between acid and base is as follows:

$$H_2PO_4^-(aq) + H_2O(l) \Leftrightarrow H_3O^+(aq) + HPO_4^{2-}(aq) \tag{4.58}$$

The pK_a for this reaction, which is based on measurement, can be found in standard tables and is 7.21. Plugging all the numbers into Eqn. 4.57 gives

$$pH = 7.21 - \log(0.2/0.1) = 6.91 \qquad (4.59)$$

It is clear why potassium phosphate is often used by biochemists to buffer biochemical reactions.

Despite the neat appearance of the calculation we've just done, the situation with phosphate buffer is rather more complicated. This is because, unlike a simple acid like HCl, phosphoric acid is **polyprotic**. That is, phosphoric acid can donate more than one proton, so it has more than one pK_a. One of the other two pK_as is much higher than pH 7, the other is much lower. Neither therefore makes a large contribution to the acid-base equilibrium at pH 7, and we can justify ignoring them in the neutral pH range. One should not, however, neglect the additional terms of a polyprotic acid if it has acidity constants that differ by less than about two pH units! We shall see why in Chapter 6. Fig. 4.12 shows the how the net charge varies with pH for two polyprotic acids, phosphoric acid (panel (A)), in which there are three, well-separated ionization constants, and the protein ribonuclease (panel (B)), in which there are numerous ionizable groups of similar value.

L. | Chemical coupling

The Gibbs free energy is a state function. As we have seen, this means that individual contributions to an overall free energy change are additive (Fig. 2.3). An extremely important consequence of this for biology is that a thermodynamically unfavorable, endergonic reaction ($\Delta G > 0$) can be 'powered' by an exergonic reaction ($\Delta G < 0$), if the two reactions are *chemically* 'coupled' and the overall free energy change under the same conditions is negative. Here's an analogy. A familiar example of *mechanical* coupling is the use of the downhill flow of a stream to turn a waterwheel that could, by means of suitable gears, be used to grind grain into flour. Another type of mechanical coupling is depicted in Fig. 4.13. In panel (A), each weight represents an energetically favorable situation: both weights will fall to the ground spontaneously. The reverse reactions, which involve an increase in gravitational potential energy, will not occur spontaneously! If, however, the pulleys are coupled, as in panel (B), the more massive weight can be used to do work on the lighter one. Experimental studies have shown that coupling of biochemical reactions is an essential thermodynamic principle for the operation of

Fig. 4.12 Acid–base titration curves. Panel (A), the titration of sodium phosphate at 25 °C. Phosphoric acid has three dissociable protons and therefore three pK_a values. The pK_as of phosphate are well-separated on the pH axis. This means that titration of the first site is effectively complete before titration of the second site begins, and so forth. Panel (B), the titration of ribonuclease at 25 °C. Ribonuclease is an extremely complex molecule. Not only are there several different types of titratable group (the different types of ionizable amino acids), but the specific chemical environment of a dissociable proton can result in a substantial shift of its pK_a relative to the value for the free amino acid (in some cases by more than 2 pH units!). This makes the titration curve of a protein very complex. Panel (B) is based on Tanford & Hauenstein (1956).

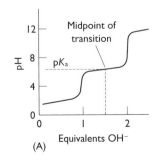

(A)

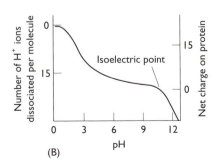

(B)

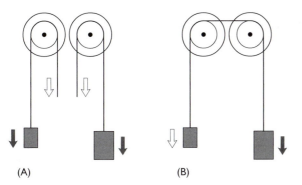

(A)   (B)

metabolic pathways, including for instance the citric acid cycle (Chapter 5).

As an example of chemical coupling, consider the following two-step process:

$$A + B \Leftrightarrow C + D \qquad \Delta G_1 \qquad (4.60)$$

$$D + E \Leftrightarrow F + G \qquad \Delta G_2 \qquad (4.61)$$

Reaction 4.60 will not occur spontaneously unless $\Delta G_1 < 0$. But let's suppose that $\Delta G_1 > 0$, so that the reaction does not occur on its own. Let's also suppose that the reaction of Eqn. 4.61 is spontaneous. Then the second reaction can be used to drive the first one if two conditions are met: the reactions involve a common compound (in this case, D), and the overall free energy change ($\Delta G_1 + \Delta G_2$) is negative. When both conditions are satisfied, the overall reaction proceeds spontaneously, even if the amount of compound D formed by Reaction 4.60 is very small. These two reactions are said to be **coupled**, and substance D is called a **common intermediate** or **energy transducer**.

Now let's look at a specific biochemical example. Transport of glucose into the cell is accompanied by phosphorylation. The negative charge on glucose prevents the diffusion of this valuable source of free energy back out of the cell. This something like a farmer making sure that all the wheat he's harvested is stored in a building that won't allow the wind to blow it away. The interior of the plasma membrane is made of hydrocarbon and, like the interior of a protein, is a region a low dielectric. It is therefore energetically unfavorable for something charged to pass into a membrane from bulk aqueous solution. It makes very good sense from an energetic point of view for phosphorylation of glucose to occur *after* it's entered the cell. The overall coupled reaction described here is one in which a phosphoryl group is transferred from ATP to glucose, but it should be noted that neither of the relevant half-reactions (ATP hydrolysis, glucose phosphorylation) *obligatorily* drives the other reaction.

Let's look at some details of the energetics of glucose phosphorylation:

$$\text{glucose} + P_i \Leftrightarrow \text{glucose-6-phosphate} + H_2O \quad \Delta G°' = 13.8 \text{ kJ mol}^{-1} \quad (4.62)$$

The blood concentration of glucose, the brain's primary fuel source, is ~ 5 mM. Given a cellular glucose concentration of about 300 μM, for the reaction to proceed to the right on the basis of concentration differences alone, the concentration of glucose-6-phosphate in the blood would have to be over 100 mM. (Can you prove that?) The way this actually works in our bodies is that phosphorylation is coupled to ATP hydrolysis. The overall reaction can be written as

$$\text{glucose} + \text{ATP} \Leftrightarrow \text{glucose-6-phosphate} + \text{ADP} \tag{4.63}$$

This coupled reaction, which has $\Delta G^{\circ\prime} = -17.2$ kJ mol^{-1}, is clearly energetically favorable and will proceed to the right spontaneously.

Examination of Eqn. 4.63 can give a good sense of how reaction coupling works on the molecular level. Minimizing the concentration of P_i would promote the forward reaction (by mass action), so an enzyme machine we might design to carry out the phosphorylation reaction should avoid the accumulation of P_i. Similarly, minimizing the concentration of H_2O in the vicinity of our molecule-sized 'workbench' would minimize the probability of transfer of P_i from ATP to water; we want the phosphate group to be transferred to glucose! Crystallographic studies of the enzyme hexokinase have revealed much about how the process probably occurs. Binding of glucose induces a conformational change in the enzyme that increases its affinity for ATP by 50-fold and excludes water from the catalytic site. The functional groups of the amino acid side chains involved in catalysis move into proper alignment and a phosphoryl group is transferred from ATP to glucose. The low activity (concentration) of water in the active site of the enzyme is crucial to the reaction. Measurements have shown that the conformational change induced in the enzyme upon glucose binding results in the release of about 60 water molecules into bulk solution. This contributes a substantial increase in entropy to the overall energetics of the reaction, offsetting the highly unfavorable entropy change of bringing glucose and ATP simultaneously into an orientation that permits phosphoryl transfer to the appropriate hydroxyl group on glucose. Hexokinase is an amazing molecular machine! Finally, it should be noted that the glucose example of coupling is less general than it might be: P_i is transferred to glucose directly from ATP, not from free P_i, and ATP hydrolysis is essentially irreversible.

M. | Redox reactions

Biologically advantageous release of the free energy of glucose and other nutrients consumed by eukaryotic organisms is controlled in cells by means of **oxidation–reduction reactions**, or **redox reactions**. Some of these reactions occur in organelles called mitochondria, the 'power-houses' of the cell. Redox reactions are of such great and general importance that we could easily devote more than one section of one chapter of this book to them; what we aim for here is a sound introduction to the topic and not a comprehensive treatment.

In redox reactions, electrons are transferred from one molecule to another. Electron transfer can be accompanied by the transfer of an atom or ion, but our main concern at the moment is electrons and changes in

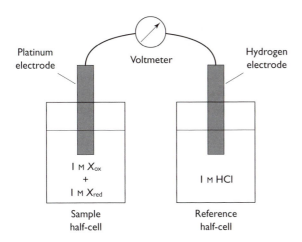

Fig. 4.14 Measurement of the standard redox potential. This device can be used to measure the standard redox potential of a compound. The hydrogen electrode in the reference cell contains 1 atm $H_2(g)$.

oxidation. The electron donor is called the **reductant** and the acceptor is called the **oxidant**. Reductant and oxidant work in pairs, or couples.

Any redox reaction can be expressed as the difference of two reduction **half-reactions**. Though only conceptual in nature, half-reactions facilitate calculations by showing the electrons explicitly. An example of a half-reaction is

$$\text{oxidized compound } (A_{ox}) + ne^- \rightarrow \text{reduced compound } (A_{red}) \qquad (4.64)$$

where n electrons are transferred to the oxidized compound, giving the reduced compound. A more complicated redox reaction is

$$B_{ox} + ne^- + mH^+ \rightarrow B_{red}mH \qquad (4.65)$$

where n is not necessarily equal to m. Note that proton (H^+) transfer is involved here. An example of a full redox reaction, where both redox pairs are given, is

$$A_{red} + B_{ox} \rightarrow A_{ox} + B_{red} \qquad (4.66)$$

An extremely important physiological redox reaction in humans is the reduction of oxygen to water by cytochrome c:

$$4\text{cyt}c^{2+} + 4H^+ + O_2 \rightarrow 2H_2O + 4\text{cyt}c^{3+} \qquad (4.67)$$

The transferred electrons are *not* shown explicitly here, but they are none the less integral to the reaction. Note that water is a product of this electron transfer.

The **standard redox potential**, $\Delta V^{\circ\prime}$, represents the ability of a redox pair to exchange electrons, the 'electron transfer potential.' A redox potential is measured as an **electromotive force** (e.m.f.), or **voltage**, of a half-cell consisting of both members of the redox couple (Fig. 4.14). (The volt, the SI unit of electrical potential, gets its name from Count Alessandro Giuseppe Antonio Anastasio Volta (1745–1827), the Italian physicist.) Just as the Gibbs free energy (a *thermodynamic* potential) must be measured with respect to some arbitrarily chosen reference state, so the redox voltage measurement must made with respect to some standard value. This is the **hydrogen electrode**, the voltage of which is set to 0.0 V at pH 0.0. The redox potential is like the Gibbs free energy in that its magnitude depends on the concentrations of the reacting species.

Table 4.6 Standard redox potentials of some important biochemical substrates

Electrode equation	n	$\Delta V^{\circ\prime}$ (V)
Acetate $+ 2H^+ + 2e^- \Leftrightarrow$ acetaldehyde	2	-0.58
$2H^+ + 2e^- \Leftrightarrow H_2$	2	-0.42
$NAD^+ + 2H^+ + 2e^- \Leftrightarrow NADH + H^+$	2	-0.32
Pyruvate $+ 2H^+ + 2e^- \Leftrightarrow$ lactate	2	-0.19
Cytochrome c (Fe^{3+}) $+ e^- \Leftrightarrow$ cytochrome c (Fe^{2+})	1	$+0.22$
$1/2O_2 + 2H^+ + 2e^- \Leftrightarrow H_2O$	2	$+0.82$

To find the concentration-dependence of the redox potential, we note that work is done when an electric charge, q, is moved through a difference in electrical potential, ΔV. The formula is

$$w' = -q\Delta V \tag{4.68}$$

where the prime indicates *non-pV* work. This is the electrical work we mentioned in passing in Chapter 2. As before, work is the product of an 'intensity factor' (q) and a 'capacity factor' (ΔV). The value of w' is negative if work is done *by* the system (energy leaves the system). We multiply the right-hand side of Eqn. 4.68 by F, the Faraday constant (96.494 kJ V^{-1} mol^{-1}), and by n, the number of electrons transferred, to put things on a molar basis. (The Faraday constant represents the electronic charge on 1 mole of electrons and is named after the renowned British physical chemist Michael Faraday (1791–1867), the first person to quantify the relationship between a chemical reaction and electric current.) This gives

$$\Delta\mu = -nF\Delta V \tag{4.69}$$

This equation shows that the free energy change will be favorable (the reaction will be spontaneous) if $\Delta V > 0$. The larger ΔV, the larger $|\Delta\mu|$ for a given value of n. It also tells us that the greater the number of electrons transferred against a given potential difference, the greater the work done. Rearranging Eqn. 4.69 in terms of ΔV gives

$$\Delta V = \frac{-\Delta\mu}{nF} \tag{4.70}$$

$$\Delta V = -\frac{\Delta\mu^{\circ\prime} + RT \ln\dfrac{[products]}{[reactants]}}{nF} = \Delta V^{\circ\prime} - \frac{RT}{nF} \ln\frac{[products]}{[reactants]} \tag{4.71}$$

where the standard state potential, $-\Delta\mu^{\circ\prime}/nF$, is symbolized as $\Delta V^{\circ\prime}$. (Standard state conditions are pH 7.0, 1 atm, and 25 °C.) Eqn. 4.71 is known as the **Nernst equation**. It is named after Walther Hermann Nernst (1864–1941), the German physicist and chemist. Nernst was awarded the Nobel Prize in Chemistry in 1920. As we shall see below, the Nernst equation has many applications in biological systems. At equilibrium, there is no driving force for change, so $\Delta V = 0$, and

$$\Delta V^{\circ\prime} = -\frac{RT}{nF} \ln K'_{eq} \tag{4.72}$$

This relates the standard cell e.m.f. to the equilibrium constant. Some biologically important standard redox potentials are given in Table 4.6.

Eqn. 4.65 can be split into two parts,

$$A_{ox} + ne^- \rightarrow A_{red} \tag{4.73}$$

and

$$B_{ox} + ne^- \rightarrow B_{red} \tag{4.74}$$

The half-cell reduction potentials here are

$$\Delta V_A = \Delta V_A^{\circ\prime} - \frac{RT}{nF} \ln \frac{[A_{red}]}{[A_{ox}^{n+}]} \tag{4.75}$$

and

$$\Delta V_B = \Delta V_B^{\circ\prime} - \frac{RT}{nF} \ln \frac{[B_{red}]}{[B_{ox}^{n+}]} \tag{4.76}$$

Note that the form of Eqns 4.75 and 4.76 is exactly the same as that of Eqn. 4.7. The redox potential of any two half-reactions is

$$\Delta V = \Delta V_{e-acceptor} - \Delta V_{e-donor} \tag{4.77}$$

An example will help to illustrate how these equations are put into practice. In glycolysis, a crucially important reaction is the reduction of pyruvate to lactate by nicotinamide adenine dinucleotide (NADH, reduced form). This is catalyzed with absolute stereospecificity by the enzyme lactate dehydrogenase:

$$\text{pyruvate} + \text{NADH} + \text{H}^+ \Leftrightarrow \text{lactate} + \text{NAD}^+ \tag{4.78}$$

The $\text{NAD}^+ - \text{NADH}$ pair is oxidized by the pyruvate–lactate pair (the reaction proceeds spontaneously to the right); the respective half-cell potentials, $\Delta V^{\circ\prime}$, $+0.32\,\text{V}$ and $-0.19\,\text{V}$ (Table 4.6), sum to a value *greater* than 0. Oxidation of NADH 'contributes' about 1/3 of a volt, while reduction of pyruvate 'costs' only about 1/5 of a volt. The forward reaction is therefore more probable than the reverse one, and we should expect the spontaneous oxidation of NADH by pyruvate. We note that, although they do not appear explicitly in Eqn. 4.78, two electrons are transferred in this reaction. Because $\Delta V^{\circ\prime} > 0$, $\Delta G^{\circ\prime} < 0$, and by Eqn. 4.69, $\Delta\mu^{\circ\prime} = 2 \times 96.5\ \text{kJ V}^{-1}\ \text{mol}^{-1} \times 0.13\ \text{V} = 25.1\ \text{kJ mol}^{-1}$. Nicotinamide adenine dinucleotide phosphate (NADPH, reduced form), a closely related molecule to NADH, plays a key role in the cellular redox reactions that enable the synthesis of compounds that are thermodynamically less stable than glucose, a starting material.

What if we change the concentrations of reactants and products? What effect will this have on the spontaneity of electron transfer? As we have seen, the standard state redox potential of the $\text{NAD}_{ox}/\text{NAD}_{red}$ couple is $-0.32\,\text{V}$. Now we wish to calculate the potential of this half-cell reaction under non-standard state conditions. Let the couple be 75% reduced and $T = 70\,°\text{C}$. By Eqn. 4.71,

$$\Delta V = -0.32\text{V} + \frac{8.314\,\text{J mol}^{-1}\,\text{K}^{-1} \times 343\,\text{K}}{2 \times 962\ 494\,\text{J V}^{-1}\,\text{mol}^{-1}} \ln\left(\frac{25}{75}\right) = -0.34\text{V} \tag{4.79}$$

There is relatively little deviation from the standard state potential at 75% reduction. At 50% reduction, $\Delta V = \Delta V^{\circ\prime}$. Marked deviations from the standard state potential only occur for extremes of temperature or extremes of the ratio of the concentration of oxidant to concentration of reductant. (Prove this.)

Let us return briefly to photosynthesis. This process makes sugar, a reduced form of carbon, using water, a reducing agent. The chemical reaction can be summarized as

$$6\,CO_2 + 6\,H_2O + light \rightarrow C_6H_{12}O_6 + 6\,O_2 + energy \qquad (4.80)$$

Water has a reduction potential of $+820$ mV and is therefore a poor reductant; energy is required to separate electrons from water. The energy is provided by photons absorbed by chlorophylls a and b in photosystems I and II. The energy trapped by chlorophyll is transferred to the photosynthetic reaction center, whence electrons are transferred to pheophytin and plastoquinone. To regenerate the reaction center, replacement electrons are transferred from water, releasing oxygen and producing a proton gradient across the membrane. We'll take a closer look at photosynthesis and glycolysis in Chapter 5.

N. | References and further reading

Arrhenius, S. (1912) *Theories of Solution*. New Haven: Yale University Press.

Atkins, P. W. (1998). *Physical Chemistry*, 6th edn, cc. 10.2 & 23.2. Oxford: Oxford University Press.

Atkins, P. W. (1994). *The Second Law: Energy, Chaos, and Form*, p. 167. New York: Scientific American.

Bada, J. L. Miller, S. L. (1968). Equilibrium constant for the reversible deamination of aspartic acid. *Biochemistry*, **7**, 3403–3408.

Banks, B.E.C. & Vernon, C. A. (1978). Biochemical abuse of standard equilibrium constants. *Trends in Biochemical Sciences*, **3**, N156–158.

Bates, R.G. (1973). *Determination of pH; Theory and Practice*. New York: John Wiley.

Bergethon, P. R. (1998). *The Physical Basis of Biochemistry: the Foundations of Molecular Biophysics*, cc. 2.17, 13 & 21.4–21.6. New York: Springer-Verlag.

Blandamer, M. J. (1992). *Chemical Equilibrium in Solution*. Hemel Hempstead: Ellis Horwood/Prentice-Hall.

Bradley, J. (1990). Teaching electrochemistry. *Education in Chemistry*, **27**, 48–50.

Bretscher, M. S. (1985). The molecules of the cell membrane. *Scientific American*, **253**, no. 4, 86–90.

Christensen, H. N. & Cellarius, R. A. (1972). *Introduction to Bioenergetics: Thermodynamics for the Biologist: A Learning Program for Students of the Biological and Medical Sciences*. Philadelphia: W.B. Saunders.

Creighton, T. E. (1993). *Proteins: Structures and Molecular Properties*, 2nd edn, ch. 7. New York: W. H. Freeman.

Dawes, E. A. (1962). *Quantitative Problems in Biochemistry*, 2nd edn, ch. 1. Edinburgh: E. & S. Livingstone.

Dawson, R. M. C., Elliott, D. C., Elliott, W. H. & Jones, K. M. (1986) *Data for Biochemical Research*, 3rd edn. Oxford: Clarendon Press.

DeKock, R. L. (1996). Tendency of reaction, electrochemistry, and units. *Journal of Chemical Education*, **73**, 955–956.

DeVries, A. L. (1982). Biological antifreeze agents in coldwater fishes. *Comparative Biochemistry and Physiology,* **73A**, 627–640.

Encyclopædia Britannica CD98, 'Activity Coefficient,' 'Arrhenius Theory,' 'Boiling Point,' 'Debye-Hückel Equation,' 'Equilibrium,' 'Free Energy,' 'Freezing Point,' 'Ideal Solution,' and 'Life,' 'Melting Point,' and 'Phase.'

Feeney, R. E. (1974). A biological antifreeze. *American Scientist,* **62**, 712–719.

Feiner, A. S. & McEvoy, A. J. (1994). The Nernst equation. *Journal of Chemical Education,* **71**, 493–494.

Franzen, H. F. (1988). The freezing point depression law in physical chemistry: is it time for a change? *Journal of Chemical Education,* **65**, 1077–1078.

Fruton, J. S. (1999). *Proteins, Enzymes, Genes: the Interplay of Chemistry and Biology.* New Haven: Yale University Press.

Gennix, R. B. (1989). *Biomembranes: Molecular Structure and Function.* New York: Springer-Verlag.

German, B. & Wyman, J. (1937). The titration curves of oxygenated and reduced hemoglobin. *Journal of Biological Chemistry,* **117**, 533–550.

Gillispie, Charles C. (ed.) (1970). *Dictionary of Scientific Biography.* New York: Charles Scribner.

Good, N. E., Winter, W., Connolly, T. N., Izawa, S. & Singh, R. M. M. (1966). Hydrogen ion buffers for biological research. *Biochemistry,* **5**, 467–476.

Green, R. L. & Warren, G. J. (1985). Physical and functional repetition in a bacterial ice nucleation gene. *Nature,* **317**, 645–648.

Gurney, R. W. (1953). *Ionic Processes in Solution.* New York: McGraw-Hill.

Harned, H. S. & Owen, B. B. (1958). *The Physical Chemistry of Electrolytic Solutions.* Princeton: Van Nostrand-Reinhold.

Harold, F. M. (1986). *The Vital Force: a Study of Bioenergetics,* cc. 1 & 2. New York: W. H. Freeman.

Harris, D. A. (1995). *Bioenergetics at a Glance,* ch. 1. Oxford: Blackwell Science.

Haynie, D. T. (1993). *The Structural Thermodynamics of Protein Folding,* ch. 4. Ph.D. thesis, The Johns Hopkins University.

Hodgman, C. D. (ed.) (1957). *C.R.C. Standard Mathematical Tables,* 11th edn. Cleveland, Ohio: Chemical Rubber Company.

Howlett, G. J., Blackburn, M. N., Compton, J. G. & Schachman, H. K. (1977). Allosteric regulation of aspartate transcarbamoylase. Analysis of the structural and functional behavior in terms of a two-state model. *Biochemistry,* **126**, 5091–5099.

Hubbard, R. (1966). The stereoisomerization of 11-*cis*-retinal. *Journal of Biological Chemistry,* **241**, 1814–1818.

Kemp, H. R. (1987). The effect of temperature and pressure on equilibria: a derivation of the van't Hoff rules, *Journal of Chemical Education,* **64**, 482–484.

Klotz, I. M. (1986). *Introduction to Biomolecular Energetics,* cc. 3–7. Orlando: Academic Press.

Kondepudi, D. & Prigogine, I. (1998). *Modern Thermodynamics: from Heat Engines to Dissipative Structures,* cc. 7.5, 8.2 & 8.3. Chichester: John Wiley.

Koryta, J. (1992). *Ions, Electrodes, and Membranes.* New York: John Wiley.

Lodish, H., Baltimore, D., Berk, A., Zipursky, S. L., Matsudaira, P. & Darnell, J. (1995). *Molecular Cell Biology,* 3rd edn, ch. 2. New York: W. H. Freeman.

McPartland, A. A. & Segal, I. H. (1986). Equilibrium constants, free energy changes and coupled reactions: concepts and misconceptions. *Biochemical Education,* **14**, 137–141.

Millar, D., Millar, I., Millar, J. & Millar, M. (1989). *Chambers Concise Dictionary of Scientists.* Cambridge: Chambers.

Nicholls, D. G. & Ferguson, S. J. (1992). *Bioenergetics 2*. London: Academic Press.

Ochs, R. S. (1996). Thermodynamics and spontaneity. *Journal of Chemical Education*, **73**, 952–954.

Ostro, M. J. (1987). Liposomes. *Scientific American*, **256**, no. 1, 102–111.

Peusner, L. (1974). *Concepts in Bioenergetics*, cc. 3, 5, 6 & 7. Englewood Cliffs: Prentice-Hall.

Price, G. (1998). *Thermodynamics of Chemical Processes*, ch. 4. Oxford: Oxford University Press.

Schrake, A., Ginsburg, A. & Schachman, H. K. (1981). Calorimetric estimate of the enthalpy change for the substrate-promoted conformational transition of aspartate transcarbamoylase from *Escherichia coli*. *Journal of Biological Chemistry*, **256**, 5005–5015.

Schultz, S. G. (1980). *Basic Principles of Membrane Transport*. Cambridge: Cambridge University Press.

Segal, I. H. (1976). *Biochemical Calculations: How to Solve Mathematical Problems in General Biochemistry*, 2nd edn, ch. 3. New York: John Wiley.

Smith, C. A. & Wood, E. J. (1991). *Energy in Biological Systems*, cc. 1.3 & 1.4. London: Chapman & Hall.

Snell, F. M., Shulman, S., Spencer, R. P. & Moos, C. (1965). *Biophysical Principles of Structure and Function*. Reading, Massachusetts: Addison-Wesley.

Spencer, J. N. (1992). Competitive and coupled reactions, *Journal of Chemical Education*, **69**, 281–284.

Tanford, C. & Hauenstein, J. D. (1956). Hydrogen ion equilibria of ribonuclease. *Journal of the American Chemical Society*, **78**, 5287–5291.

Tombs, M. P. & Peacocke, A. R. (1974). *The Osmotic Pressure of Biological Macromolecules*. Oxford: Clarendon Press.

Tydoki, R.J. (1995). Spontaneity, accessibility, irreversibility, 'useful work': the availability function, the Helmholtz function and the Gibbs function, *Journal of Chemical Education*, **72**, 103–112.

Tydoki, R.J. (1996). The Gibbs function, spontaneity, and walls. *Journal of Chemical Education*, **73**, 398–403.

van Holde, K. E. (1985). *Physical Biochemistry*, 2nd edn, cc. 2.1, 2.3, 2.4 & 3. Englewood Cliffs: Prentice-Hall.

Voet, D. & Voet, J. G. (1995). *Biochemistry*, 2nd edn, cc. 2-2, 3, 4, 11-2, 15-4–15-6, 16, 18-1, 19-1, 28-3 & 34-4B. New York: Wiley.

Williams, T. I. (ed.) (1969). *A Biographical Dictionary of Scientists*. London: Adam & Charles Black.

Wood, S.E. & Battino, R. (1996). The Gibbs function controversy, *Journal of Chemical Education*, **73**, 408–411.

Wrigglesworth, J. (1997). *Energy and Life*, cc. 3, 5.7.2, 7.1, 7.3 & 7.5.1. London: Taylor & Francis.

Youvan, D. C. & Marrs, B. L. (1987). Molecular mechanisms of photosynthesis. *Scientific American*, **256**, no. 6, 42–48.

O. Exercises

1. State whether the following phrases pertain to (A) the expansion of a gas into a vacuum, (B) two objects coming to thermal equilibrium, (C) both of these processes, or (D) neither of these processes.

   (1) Involves a change in enthalpy.

(2) Involves an increase in entropy.

(3) Involves a decrease in Gibbs free energy.

(4) Can be made to proceed in the opposite direction.

2. State whether the following phrases pertain to (A) spontaneity, (B) reversibility, (C) both spontaneity and reversibility, or (D) neither spontaneity nor reversibility.

(1) Established for $\Delta G < 0$ at constant T.

(2) Established for $\Delta S < 0$.

(3) Established for a process in which the work done is a maximum.

(4) Illustrated by the migration of a solute from a region of high concentration to low concentration.

(5) Required for determination of ΔS by the heat transferred.

(6) Implies that the coupling of coupled reaction is very efficient.

3. State whether the following phrases pertain to (A) ΔU, (B) ΔG, (C) both ΔU and ΔG, or (D) neither ΔU nor ΔG.

(1) Does not depend on pathway during a change of state.

(2) Consists of the heat transferred and the work done.

(3) Must be negative if an isothermal and isobaric process is spontaneous.

(4) Measures the degree of disorder of a system.

(5) Is zero at equilibrium for an isothermal and isobaric process.

(6) Used to determine whether one reaction can drive another by coupling.

(7) Includes only energy that can do work, at constant temperature and pressure.

4. State whether the following phrases pertain to (A) ΔG, (B) $\Delta G°$, (C) both ΔG and $\Delta G°$, or (D) neither ΔG nor $\Delta G°$.

(1) Equals $-RT \ln K_{eq}$.

(2) Equals $-nF\Delta V°'$.

(3) Is zero if the change in the state of the system is spontaneous.

(4) Equals $\Delta H - T\Delta S$ at constant T.

(5) Is given for one mole of the reaction for a given reaction.

(6) Is equal to the sum of the chemical potentials of the products minus the chemical potentials of the reactants, with each chemical potential multiplied by the number of moles involved in the reaction.

(7) Is independent of the concentration of the components of a reaction.

5. State whether the following phrases pertain to (A) a, (B) μ, (C) both a and μ, or (D) neither a nor μ.

(1) Equals the concentration times the activity coefficient.

(2) Needed to calculate ΔG if $\Delta G°$ is known under specified conditions for all components of a reaction.

(3) Used to calculate ΔG for process after multiplication by the number of moles of that component involved in the process.

6. In Chapter 1 we said that all living organisms depend on the Sun in order to meet the energy requirements of life. This is only partially true of the chemosynthetic bacteria that live at the bottom of the ocean. Explain the energy requirements for life in completely

general terms. Although it may be that the Sun played an indispensable role in the formation of life as we know it, is the Sun absolutely necessary for life? Why or why not?

7. What are the units of K_{eq}? Explain.

8. Calculate $\Delta G°(25\,°C)$ for $K_{eq} = 0.001, 0.01, 0.1, 1, 10, 100$ and 1000.

9. The multi-component enzyme aspartate transcarbamoylase catalyzes the formation of N-carbamoylaspartate from carbamoyl phosphate and aspartate. Arthur Pardee has demonstrated that this reaction is the first step unique to the biosynthesis of pyrimidines, including cytosine, thymine and uracil, major components of nucleic acids. Aspartate transcarbamoylase has at least two stable folded conformations, known as R (high substrate affinity) and T (low substrate affinity). Interestingly, the relative stability of the T and R states is affected by the binding of ATP (a purine) to R and CTP (a pyrimidine) to T, a topic covered in Chapter VII. Measurement of the standard state free energy difference between R and T in the absence of ATP and CTP yielded the value 3.3 kcal mol^{-1}. Calorimetric determination of $\Delta H°$ for the transition was -6 kcal mol^{-1}. Calculate the standard state entropy change for the T→R transition.

10. When a photon in the visible range is absorbed in the retina by rhodopsin, the photoreceptor in rod cells, 11-*cis*-retinal is converted to the all-*trans* isomer. Light energy is transformed into molecular motion. The efficiency of photons to initiate the reaction is about 20% at 500 nm (57 kcal mol^{-1}). About 50% of the absorbed energy is available for the next signaling step. This process takes about 10 ms. In the absence of light, spontaneous isomerization of 11-*cis*-retinal is *very* slow, on the order of 0.001 yr^{-1}! Experimental studies have shown that the equilibrium energetics of retinal isomerization are $\Delta S° = 4.4$ cal mol^{-1}K^{-1} and $\Delta H° = 150$ cal mol^{-1}. Calculate the equilibrium constant for the reaction.

11. Which one of the following equations is used to evaluate free energy changes in cells under physiological conditions? What makes it appropriate?
    A. $\Delta G = RT \ln K_{eq}'$
    B. $\Delta G = \Delta G°' + RT \ln[\text{products}]/[\text{reactants}]$
    C. $\Delta G = RT \ln[\text{products}]/[\text{reactants}]$
    D. $\Delta G = \Delta H - T\Delta S$
    E. $\Delta G = \Delta G°' + RT [\text{products}]/[\text{reactants}]$

12. The direction of a reaction with a very large or very small value of K_{eq} is difficult, though not impossible, to alter by changing the mass action ratio. Explain.

13. Show that, for a reaction at 25 °C which yields 1 mol of H_2O, $\Delta G°' = \Delta G° + 9.96$ kJ mol^{-1}.

14. Calculate K_{eq} for the hydrolysis of the following compounds at neutral pH and 25 °C: phosphoenolpyruvate ($\Delta G°' = -61.9$ kJ mol^{-1}), pyrophosphate ($\Delta G°' = -33.5$ kJ mol^{-1}), and glucose-1-phosphate ($\Delta G°' = -20.9$ kJ mol^{-1}). These compounds are involved in the glycolytic pathway.

15. $\Delta G^{\circ\prime}$ for the conversion of fructose-1,6-bisphosphate (FBP) into glyceraldehyde-3-phosphate (GAP) and dihydroxyacetone phosphate (DHAP) is $+22.8$ kJ mol^{-1}. This reaction is step four of the glycolytic pathway and is catalyzed by aldolase. In the cell at $37\,^\circ$C the mass action [DHAP]/[GAP] $= 5.5$. What is the equilibrium ratio of [FBP]/[GAP] when [GAP] $= 2\times 10^{-5}$ M? When [GAP] $= 1\times 10^{-3}$ M?

16. Calculate ΔG when the concentrations of glucose-1-phosphate and glucose-6-phosphate are maintained at 0.01 mM and 1 mM, respectively. Compare the sign of ΔG with what was obtained in the worked example above. Suggest how this might be significant in metabolism.

17. Lactate dehydrogenase (LDH) catalyzes the oxidation of pyruvate to lactate and NADH to NAD$^+$ in glycolysis, the pathway by which glucose is converted to pyruvate with the generation of 2 mol of ATP mol^{-1} of glucose. The reaction is particularly important during strenuous activity, when the demand for ATP is high and oxygen is depleted. The relevant half-reactions and their standard reduction potentials are given in Table 4.6. Calculate ΔG for the reaction under the following conditions: [lactate]/[pyruvate] $=$ [NAD$^+$]/[NADH] $= 1$; [lactate]/[pyruvate] $=$ [NAD$^+$]/[NADH] $= 160$; [lactate]/[pyruvate] $=$ [NAD$^+$]/[NADH] $= 1000$. What conditions are required for the reaction to spontaneously favor oxidation of NADH? [NAD$^+$]/[NADH] must be maintained close to 10^3 in order for the free energy change of the glyceraldehyde-3-phosphate reaction to favor glycolysis. This function is performed by LDH under anaerobic conditions. What is the largest [lactate]/[pyruvate] can be in order for the LDH reaction to favor the production of NAD$^+$ and maintain [NAD$^+$]/[NADH] $= 10^3$?

18. The citric acid cycle is the common mode of oxidative degradation in eukaryotes and prokaryotes (Chapter V). Two components of the citric acid cycle are α-ketoglutarate and isocitrate. Let [NAD$_{ox}$]/[NAD$_{red}$] $= 8$; [α-ketoglutarate] $= 0.1$ mM; [isocitrate] $= 0.02$ mM. Assume $25\,^\circ$C and pH 7.0. Calculate ΔG. Is this reaction a likely site for metabolic control? Explain.

19. Mixing. Refer to Fig. 4.4. In the text we noted that at first entropy is at a minimum and free energy is at a maximum. Later, ... If the two liquids are the same, what are ΔS and ΔG of mixing?

20. Refer to Fig. 3.2. The stopcock is closed. In the bulb on the left there is an inert gas at concentration x. In the bulb on the right there is the same inert gas at concentration x. What are the entropy and free energy differences between this state and the equilibrium state obtained by opening the stopcock?

21. Cytochromes are redox-active proteins that occur in all organisms except a few types of obligate anaerobes. These proteins contain heme groups, the iron atom of which reversibly alternates between the Fe(II) and Fe(III) oxidation states during electron transport. Consider the reaction

$$\text{Cyt } c\,(\text{Fe}^{2+}) + \text{Cyt } f\,(\text{Fe}^{3+}) \Leftrightarrow \text{Cyt } c\,(\text{Fe}^{3+}) + \text{Cyt } f\,(\text{Fe}^{2+})$$

involving cytochromes c and f. If $V°' = 0.365$ V for electron transfer to Cyt f (Fe^{3+}), and $V°' = 0.254$ V for electron transfer to Cyt c (Fe^{3+}), can ferrocytochrome c ($2+$ oxidation state) reduce ferricytochrome f ($3+$ oxidation state) spontaneously?

22. Calculate ΔV in Eqn. 4.79 when the couple is 99% reduced and the temperature is $37\,°C$.

23. Table 4.1 presents thermodynamic properties of water. On the basis of these data, rationalize the suitability, or lack thereof, of each thermodynamic function as an index of spontaneous change.

24. Cholesterol increases membrane rigidity. What effect will it have on the character of the lipid bilayer order–disorder transition? Why?

25. Some organisms are able to tolerate a wide range of ambient temperatures, for instance bacteria and poikilothermic (cold-blooded) animals such as fish. The membrane viscosity of *E. coli* at its growth temperature is approximately constant over the range $15–43\,°C$. Knowing aspects of the physical basis of the solid–gel transition in lipid bilayers, suggest how bacteria and fish might cope with changes of temperature.

26. Use your knowledge of the physical properties of lipids to outline several design characteristics of a liposome-based drug delivery system. A liposome is a bilayer structure that is self-enclosing and separates two aqueous phases.

27. The reversible deamination of aspartate yields ammonium and fumarate. Fumarate is a component of the citric acid cycle. Aspartate deamination is catalyzed by the enzyme aspartase. Experimental studies on the deamination reaction have shown that

$$\log K_{eq} = 8.188 - (2315.5/T) - 0.010\,25T$$

where T is in degrees kelvin. Note that the units of the coefficient of $1/T$ (i.e. 2315.5) must be K, while those of 0.010 25 are K^{-1}. Calculate $\Delta G°$ at $25\,°C$. Remember that $K = 10^{\log K}$ and $2.303 \log x \approx \ln x$. Follow the development leading up to Eqn. (4.50) to show that $\Delta H° = 2.303 \times R \times (2315.5 - 0.010\,25T^2)$. Calculate $\Delta H°$ and $\Delta S°$ at $25\,°C$. From Chapter 2, $\Delta C_p = \Delta(\Delta H°)/\Delta T$. Use this to show that $\Delta C_p = -2.303 \times R \times 0.0205T$. Evaluate ΔC_p at $25\,°C$.

28. State whether the following phrases pertain to (A) chemical potential of the solute, (B) chemical potential of the solvent, (C) both of these chemical potentials, or (D) neither chemical potential.
    (1) Equal $RT \ln a$.
    (2) Equals $\mu° + RT \ln a$.
    (3) At equilibrium, its value is the same on both sides of a membrane.
    (4) Is proportional to the osmotic pressure.
    Note: Osmotic pressure is treated as an application of the Gibbs free energy in chp. 5.

29. Calculate the value of x for which the approximation $\ln(1 + x) \approx x$ gives an error of 5%.

30. State whether the following phrases pertain to (A) ΔG, (B) ΔV, (C) both ΔG and ΔV, or (D) neither ΔG nor ΔV.
    (1) Indicates whether an oxidation–reduction reaction is spontaneous.

(2) Standard value for a reaction is determined with all components is their standard states.

(3) Is positive for a spontaneous reaction.

(4) Is called the standard electrode reduction potential.

(5) Can be used to calculate the equilibrium constant for a reaction for a known set of concentrations of all components of a reaction at a given temperature.

31. Chemical coupling. The equilibrium constant for $Glu^- + NH_4^+ \Leftrightarrow Gln + H_2O$ is $0.00315\ M^{-1}$ at pH 7 and 310 K; the reaction lies far to the left. The synthesis of Gln from Glu is made energetically favorable by coupling it to hydrolysis of the terminal phosphodiester bond of ATP. The products of ATP hydrolysis are ADP and P_i. The equilibrium constant for the coupled reaction, which is known from experiments with glutamine synthase, is 1200. Calculate the phosphate bond energy in ATP at pH 7 and 310 K.

32. What is the pH value of 0.001 M HCl solution?

33. Calculate the hydrogen ion concentration of solution of pH 6.0.

34. Calculate the ionic strength of a 0.35 molal aqueous solution of $MnCl_2$. Assume that dissociation of the salt into ions is complete at this concentration.

35. Calculate the percentage non-ionized for an acid with a pK_a of 4 in an environment of pH 1.

36. Calculate the ionic strength of 0.01 N acetic acid if the dissociation constant of the acid is 1.8×10^{-5}.

37. Calculate the activity coefficient and activities of the ions in aqueous solution of (a) 5 mM H_2SO_4 and (b) 2 mM NaCl.

38. The following data were obtained by German & Wyman for horse hemoglobin, an oxygen-binding blood protein, in 0.333 M NaCl.

Deoxygenated hemoglobin		Oxygenated hemoglobin	
Acid (−) or base (+) per gram Hb	pH	Acid (−) or base (+) per gram Hb	pH
−0.514	4.280	−0.514	4.280
−0.452	4.415	−0.453	4.410
−0.419	4.525	−0.420	4.525
−0.390	4.610	−0.392	4.618
−0.323	4.842	−0.324	4.860
−0.258	5.160	−0.259	5.188
−0.224	5.320	−0.225	5.430
−0.172	5.590	−0.173	5.800
−0.130	6.072	−0.130	6.055
−0.064	6.541	−0.063	6.430
0.0	6.910	+0.001	6.795
+0.070	7.295	+0.072	7.130
+0.131	7.660	+0.133	7.510
+0.171	7.860	+0.172	7.725
+0.208	8.140	+0.209	8.043

Deoxygenated hemoglobin		Oxygenated hemoglobin	
Acid (–) or base (+) per gram Hb	pH	Acid (–) or base (+) per gram Hb	pH
+0.254	8.545	+0.254	8.450
+0.288	8.910	+0.288	8.890
+0.311	9.130	+0.292	8.990
+0.331	9.350	+0.311	9.130
+0.350	9.410	+0.331	9.355
+0.357	9.465	+0.350	9.410
+0.407	9.800	+0.357	9.480
		+0.407	9.800

Plot the titration data to find which form of hemoglobin is the stronger acid. The stronger an acid, the more readily it gives up protons. We shall study hemoglobin in considerably greater depth in Chapters 5 and 7.

39. The history of science is full of 'partly true' ideas pursued with vigor until they no longer became tenable. As we have seen in Chapter 2 Galileo's assumption about the shapes of planetary orbits, which was based on the speculations of thinkers of classical antiquity, was eventually superseded by the very detailed measurements of the Danish astronomer Tycho Brahe (1546–1601) and analysis of Johannes Kepler. Similarly, Galileo's work on the relative motion of bodies was a great addition to the physics of his day (mostly that of Aristotle), and it prepared the way for Newton; but in the twentieth century, Galilean (Newtonian) relativity was seen to be a limiting case of the more general view proposed by Einstein. In the nineteenth century research in thermochemistry was motivated in part by the belief that the heat of a reaction measured its 'affinity:' the greater the energy liberated, the greater the affinity of the reactants for each other. This view became untenable by the discovery of spontaneous endothermic reactions. Explain.

40. ΔG cannot generally be equated with $\Delta G°$. To a very good first approximation ΔH can be equated with $\Delta H°$. Explain.

41. Tris, a base, is a popular buffer for biochemical research. Its pK_a is strongly dependent on temperature. Would it make a very good buffer for a scanning calorimetry experiment? Why or why not? Assuming that $K_a = 8.3 \times 10^{-9}$ M, calculate the ratio of acid to base at pH 8.0. Let the total concentration of Tris be 150 mM, and divide the stock into to two parts. To one, add 10 mM HCl. How does the pH change? Hint: assume complete dissociation of HCl. To the other, reduce the concentration to 30 mM. How does the pH change?

42. The concentration of creatine in urine is c. 40-fold greater than in serum. Calculate the free energy change per molecule required for transfer of creatine from blood to urine at 37 °C.

43. Which of the following redox pairs is the strongest reducing agent?

Redox pair	$V^{\circ\prime}$ in volts
Oxidized ferrodoxin/reduced ferrodoxin	−0.43
NADP/NADPH	−0.32
Oxidized glutathione/reduced glutathione	−0.23
Pyruvate/lactate	−0.19
Ubiquinone/hydroquinone	0.10

For solutions, see http://chem.nich.edu/homework

Chapter 5

Gibbs free energy – applications

A. Introduction

The Gibbs free energy is so important in biology research because it enables one to predict the direction of spontaneous change for a system under the constraints of constant temperature and pressure. These constraints generally apply to all living organisms. In the previous chapter we discussed basic properties of the Gibbs free energy, showed how its changes underlie a number of aspects of physical biochemistry, and touched on what the biological scientist might do with such knowledge. Here we build on that introduction to the basics and show how the elementary concepts can be applied to a wide variety of topics that are specifically biological or biochemical. We wish to illustrate, by way of a range of examples, when, where, why, and how the Gibbs free energy is such a useful concept. We shall discuss the energetics of very different types of biological structure, including small organic molecules, membranes, nucleic acids and proteins. This will help to give a deeper sense of the relatedness of the seemingly very different topics one encounters in biochemistry.

B. Photosynthesis, glycolysis, and the citric acid cycle

This section is a 'low-resolution lens' on the energetics of photosynthesis, glycolysis, and the citric acid cycle. The details we shall omit are certainly important (entire books have been written on each subject!), but our main aim here is to consider biological energy in a global, qualitative way; to see 'the big picture.' The names of specific enzymes, chemical intermediates and other pertinent details we shall leave out are described in all good biochemistry textbooks.

Over 99% of the free energy in our biosphere comes from the Sun. Green plants, certain unicellular organisms like diatoms, cyanophytes (blue-green algae), and various kinds of bacteria are known as photoautotrophs. These organisms convert the light energy of the Sun and CO_2 into the chemical energy of bonding electrons in sugar molecules by a process called **photosynthesis**. The remaining less than 1% of the free energy in the biosphere comes from the oxidation of inorganic matter, mainly hydrogen and sulfur, by microorganisms called chemolithotrophs. Whether photoautotrophs preceded or followed chemolithotrophs in the flowering of life on Earth is an intriguing open question (see Chapter 9).

The overall chemical reaction of photosynthesis is:

$$CO_2 + H_2O + light \rightarrow (CH_2O) + O_2 \tag{5.1}$$

CO_2 and H_2O are reduced to sugar and oxygen in this redox reaction. The process carried out in photosynthetic protists and cyanophytes resembles that in green plants, though compounds other than water serve as a reactant in photosynthetic bacteria and oxygen is not produced. All known photosynthetic organisms (excluding the halobacteria that thrive in the high salt environment of the Dead Sea but including all other types of photosynthetic prokaryotes) contain the light-absorbing pigment chlorophyll (Fig. 1.3). This molecule plays a key role in the transformation of light energy to chemical compounds. Chlorophyll, like the heme group of the vertebrate oxygen transport protein hemoglobin (see below) and the heme group of the electron transfer protein cytochrome c, is derived from protoporphyrin IX, a complex ring structure synthesized from glycine and acetate (Fig. 5.1).

Figure 5.2 depicts the energetics of photosynthesis in schematic form. Absorption of photons results in the ejection of electrons from P680, the reaction center chlorophyll of photosystem II (680 nm is the wavelength of the absorption maximum of the reaction center chlorophyll). Each electron passes through a chain of electron carriers to plastoquinone, giving plastoquinol. By means of a series of redox reactions, the electrons are delivered to plastocyanin, which regenerates photooxidized P700, the reaction center chlorophyll of photosystem I. The electron ejected from P700 then passes *via* a chain of electron carriers to the

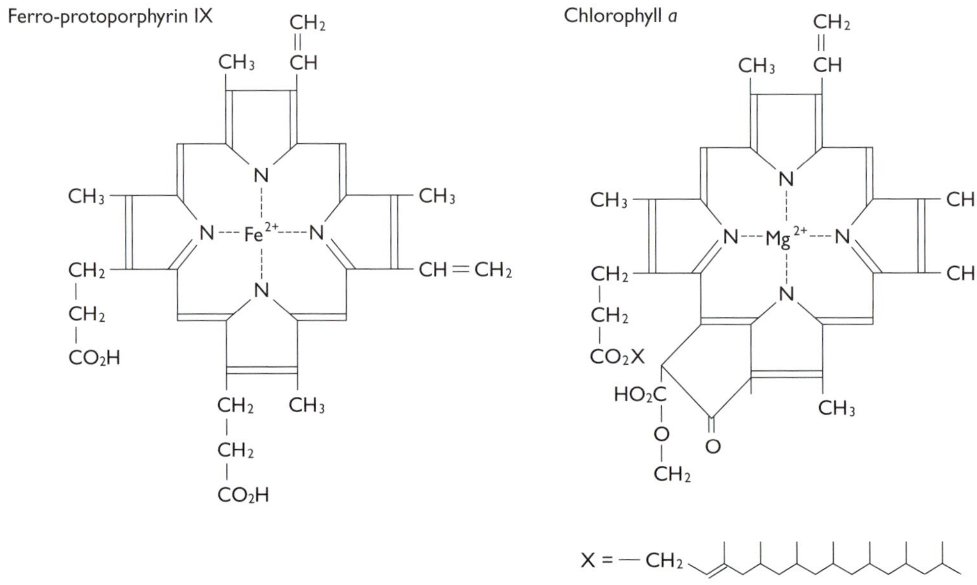

Fig. 5.1 Molecular formulae of the heterocyclic ring systems of ferro-protoporphyrin IX (heme) and chlorophyll a. The iron bound to heme is normally in the Fe(II) (ferrous, 2+) oxidation state regardless of whether oxygen is bound. Heme plays a key role in oxygen transport in a broad range of organisms, notably humans. The structure of chlorophyll b is nearly identical to that of chlorophyll a: a formyl group is found in place of a methyl group. Bacteriochlorophylls a and b, which are important in photon capture in photosynthetic bacteria, are also very similar in structure to chlorophyll a. The long aliphatic tail of chlorophyll probably helps to increase its solubility in nonpolar environments. It is a remarkable indication of the unity of all known living things that such similar ring structures should play important roles in biological energetics in organisms as different as bacteria and humans.

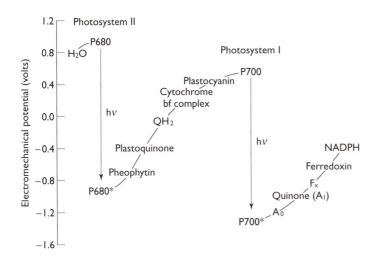

Fig. 5.2 Schematic diagram ('Z-scheme') of the energetics of electron transport in photosynthesis. The electrochemical potential (free energy) is measured in volts. Electrons tend to flow spontaneously from a state of higher to lower free energy. In terms of electrochemical potential, electrons migrate spontaneously from a more negative to a more positive reduction potential. PSII is coupled to PSI via the quinone Q and plastocyanin.

oxidized form of nicotinamide adenine dinucleotide phosphate ($NADP^+$), an intracellular electron carrier. Photosynthetic electron transport drives the formation of a proton (pH) gradient, a difference in the concentration of protons on opposite sides of a membrane (in plants, the thylakoid membrane in chloroplasts). Movement of protons from a region of high chemical potential to low chemical potential powers the synthesis of ATP in plants in manner that closely resembles oxidative phosphorylation, the endergonic synthesis of ATP from ADP and P_i in mitochondria in animal cells (see below). Plants also use light energy to make cellulose and other sugar molecules.

Glucose is the six-carbon sugar that is quantitatively the most important source of energy for cellular processes in all known organisms. **Glycolysis**, the metabolism of glucose, is a sequence of biochemical reactions by which one molecule of glucose is oxidized to two molecules of pyruvate, a three-carbon molecule (Fig. 5.3). The word 'metabolism' comes from the Greek, *metabolikon*, disposed to cause or suffer change; it was coined by the German biologist Theodor Schwann (1810–1882). Pyruvate is converted by a series of reactions to carbon dioxide and water. In combination with other aspects of oxidative carbohydrate metabolism, glycolysis is essentially the reverse process of photosynthesis. The *overall* chemical reaction for glucose metabolism is

$$C_6H_{12}O_6 + 6O_2 \rightarrow 6CO_2 + 6H_2O \tag{5.2}$$

Compare Eqn. 5.2 to Eqn. 5.1. The free energy change for the *complete* redox reaction is $\Delta G^{\circ\prime} = -2823$ kJ mol^{-1}, and 24 electrons are transferred in the process. The standard state free energy change ($\Delta G^{\circ\prime}$) for glycolysis alone is -43.4 kJ mol^{-1}, while the physiological free energy change (ΔG) for glycolysis, which includes the synthesis of 2 moles of ATP, is -74 kJ mol^{-1}. Figure 5.4 depicts the physiological energetics of glycolysis in schematic form.

Glycolysis occurs by similar means in all organisms. Once glucose has entered the cell, it is immediately phosphorylated at the expense of one molecule of ATP. (It is interesting that ATP is invested in a process

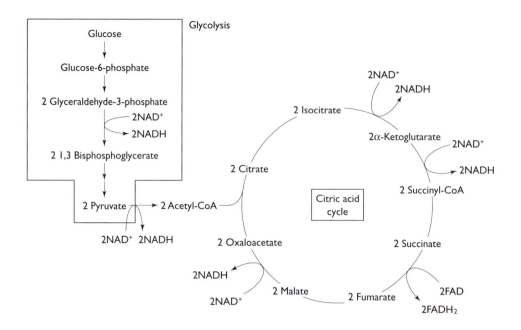

Fig. 5.3 Schematic diagram of glycolysis and the citric acid cycle. The figure shows the points at which the electron carriers NAD$^+$ and FAD are reduced by electron transfer to form NADH and FADH$_2$.

which, as we shall see, leads to ATP production.) Phosphorylation of glucose is an essentially irreversible reaction, because the free energy change of removal of the phosphoryl group from ATP is large and negative. Phosphorylation of glucose ensures that once it has entered the cell, the chemical energy of glucose stays trapped there. The fate of pyruvate depends on the organism, tissue and conditions. For example in stressed, oxygen-depleted skeletal muscle, pyruvate is converted to lactate (the conjugate base of lactic acid) and one molecule of ATP is produced. Fermentation of yeast, a process integral to making bread, beer, and wine, involves the conversion of pyruvate to ethanol and CO_2. In the presence of oxygen, however, the three carbons of pyruvate are completely oxidized to CO_2.

In the previous chapter we said that extreme values of K_{eq} correspond to mass action ratios that are difficult to shift by changes in the concentrations of reactants or products alone. The unfavorability of a process, however, can be overcome by the cell's maintaining concentrations that promote the reaction. One such reaction occurs in glycolysis. Fructose-1,6-bisphosphate (FBP) is cleaved by aldolase into two triose phosphates, dihydroxyacetone phosphate and glyceraldehyde phosphate (GAP). (Note that both trioses are phosphorylated, preventing escape from the cell!) Cleavage of the C–C bond is highly endergonic; $\Delta G°$ is large and positive. In order for the reaction to occur, $\ln([GAP]^2/[FDP])$ must be negative; in other words, the mass action ratio must be much less than 1. This step of glycolysis occurs only because the cellular concentrations of the products are kept below 1 μM; the mass action ratio is less than 1 for concentrations of FBP greater than 1 pM!

The **citric acid cycle** (Fig. 5.3) is the terminal stage of the chemical processes by which the major portion of carbohydrate, fatty acid and amino acids are converted into a more useful form of chemical energy.

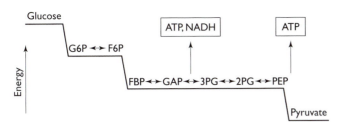

Fig. 5.4 Control by phosphofructokinase (PFK) of the flux of glycolysis breakdown products. Abbreviations: G6P, glucose-6-phosphate; F6P, fructose-6-phosphate; FBP, fructose-bisphosphate; GAP, glyceraldehyde-3-phosphate; 3PG, 3-phosphoglycerate; 2PG, 2-phosphoglycerate; PEP, phosphoenolpyruvate. The physiological free energy changes (in kJ mol^{-1}) are: -27.2, -1.4, -25.9, -5.9, $+3.3$, -0.6, -2.4, and -13.9 (from Table 16-1 in Voet & Voet, 1995). There are three 'irreversible' steps in the metabolism of glucose to pyruvate. These occur between glucose and G6P, F6P and FDP and PEP and pyruvate. The irreversibility of these reactions is extremely important for cellular function. For only at an irreversible step of a process can control be exerted; irreversibility permits regulation of the speed of the reaction. Since regulation of a pathway at a particular point affects all reactions that occur downstream, PFK controls the flux of glycolysis. Adapted from Fig. 1.3 of Harris (1995).

The cycle is the common mode of oxidative degradation in cells in animals, plants, microorganisms, and fungi; it is a main feature of cellular chemistry that is shared by all known forms of life. One complete cycle yields two molecules of carbon dioxide, one molecule of ATP, and numerous biosynthetic precursors. Because the cycle is used twice in the oxidation of a single glucose molecule (one glucose gives two pyruvates), it produces six molecules of nicotinamide adenine dinucleotide (NADH, a close relative of NADPH) and two molecules of flavin adenine dinucleotide (FADH$_2$) per glucose molecule, by way of redox reactions. The electron carriers NADH and FADH$_2$ (Fig. 5.5), which incidentally are synthesized from vitamin precursors, are of great importance to ATP production in oxidative phosphorylation (see next section). The citric acid cycle was first proposed in 1937 by Sir Hans Adolf Krebs (1900–1981), a biochemist who emigrated from Germany to England in 1933. Krebs shared the 1953 Nobel Prize in Medicine or Physiology with the American Fritz Albert Lipmann (1899–1986).

C. | Oxidative phosphorylation and ATP hydrolysis

The NADH and FADH$_2$ molecules generated by the citric acid cycle play a central role in **oxidative phosphorylation**, the complex process by which ADP and inorganic phosphate (P$_i$) are combined to make ATP. From a quantitative point of view, oxidative phosphorylation is the most important means by which a cell generates ATP: complete metabolism of one mole of glucose *via* the citric acid cycle yields a maximum of 38 moles of ATP (two from glycolysis, two from the citric acid cycle, and 34 from reoxidation of NADH and FADH$_2$). ATP is the most commonly utilized form of energy in a cell.

The term *bioenergetics* refers the way the in which cells generate energy from foodstuffs. The central concept of **bioenergetics** is the chemiosmotic theory, which states that energy stored as a proton gradient across biological membranes (the so-called proton-motive force) is converted to useful chemical energy in the form of ATP. One of the key contributors to the understanding of biological energy transfer was the British biological chemist Peter Dennis Mitchell (1920–1992), who was awarded the Nobel Prize in Chemistry for his work in 1978. The **proton motive force** is generated across the inner membrane of mitochondria in animals, the inner membrane of chloroplasts in plants, and the plasma membrane of aerobic bacteria (Fig. 5.6). The proton gradient comes about by the harnessing of energy released

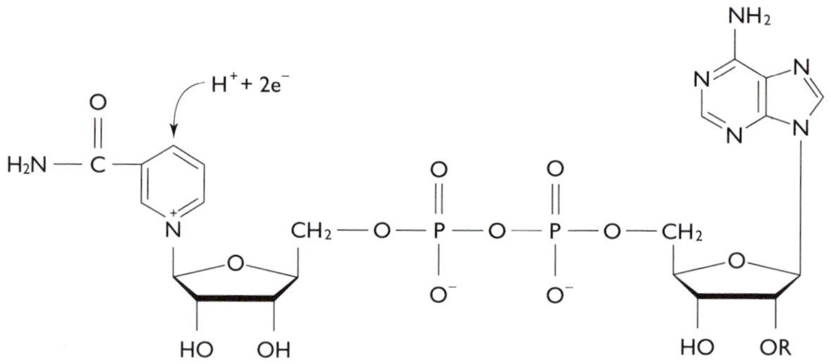

(A) Nicotinamide adenine dinucleotide (oxidized).

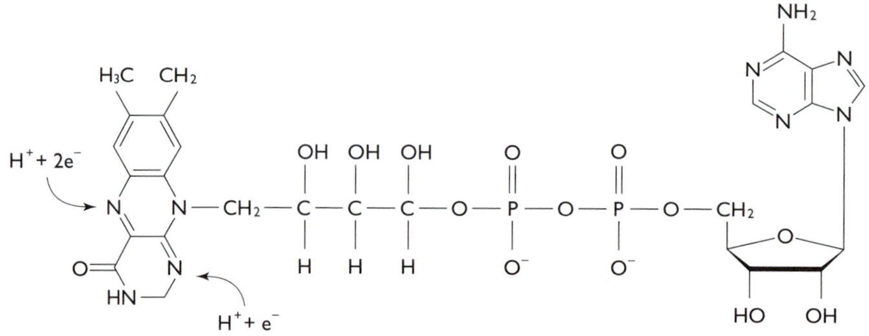

(B) Flavin adenine dinucleotide (oxidized).

Fig. 5.5 Electron carriers in metabolism. NAD is a major soluble redox intermediate in metabolism. It is closely related to NADP, another redox intermediate. NAD and NADP differ in that the latter is phosphorylated on the adenylate ribose (R = phosphate in NADP, R = H in NAD). NADH shuttles electrons to electron transfer chains, NADPH provides electrons for biosynthesis. Neither NADH nor NADPH can form a stable one-electron intermediate. In contrast, FAD, a protein-bound cofactor, can form a one-electron semiquinone. Both NAD^+ and FAD comprise AMP and are synthesized from ATP (see Fig. 5.7).

from electron-transfer events in membrane-bound proteins. Lest you think that these electron-transfer reactions are not all that important, the mechanism underlying the toxicity of the highly poisonous cyanide ion involves binding to and inhibition of the cytochrome a–cytochrome a_3 complex (cytochrome oxidase) in mitochondria. As we have seen above, the chain of electron transfers terminates in the reduction of oxygen to water and the otherwise thermodynamically unfavorable pumping of protons across the membrane against a proton gradient. The movement of protons down their gradient through ATP synthase, the most complex structure in the inner mitochondrial membrane, results in the synthesis of ATP from ADP and inorganic phosphate. The difference in proton concentration across the membrane can be measured as a difference in pH. The role of mitochondria in coupling the phosphorylation of ADP to the electron transfer from reduced NAD to oxygen was shown by Albert Lester Lehninger (1917–1986) and his associates at Johns Hopkins. The poison sodium azide, which is added to protein solutions to inhibit the growth of bacteria, inhibits cytochrome c oxidase and ATP synthase.

Once synthesized, ATP is used in many ways (Fig. 1.6). Perhaps the most important of these is the use of the free energy change of ATP hydrolysis to power a tremendous variety of otherwise thermodynamically unfavorable biochemical reactions. The essence of what ATP does in this context is provide free energy on the loss of its terminal phosphate group by hydrolysis of the **phosphoanhydride bond** (Fig. 5.7). Chemical

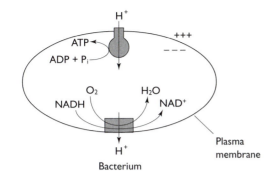

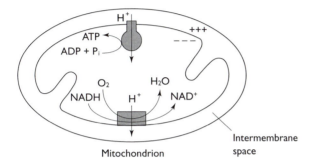

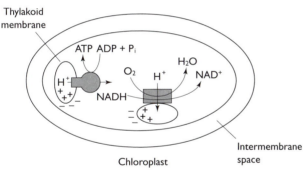

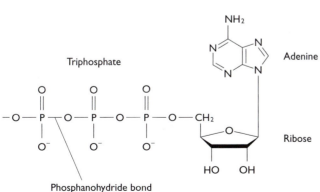

Fig. 5.6 Proton movement in bacteria, mitochondria, and chloroplasts. Note the similarities and differences in membrane orientation and direction of proton movement. In bacteria, mitochondria, and chloroplasts, the protein complex in which ATP is synthesized is situated on the cytosolic face of the membrane. Electron transport results in translocation of protons from the cytosolic side to the exoplasmic side of the membrane, creating a pH gradient. This is used to generate ATP as protons move down the pH gradient into cytoplasmic side. The similarities in ATP generation in bacteria, mitochondria, and chloroplasts point to the profound unity of all known living organisms. Adapted from Fig. 17-14 of Lodish *et al.* (1995).

Fig. 5.7 The structure of adenosine triphosphate. There are three main components: adenine, a base found in RNA and DNA; ribose, a sugar; and triphosphate. In most biochemical reactions in which it is involved, ATP is hydrolyzed to ADP and inorganic phosphate. The bond broken in this reaction is a phosphoanhydride bond. The pK_as of the dissociable protons are different (see Chapter 4).

coupling of ATP hydrolysis then 'energizes' metabolic reactions which on their own would not occur spontaneously (Chapter 4). ATP is a common intermediate of energy transfer during anabolism, cellular processes by which energy is used to synthesize complex molecules from simpler ones.

Table 5.1 | ATP requirements of macromolecule formation

Macromolecule	Subunit type	ATP expenditure per monomer added (mol mol^{-1})
Polysaccharide	Sugar	2
Protein	Amino acid	4
Lipid	C$_2$ from acetic acid	1
DNA/RNA polymerization	Nucleotide	2

In certain specialized cells or tissues, the chemical energy of ATP is used to perform other types of chemical work, for instance the mechanical work of muscle contraction and cell movement (Chapter 8). ATP is required for osmotic work, the transport of ions or metabolites through a membrane against a concentration gradient (more on this below). ATP is also a major energy source in the synthesis of macromolecules from monomers, e.g. polysaccharides from sugars and polypeptides from amino acids (Table 5.1). In respect of all this, ATP is known as the 'universal biochemical energy currency.' There are many possible ways in which the free energy of a single glucose molecule can be distributed throughout a cell! The importance of ATP in metabolism was first recognized by Fritz Lipmann and Herman Kalckar in 1941.

The hydrolysis of ATP to ADP and P$_i$ can be symbolized as

$$ATP + H_2O \Leftrightarrow ADP + P_i + H^+ \tag{5.3}$$

Using Eqns. 4.32 and 4.38, the free energy change for this reaction can be expressed as

$$\Delta G = \Delta G^{\circ\prime} + RT \ln[ADP][P_i]/[ATP] \tag{5.4}$$

To keep things simple, we assume ideal behavior. Note that [H$^+$] and [H$_2$O], which are practically independent of the concentrations of the other species, are not included explicitly in Eqn 5.4 (refer to the previous chapter if you are not sure why!). $\Delta G^{\circ\prime}$ for Eqn. 5.4 is about -7 kcal mol^{-1}. A question we might ask is whether this is relevant to the cell, where conditions are of course very different from the standard state. Assuming that the concentration of each species is 10 mM in the cell, a *very rough* estimate, Eqn. 5.4 says that $\Delta G = -7$ kcal mol^{-1} + (1.987 cal mol^{-1}K^{-1} × 298 K × ln(0.010) = -7 kcal mol^{-1} $-$ 2.7 kcal mol^{-1} ≈ -10 kcal mol^{-1} ≈ -42 kJ mol^{-1} (in skeletal muscle, [ATP] is ~50 times [AMP] and ~10 times [ADP]; using these values, ΔG is even more exergonic, possibly as large as -60 kJ mol^{-1}). That's a 40% increase in the driving force for hydrolysis over standard state conditions! In other words, the equilibrium in Eqn. 5.3 has shifted heavily towards the products when the solution is dilute. And according to the Second Law, if ATP hydrolysis releases about 10 kcal mol^{-1} at cellular concentrations, *at least* that much energy must have been consumed to synthesize ATP in the first place.

In addition to being just that much more amazed by how the world is put together, we can draw some practical lessons from this discussion. Hydrolysis of ATP is clearly spontaneous in aqueous solution, and the reaction occurs relatively rapidly at 25 °C. (*In vitro*, the half-life of ATP is on the order of days at this temperature, and in the cell, where it is needed for metabolism, its half-life is less than 1 s. If the ratio of the *in vitro* half-life to the *in vivo* one were not large, ATP would be less a useful energy storage molecule than it is.) The turnover rate of ATP and its dependence of concentration requires that ATP-containing buffers be made up fresh or stored in the cold. For the same reason solutions of the free nucleotides used in the polymerase chain reaction (PCR) are usually stored frozen at −20 °C and thawed immediately before use. We'll come back to PCR later in this chapter.

Measurement of the enthalpy change of ATP hydrolysis has shown that $\Delta H° = -4$ kcal mol^{-1}. That is, hydrolysis of one mole of ATP at 25 °C results in about 4 kcal being transferred to the solution in the form of heat and about 3 kcal remaining with ADP and P$_i$ in the form of increased random motion. We can combine our knowledge of the free energy and enthalpy changes to calculate the entropy change of ATP hydrolysis. Solving Eqn. 4.2 for $\Delta S°$ when $\Delta T = 0$ gives $\Delta S° = (\Delta H° - \Delta G°)/T$. At 310 K, $\Delta S° = ((-4 \text{ kcal mol}^{-1} - (-7 \text{ kcal mol}^{-1}))/(310 \text{ K}) = 10$ cal mol^{-1} K^{-1}. This is *roughly* the amount of entropy your body generates every time an ATP molecule is hydrolyzed. No matter how much you might feel like you're descending into a permanent vegetative state at the moment, you are nevertheless doing a splendid job of degrading the useful energy of the universe!

A couple of other points can be made here. One is that three of the four phosphate hydroxyl groups of ATP have pK$_a$ values around 1.5. These are effectively completely ionized at neutral pH. In contrast, the fourth one has a pK$_a$ of 6.5. This suggests that the net charge on a given ATP molecule might have a large impact on its cellular function. The second point is that the free energy difference between ATP and ADP + P$_i$ is *not* the same as that between the plus-phosphate and minus-phosphate forms of other biomolecules. Glucose-6-phosphate, for instance, an important molecule in glycolysis, transfers its phosphate group to water with a standard state free energy of about −3 kcal mol^{-1}. This is a substantially smaller energy change than for hydrolysis of ATP. The driving force for the chemical transfer of a phosphoryl group is known as phosphoryl **group-transfer potential**. ATP has a higher phosphoryl group-transfer potential to water than does glucose-6-phosphate. One might guess that ATP has the *highest* standard free energy of hydrolysis of all biologically important phosphates, but in fact that's not so. Instead, ATP occupies a position about midway between extremes in tables of the standard free energy of hydrolysis of phosphate compounds (Table 5.2).

Before leaving this section, we wish to touch on a few other aspects of the cellular role of ATP: activity of glycogen synthase, synthesis of cyclic AMP, binding of ATP to hemoglobin, and inhibition of thermogenin in heat generation. Glycogen is a polymeric form of glucose that can be metabolized readily in times of need. Synthesis of glycogen involves

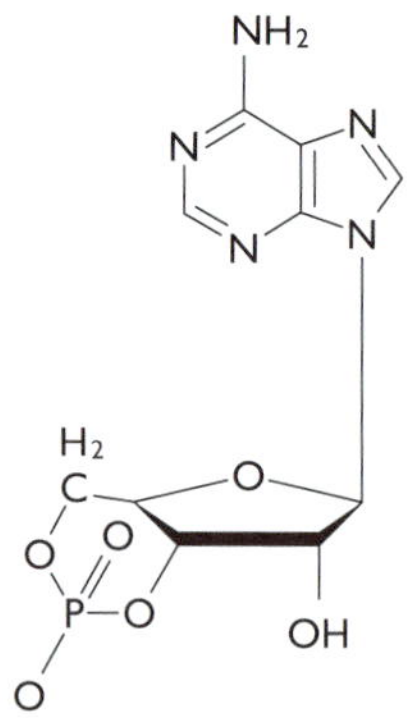

Table 5.2 | Standard free energy changes of hydrolysis of some phosphorylated compounds

Compound	$\Delta G^{\circ\prime}$ (kJ mol^{-1})
Glucose-1-phosphate	−20.9
Glucose-6-phosphate	−13.8
Fructose-6-phosphate	−13.8
ATP → ADP + P$_i$	**−30.5**
ATP → AMP + P$_i$	−32.5
Phosphocreatine	−43.1
Phosphoenolpyruvate	−61.9

Source: data are from Jencks, W.P., in Fasman, G.D., ed., *Handbook of Biochemistry and Molecular Biology*, 3rd edn, Physical and Chemical Data, Vol. I, pp. 296–304 (Boca Raton: CRC Press, 1976).

Fig. 5.8 The structure of cAMP. This molecule is synthesized from ATP and plays an important role in a variety of cellular processes. Principal among these is the control of glycogen metabolism in muscle. Glycogen is the highly branched high molecular mass glucose polysaccharide that higher animals synthesize to protect themselves from potential fuel shortage. The corresponding polymer in plants is starch (Fig. 1.1). Glycogen synthesis involves glycogen synthase. This enzyme catalyzes the transfer of the glucosyl unit of UDP-glucose (itself synthesized from glucose-1-phosphate and UTP) to glycogen. The nucleotides of information storage play an important role in energy storage and utilization in all known living organisms. In glycogen breakdown, the enzyme glycogen phosphorylase cleaves the glycosidic bond linking glucose monomers by the substitution of a phosphoryl group. The products are a slightly smaller glycogen molecule and one molecule of glucose-1-phosphate (G1P). G1P converted to glucose-6-phosphate by phosphoglucomutase. cAMP activates a protein kinase which activates phosphorylase kinase which, through phosphorylation, activates glycogen phosphorylase and inactivates glycogen synthase. The cellular concentration of cAMP is increased by adenylate cyclase, which is activated by the binding of glucagon or epinephrine to its receptor in the plasma membrane. When the hormone insulin binds to its receptor, glycogen phosphorylase is inactivated and glycogen synthase is activated.

the transfer of the glycosyl unit of uridine diphosphate glucose (UDPG) to an existing carbohydrate chain. UDPG is synthesized from glucose-6-phosphate and uridine triphosphate (UTP), a molecule involved in the synthesis of mRNA. Note the extraordinary close 'coupling' between energy storage and metabolism and information storage and expression. Replenishment of UTP occurs by means of a phosphoryl transfer reaction mediated by nucleotide diphosphate kinase. This enzyme catalyzes the transfer of a phosphoryl group from ATP to UDP, yielding ADP and UTP. Replenishment of ATP occurs by means of a phosphoryl reaction mediated by ATP synthase and a proton gradient. Replenishment of the proton gradient occurs by means of oxidation of glucose ….

ATP is a precursor in the synthesis of 3′,5′-cyclic AMP (cAMP), an important intracellular signaling molecular known as a **second messenger** (Fig. 5.8). This term was introduced in 1964 by Earl Sutherland (1915–1974), the American discoverer of cAMP. The concentration of cAMP in the cell increases or decreases in response to the tight and specific binding of an extracellular molecule to a cell-surface receptor. For instance, [cAMP] goes up when a specific odorant receptor on a cell in the olfactory epithelium binds an odorant molecule, e.g. an aromatic amine. Binding induces a conformational change in the receptor, and a protein that interacts with the receptor activates adenylyl cyclase, the membrane-bound enzyme responsible for synthesis of cAMP from ATP. cAMP then moves throughout the cytoplasm, interacting with a wide range of proteins. In this way, cAMP mediates the response of the cell to the ligand, be it an odorant molecule, hormone, or neurotransmitter. Again, we have a connection between energy and information, in that the small energy molecule ATP is involved in the communication throughout the cell of a signal received at the cell membrane. Later in this chapter we shall look at an example of the mechanics of binding interactions, and the subject will be covered in considerable detail in Chapter 7.

The physiological role of ATP does not always involve hydrolysis or chemical conversion into a electron carrier or second messenger. In fish and most amphibians, ATP binds tightly to deoxygenated hemoglobin but only weakly to oxygenated hemoglobin. The protein hemoglobin plays a crucial role in respiration by transporting oxygen to cells for oxidative phosphorylation. Binding to ATP regulates the function of hemoglobin by reducing its affinity for oxygen (see Section K below and Chapter 7 for more on hemoglobin).

Above we saw how the proton gradient in mitochondria can be coupled to the membrane protein ATPase and used to synthesize ATP. In brown adipose tissue, which contains large amounts of triacylglycerols (fatty acid triesters of glycerol, or fats, see Table 1.2) and many mitochondria, the proton gradient can be uncoupled from ATP synthesis by means of a channel protein called thermogenin. Dissipation of the proton gradient in the absence of ATP generation enables brown adipose tissue to act as a 'built-in heating pad.' Thermogenin is particularly plentiful in cold-adapted animals. The activity of thermogenin is under hormonal control. The adrenal hormone norepinephrine binds to its receptor (see Section J below) and activates adenylate cyclase. Then cAMP activates a kinase that phosphorylates a lipase, and the activated lipase hydrolyzes triacylglycerols to free fatty acids. When the concentration of free fatty acids is sufficiently high, thermogenesis is activated, and thermogenin changes the permeability of the inner mitochondrial membrane to protons and allows them back into the mitochondrial matrix without ATP production. Proton flow under the control of thermogenin is inhibited by ATP, GTP, and the diphosphate forms of these nucleotides. Energy, information, life: biological thermodynamics.

D. | Substrate cycling

The reaction catalyzed by the glycolytic enzyme phosphofructokinase is highly exergonic (Fig. 5.4). Under physiological conditions,

$$\text{fructose-6-phosphate} + \text{ATP} \rightarrow \text{fructose-1,6-bisphosphate} + \text{ADP} \quad (5.5)$$

with $\Delta G = -25.9$ kJ mol^{-1}. This reaction is so favorable that it is essentially irreversible. But this is not to say that the reverse reaction cannot occur! It just won't do so on its own. In fact, the enzyme fructose-1,6-bisphosphatase is present in many mammalian tissues, and it catalyzes the removal of a phosphate group from fructose-1,6-bisphosphate as follows:

$$\text{fructose-1,6-bisphosphate} + \text{H}_2\text{O} \rightarrow \text{fructose-6-phosphate} + \text{P}_\text{i} \quad (5.6)$$

This does not occur spontaneously since $\Delta G = -8.6$ kJ mol^{-1}. The net reaction (Eqn. 5.5 + Eqn. 5.6), however, is simply ATP hydrolysis, and $\Delta G = -34.5$ kJ mol^{-1}. Note that, although the overall free energy change is negative, this coupled reaction is less favorable than transfer of the terminal phosphoryl group of ATP to water. The opposing reactions of Eqns. 5.5 and 5.6 are known as a **substrate cycle**.

Substrate cycles might seem to serve no useful purpose, because all they do is consume energy. But nature is almost always subtler than an initial guess might suggest. Far from being futile, the reverse reaction constitutes a means of regulating the generation of product by the forward reaction, as the activities of the enzymes themselves are regulated (Chapter 7). In cases where a substrate cycle is operative, metabolic flux is not simply a matter of the activity of an enzyme, but the combined activity of the enzymes working in opposite directions. This enables exquisite regulation of a metabolic pathway, in accordance with the cell's metabolic needs. The price paid for such control is the energy lost in the combined forward and reverse reactions.

Substrate cycles also functions to produce heat, helping to maintain an organism's temperature. (Some organisms obtain a significant amount of heat from their surroundings, for instance basking lizards and snakes, which absorb heat from the Sun. Others keep warm by different means.) As we shall see in Chapter 7, one important role of heat production could be to increase the rate of enzyme activity. In bumblebees (humblebees in British English), the presence of fructose-1,6-bisphosphatase in flight muscle is thought to enable these insects to fly at temperatures as low as 10 °C, because honeybees, which do not have fructose-1,6-bisphosphatase, cannot fly when it's cold. Substrate cycling probably plays a key role in maintaining body heat in many animals, including humans. It is stimulated by thyroid hormones, which are activated upon exposure of an organism to cold.

E. | Osmosis

And now for a change of topic. Having covered the concept of chemical potential in Chapter 6, we should feel familiar enough with it to develop a topic of general importance in biochemistry: osmosis (Greek, *push*). When mineral ions are absorbed by the small intestine, water follows *by osmosis*. We treat this subject in a fair amount depth for two reasons: **osmotic work** underlies many physiological functions – nerve conduction, secretion of hydrochloric acid in the stomach, removal of water from the kidneys – and the subject involves a number of important subtleties of thermodynamics. Before looking at the mathematics of osmosis, let's first approach the subject qualitatively. This way, we'll be sure to have a general sense of what is going on before facing a page full of equations. That *osmosis* is such a nice-sounding word will help to push the discussion along!

Osmosis is an equilibrium phenomenon that involves a semi-permeable membrane, though not necessarily a biological membrane. *Semi-permeable* in this context means that there are pores in the membrane that allow small molecules like solvents, salts and metabolites to pass through but prevent the passage of macromolecules like DNA, polysaccharides and proteins. Biological membranes are semi-permeable; large solute molecules are 'impermeant.' Like freezing point depression and boiling point elevation, osmosis is a colligative property.

Suppose we have an osmometer, also called a U-tube, with arms separated by a semi-permeable membrane (Fig. 5.9). Let the temperature

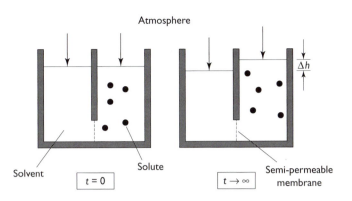

Atmosphere

Solvent
Solute
$t = 0$
Δh
Semi-permeable
membrane
$t \to \infty$

Fig. 5.9 A simple osmometer. A solute moves freely in a fraction of the total volume of solvent. The solution is separated from pure solvent by a membrane that is permeable to the solvent but not the solute. There is a net flow of solvent from the pure solvent to the solution, resulting in the development of a head of pressure. This pressure is the osmotic pressure, $\pi = \rho g \Delta h$, where ρ is density of the solution, g is gravitational acceleration, and Δh is the difference in fluid levels. As described by van't Hoff, $\pi = CV_oRT/m$, where C is the mass of solute in the volume of solvent, V_o is the partial molar volume of the solvent, and m is the molecular mass of the membrane-impermeant solute. Note that π is a linear function of C (in the first approximation). Osmotic pressure data can thus be used to measure the molecular mass of an osmotic particle.

be constant. If no solute is present the height of the solvent is the same on both sides, because the pressure of the external environment is the same on both sides. The situation changes substantially, however, upon introduction of an impermeant solute to one side. Let the solute be a large protein, say hemoglobin, and let it be freeze-dried before being added to the solvent. Freeze-dried protein occupies a relatively small volume. Initially, the height of the fluid is the same on both sides of the osmometer, just as when no solute was present. But whereas before the solute occupied a small volume on the bench top, now it is able to move freely throughout one side of the osmometer. There has been a large increase in the entropy of the solute! (If you are not sure why, see the discussion on perfume in Chapter 3.) We require that the solute particles be free to roam about the entire volume on their side of the membrane, but that they not be able pass through the membrane. What happens? There is a net movement of water from the side where no solute is present to the side where it is present. This decreases the volume of pure solvent and increases the volume of solution. Just as a confined gas pushes against the walls of its contained (Chapter 2), the solution pushes against the atmosphere and against the walls of the osmometer. How can we explain what has happened?

We know from Chapter 4 that adding a solute to a solvent reduces the chemical potential of the solvent. So, the addition of hemoglobin to one side of the osmometer has created a difference in the chemical potential of the solvent between the pure side and the impure side. This is a thermodynamically unstable situation. The impure side has a lower solvent chemical potential, and water moves down its chemical potential until equilibrium is reached. From an entropic point of view, the flow of water into the side with solute increases the entropy of the solute, making the situation more stable. How is entropy increased? Neglecting interactions between the solute and solvent, the greater the volume of solvent present, the larger the volume throughout which the solute can distribute itself. Note how similar this is to the discussion of the expansion of an ideal gas that we discussed in Section E of Chapter 3 (see Eqn. 3.22). In the context of the perfume example, if it is applied in the bathroom, the perfume molecules become distributed throughout the bathroom, but when the door is opened, the molecules begin to spread into the hall. At equilibrium, the perfume molecules will occupy the bathroom and hall, i.e. the entire accessible volume.

The flow of water from one side of the U-tube to the other must result in a change in the height of the water on the two sides. It becomes lower on the side of the pure solvent and higher on the side of the impure solvent. After enough time, the system comes to equilibrium. As we have seen, equilibrium represents a balance, so the driving force for water to move through the membrane from the pure solvent to the solution must be equal in magnitude to the hydrostatic pressure arising from the difference in height of the water in the two arms ($p_{hydrostatic} = \rho g \Delta h$, where ρ is the density of the solution). The hydrostatic pressure is the same as the *additional* pressure one would have to apply to the side of the U-tube with the solute in order to equalize the height of the water on the two sides of the membrane. This pressure is called the **osmotic pressure**, and it was first studied in the 1870s by the German botanist and chemist Wilhelm Friedrich Philipp Pfeffer (1845–1920), the son of an apothecary.

We now wish to take a more mathematical approach to osmosis. This way of thinking about the subject is not necessarily superior to the qualitative approach just because it involves more equations, but it does provide additional insight to our subject, and that's what we want. The system is now regarded as consisting of two *phases*, x and y. In x, the impermeant molecules (component 2) are dissolved in component 1 (the solvent). In y, only solvent molecules are present. This is just as before. Considering the *solvent* alone, the requirement for equilibrium between the two phases is

$$\Delta G = \mu_1^x \Delta n_1^x + \mu_1^y \Delta n_1^y = 0 \tag{5.7}$$

(See Section E of Chapter 4.) Δn_1 stands for an incremental change in the number of moles of *solvent*. It follows that

$$\mu_1^x = \mu_1^y \tag{5.8}$$

since $\Delta n_1^x = -\Delta n_1^y$, the number of molecules *gained* by phase x must be equal to the number of molecules *lost* by phase y. The ledger balances. But wait! Something funny's going on; for regardless of the amount of solvent transferred through the membrane, we can't avoid the requirement that $\mu_1 - \mu_1^\circ = \Delta\mu < 0$ (see Eqn. 4.10); the chemical potential of the solvent plus solute *must* be lower than that of the pure solvent. Nevertheless, Eqn. 5.8 does say that the chemical potentials of the solvent in the two phases *must* be equivalent. Where did we err?

We didn't! We concluded that there is a contradiction, that biological thermodynamics is illogical and therefore a waste of time, that it was a mistake to study biology or in any case to do it in a place where biological thermodynamics forms part of the curriculum, and that our best option would be to make our way to the college bar and drink ourselves to oblivion. Right? No way! Things are just starting to get interesting! Let's see if we can't crack this nut now, and think about a celebratory night out later on. But what can we do?

In Chapter 4 we showed that $\Delta G = V\Delta p - S\Delta T$. When the temperature is constant, $\Delta T = 0$ and the free energy change is proportional to the pressure change. To make the resulting expression tell us what happens when we change the number of solvent molecules present, we divide both sides by Δn_1. This gives

$$\Delta G / \Delta n_1 = \mu_1 = V_{m,1}\Delta p \qquad (5.9)$$

where $V_{m,1}$ is the molar volume of component 1. This is the 'missing' term our expression of the chemical potential! Taking into account the development leading up to Eqn. 4.12, where we saw how the chemical potential of a solvent changes when a solute is added, and adding in Eqn. 5.9, we have

$$\mu_1 - \mu_1^\circ \approx -RTC_2 V_1^\circ / M_2 + RT \ln f_1 + V_1^\circ \pi \qquad (5.10)$$

where the pressure difference has been written as π (not the ratio of the circumference of a circle to its diameter but the symbol that is traditionally used to designate the osmotic pressure; π starts with the same sound as *pressure*). It probably seems like our adding in the extra term (Eqn. 5.9) was a rather arbitrary move, but in fact it was not. In order for the equilibrium condition to be met, we must have a balance of forces, and we can write down an equation that expresses that balance only if we take into account *everything* that's going on. There was a contradiction earlier because we had assumed that the system was at equilibrium when we hadn't taken the pressure term into account. Note that in Eqn. 5.10 we have assumed $V_{m,1} \approx V_1^\circ$, the molar volume of pure solvent; this holds for dilute solutions. The pressure difference π is the pressure that must be applied to the side of the solute to make the height the same on both sides. If atmospheric pressure is pushing down on both sides, π measures the pressure head resulting from the difference in height of solvent in the two arms of the osmometer.

We can simplify Eqn. 5.10 a bit. If the solution is approximately ideal, $f_1 \approx 1$ and $RT \ln f_1 \approx 0$. At equilibrium, $\mu_1 - \mu_1^\circ = 0$, and from this it follows that

$$\pi = RTC_2 / M_2 \qquad (5.11)$$

This is the van't Hoff law of osmotic pressure for ideal dilute solutions, named in honor of the person who gave Pfeffer's work a firm mathematical foundation. Eqn. 5.11 can be used to measure the mass of an impermeant solute particle (though there are simpler and more accurate ways of doing it). Note how Eqn. 5.11 looks like Eqn. 4.12. Eqn. 5.11 also resembles the ideal gas law ($pV = nRT$ or $p = nRT/V = CRT$, where n is the number of particles and C is the concentration). C_2, the concentration of solute, is the mass of solute particles per added to a known volume of pure solvent, or n/V in the context of the gas law. What van't Hoff found was that the measured osmotic pressure was basically the pressure of n solute particles moving around in volume V, the volume of the solvent through which the solute particles are free to move!

The degree to which Eqn. 5.11 matches experimental results varies

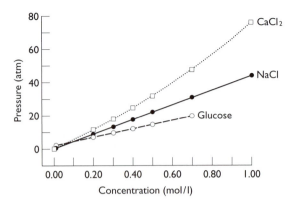

Fig. 5.10 Osmotic pressure measurements. Osmotic pressure increases with concentration of solute, as predicted by the van't Hoff law. The pressure at a given concentration of solute depends significantly on the solute. If the solute is a salt, dissociation in aqueous solution will result in greater number of particles than that calculated from the molecular mass of the salt. The van't Hoff law is exact for an ideal solution. At high solute concentrations, non-linear behavior can be detected. This can be accounted for by higher order terms in C. The data are from Table 6-5 of Peusner (1974).

with concentration and solute (Fig. 5.10). There are several different ways of trying to cope with this, but we shall concern ourselves with but one of them. We express the osmotic pressure as a series of terms (power series expansion) and check to see that the dominant term is the same as Eqn. 5.10 in the limit of low solute concentration:

$$\pi = \frac{C_2 RT}{M_2}\left(1 + B_1(T)C_2 + B_2(T)C_2^2 + \ldots\right) \tag{5.12}$$

The $B_i(T)$ terms are *constant* coefficients whose values are solute- and temperature-dependent and must be determined *empirically*. If C_2 is small, only the first term makes a significant contribution to π (convince yourself of this!), just as required. If just the first two terms contribute to π, a plot of π/C_2 will be linear in C_2 with slope $B_1(T)RT/M_2$ and intercept RT/M_2. This permits indirect measurement of M_2 and $B_1(T)$. Eqn. 5.12 can easily be generalized to account for the contribution of multiple osmotic species:

$$\pi_{\text{total}} = \pi_1 + \pi_2 + \ldots + \pi_n = \Sigma \, \pi_i \tag{5.13}$$

If a solute species is present on both sides of the membrane, and if this solute cannot pass through the membrane, it will make a contribution to the total osmotic pressure, but only if there is concentration difference. In such cases, π_i is proportional not to C_i, as in Eqn. 5.12, but to ΔC_i, the concentration difference across the membrane.

Now let's leave the wispy world of mathematics and return to a more material physical biochemistry. Osmosis can be a very strong effect. At 25 °C, a 1 M solution of a glucose, a relatively small 'osmolyte,' gives a pressure more than 25 times greater than that of the atmosphere. 1 M solutions of salts give even larger osmotic pressures (Fig. 5.10), though the ions are smaller than glucose, even when hydration is taken into account. Osmotic forces are important here because they play a key role in membrane transport of water in living organisms. Another qualitative example of osmotic pressure in action is as follows. Red blood cells are full of impermeant solute particles, mainly hemoglobin; red cells have been called 'bags of hemoglobin.' M_2 is large, about 68 000, and C_2 is high, about 0.3 M or greater. When placed in pure water, a red cell experiences a very large π, about 8 atm or higher, the approximate pressure one would feel scuba diving at a depth of 70 m! The plasma membrane of a red cell is not strong enough to withstand an osmotic pressure of this magnitude and breaks, releasing the hemoglobin. To prevent damage to red cells after separation from plasma by centrifugation, blood banks resuspend them in a sucrose solution (sucrose is membrane impermeant) of approximately the same solute concentration as blood plasma.

Note the subtle difference between the red blood cell in a hypotonic solution (low solute concentration) and impermeant particles in an osmometer. In the osmometer, there is a very real pressure difference. The presence of the impermeant particles results in the formation of a pressure head, and the solution pushes down harder than it would if the solute were not there. An osmometer need not have vertical arms, however: it could be a capillary tube oriented horizontally with a membrane in the middle. Water would still move through the membrane as before. One can think of the effect as arising from the pressure of n solute particles confined to a volume V of solvent. This suggests that red blood cells burst in hypotonic solution because the hemoglobin molecules inside bang on the cell membrane much harder than the water molecules bang on the membrane from the outside. Do they?

If the solution is at thermal equilibrium, then *all* the particles have the same average thermal energy, irrespective of their size. Big molecules like hemoglobin are relatively slow, and little ones like water are relatively fast. But these molecules do *not* have the same momentum, which is what gives rise to pressure (see Chapter 2). From physics, the kinetic energy of a particle is $mv^2/2 = \mathbf{p}^2/m$, where m is the mass, v is the velocity and $\mathbf{p} = mv$ is the momentum. Thermal energy is proportional to T, and at thermal equilibrium the kinetic energy of a particle is equal to its thermal energy. So, $\mathbf{p} \propto (mT)^{1/2}$. In other words, the more massive the particle, the greater its momentum. Note that $\mathbf{p}$ has nothing to do with particle *volume*. This means that in a hypotonic solution, where in the case of pure water there is nothing on the outside that cannot get in, the hemoglobin does bang into the membrane from all directions a lot harder than the water bangs on the membrane from all directions. But wait, is not water rushing in? Yes, in accordance with the Second Law of Thermodynamics! The driving force for this movement of water is very large, and it is mainly *entropic* in nature. The concentration of water outside is relatively large, and the concentration of water inside is relatively small (because the hemoglobin concentration is so high), and the water moves *down* its concentration gradient. It's not that water couldn't enter the cell before we put it in a hypotonic solution; it could come in and go out, but the number of water molecules per unit volume outside was identical to the number of water molecules per unit volume inside, and there was no net flow of water. In hypotonic solution, as water flows in, the red blood cell swells, creating a larger and larger volume in which the massive hemoglobin molecules can diffuse.

The actual situation with hemoglobin is not as simple as we've made it sound. This is because hemoglobin does not just float around in the water of the red blood cell, it *interacts* with it. So, when particle interactions are taken into account, the volume of the particle does matter, because the more space it takes up, the more surface it will expose to the solvent. This is the source of the higher order terms in Eqn. 5.12. It is said that thermodynamics is a difficult subject because there are so many different ways of thinking about the same thing. Osmosis is a good example of this.

In contrast to red cells, bacteria do not burst when placed in a hypotonic solution. This is because these organisms (as well as plant cells and fungi) can withstand high osmotic pressures by means of a rigid cell wall. When, however, bacteria come into contact with lysozyme, an enzyme we have already encountered several times in this book, they burst. This is caused by the biochemical activity of lysozyme, which is to cleave certain glycosidic bonds in the polysaccharides that give the bacterial cell wall its strength. It is a rather good thing that our bodies station lysozyme molecules at common points of entry of foreign microbes, e.g. the mucosal membrane in the nasal passage. People with a mutant lysozyme gene have a tougher time than most in fighting off infection, and they tend to die relatively young. You might surmise that early death results from too little lysozyme being available to make the cell walls of bacteria more susceptible to osmotic stress. Again, however, the actual situation is more complex than we've made it sound. That's because the immune system has a role to play in fighting off infection, few pathogenic bacteria are susceptible to lysozyme alone, and the mutant lysozyme proteins, which are less active than the wild-type enzyme, are also less thermostable than the wild-type enzyme and give rise to amyloid fibril formation! Not only are these mutant lysozymes less active and therefore less able to fight off infection, there is a net flow of them into rigid fibril structures where they have effectively no enzymatic activity all. To make things worse, the body has a difficult time getting rid of the fibrils, and their continued increase in size can be pathological. We'll return to protein pathology in Chapter 8.

F. | Dialysis

This section follows on directly from the previous one. There are two basic forms of dialysis in biochemistry: non-equilibrium dialysis and equilibrium dialysis. We look at both forms here; the physics involved is essentially the same for both of them. Dialysis is useful to the biochemist because it can be used to separate molecules according to size. It does this by means of a semi-permeable membrane, like the one discussed in the previous Section (Fig. 5.11). Some membranes used for dialysis are made of cellophane (cellulose acetate).

Non-equilibrium dialysis is the use of a semi-permeable membrane to change the composition of the solution in which macromolecules are dissolved. For instance, one means of purifying recombinant proteins from *E. coli* host cells is to lyse the cells in 8 M urea, a small organic compound. Urea denatures most proteins at room temperature when the concentration is above about 8 M. Once the recombinant proteins have

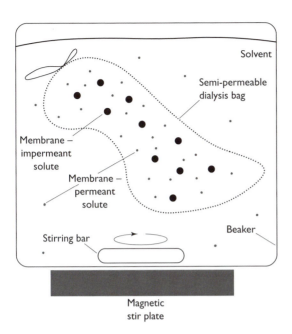

Fig. 5.11 Dialysis. A dialysis bag containing a membrane-impermeant solute is submerged in solvent in a beaker. Membrane-permeant solute appears on both sides of the membrane. The dialysis bag-solvent system is not at equilibrium. At equilibrium, the concentration of membrane-permeant solute will be the same on both sides of the membrane. A magnetic stir plate and a stirring bar are used to accelerate the approach to equilibrium; the net flow rate of membrane-permeant solute out of the dialysis bag is related to the concentration gradient of that solute.

been separated from the bacterial ones (often by a type of affinity chromatography), the recombinant protein solution is transferred to a dialysis bag. The bag is then sealed and placed in a large volume of buffer containing no urea. When equilibrium is reached several hours later, the concentration of urea inside the bag has decreased, the concentration outside has increased, and the concentration of urea is about the same on both sides of the membrane.

What causes this? There are several relevant considerations. Osmosis is occurring, because the concentration of solute particles is much higher within the dialysis bag than in the solution outside. At the same time, the chemical potential of urea is much higher in the bag and low outside. Urea will therefore diffuse out of the bag until the gradient is gone. Because the concentration of urea both inside the bag and outside changes continually until equilibrium is reached, the osmotic pressure changes continually as well. The osmotic effect can be substantial when working with such a high urea concentration, leading to a large increase in the volume of material inside the bag. So, to avoid rupture and loss of a precious protein sample, it's a good idea to leave room within the bag for the influx of buffer.

Is any work done by diffusion of urea out of the dialysis bag? No! Despite similarities to osmosis, the situation here is qualitatively different. True, urea moves down its concentration gradient, but there is no corresponding development of a pressure head, as in osmosis. In other words, nothing retards the dilution of urea, so no pV-work is done. The experiment can be carried out in isolation, so $q = 0$. By the First Law, $\Delta U = 0$. If the pressure is constant, $\Delta H = 0$. If the temperature is constant as well, then the Gibbs free energy is the appropriate thermodynamic potential. However, this process is irreversible! But if we carry out the

Fig. 5.12 Equilibrium dialysis. At the beginning of the experiment ($t = 0$), the membrane-impermeant macromolecule and membrane-permeant ligand are on opposite sides of a semi-permeable dialysis membrane. The two-chambered system is not at equilibrium. After a long time ($t \to \infty$), the concentration of *free* ligand is the same on both sides of the membrane, in accordance with the Second Law of Thermodynamics. The number of ligand molecules, however, is not the same on both sides of the membrane, as some ligands are bound to the membrane-impermeant macromolecules. The bound ligand molecules are nevertheless in equilibrium with the free ones. Measurement of the concentration of free ligand at equilibrium and the total concentration of ligand determines the amount of bound ligand at equilibrium.

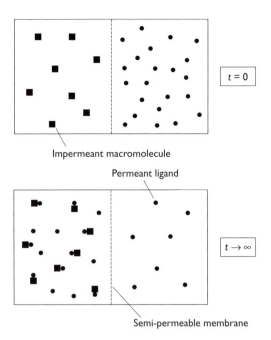

Impermeant macromolecule

Permeant ligand

Semi-permeable membrane

process very slowly, the system will pass through a series of near equilibrium states, and we can evaluate ΔG. Because G is a state function, the value of ΔG depends only on the initial and final states, and not on whether the process used to get from one to the other was reversible or irreversible. Because the diffusion of the urea is spontaneous, $\Delta G < 0$. Hence, $\Delta S > 0$.

Our final consideration in non-equilibrium dialysis is that of charge. Typically, the macromolecule in the dialysis bag will be charged, and this will affect osmosis. Similarly, the solution one dialyzes out of or into the bag will typically be a buffer of some sort, containing both charged and uncharged solute particles, and the ratio and relative abundance of these will have an impact on the migration of water through the membrane. The effect of charge on membrane equilibrium is a complicated subject; we'll come back to the subject below and treat it in some depth.

What about equilibrium dialysis? Suppose you are interested in the binding of a macromolecule to a ligand, and that the ligand is membrane-permeant. This presents an opportunity for quantitative analysis of the binding interaction. To see how, suppose we have a two-chambered device like that shown in Fig. 5.12. On the left side is a known amount of macromolecule in our favorite buffer, and on the right a known amount of ligand dissolved in the same buffer. The ligand will diffuse down its concentration gradient through the membrane, and by mass action become bound to the macromolecule. After a sufficiently long time, equilibrium will be reached. At equilibrium, the concentration of free ligand will be the same on both sides of the membrane. The amount of ligand on the side of the macromolecule, however, will be higher by an amount depending on the strength of ligand binding (i.e. the magnitude of the equilibrium constant of binding). If one then assays the amount of ligand on both sides of the membrane, the difference will be the amount bound to the macromolecule. A comparison of the concentration of 'bound' ligand to the concentration of macromolecule and the concentration of 'free' ligand enables determination of the binding constant and the number of ligand molecules bound per macromolecule. Binding is so important to the biochemist that an entire chapter is devoted to it below.

G. | Donnan equilibrium

Our discussion of dialysis merely mentioned charge effects in passing. Here we shall see just how much more complicated things are when charge is taken into account explicitly. It is important to engage with the subject with this added degree of complexity, as three of the four major classes of biological macromolecule – proteins, nucleic acids and phospholipids – are charged. Moreover, in the living organism these molecules are found not in pure water but in a saline solution.

Now, suppose we have a polyelectrolyte like DNA dissolved in a solution containing a simple salt (e.g. NaCl). Suppose further that there are two phases to our system: one (α) consisting of water, Na^+ and Cl^-, the other (β) of water, Na^+, Cl^- and DNA. The phases are separated by a semipermeable membrane, and DNA alone is impermeant. The equilibrium concentration of ions will not be the same on the two sides of the membrane except in the limit that $[DNA] \rightarrow 0$. Moreover, because DNA is anionic, we should expect the concentration of sodium to be higher on the side of the DNA. In symbols, $[Na^{+\beta}] > [Na^{+\alpha}]$. Let's see how we can obtain quantitative expressions of the ion concentrations.

At equilibrium,

$$\mu_{NaCl}{}^{\alpha} = \mu_{NaCl}{}^{\beta} \tag{5.14}$$

To simplify matters, we assume that the solution is ideal. The chemical potential of the salt is

$$\mu_{NaCl} = \mu_{NaCl}{}^{\circ} + RT \ln[Na^+][Cl^-] \tag{5.15}$$

At equilibrium, the chemical potential must be the same in both phases, so

$$[Na^{+\alpha}][Cl^{-\alpha}] = [Na^{+\beta}][Cl^{-\beta}] \tag{5.16}$$

Moreover, the net charge of each phase must be equal to zero, a condition known as **electroneutrality**. We express electroneutrality in mathematical terms as

$$[Cl^{-\alpha}] = [Na^{+\alpha}] \tag{5.17}$$

$$z[DNA^{\beta}] + [Cl^{-\beta}] = [Na^{+\beta}] \tag{5.18}$$

where z is the number of negative charges on the DNA. With a bit of algebra, these equations can be combined to give

$$[Na^{+\beta}] = [Na^{+\alpha}]\left(1 + \frac{z[DNA^{\beta}]}{[Cl^{-\beta}]}\right)^{1/2} \tag{5.19}$$

$$[Cl^{-\beta}] = [Cl^{-\alpha}]\left(1 - \frac{z[DNA^{\beta}]}{[Na^{+\beta}]}\right)^{1/2} \tag{5.20}$$

As expected, $[Na^{+\beta}] > [Na^{+\alpha}]$, and this neutralizes the charge on DNA in phase β. Similarly, $[Cl^{-\beta}] > [Cl^{-\alpha}]$, though the minus sign tells us that the difference is not as great as for the DNA counterions. In such situations, the osmotic pressure is produced by both the impermeant macromolecule and the asymmetric distribution of small ions. This effect, called the **Donnan equilibrium**, was first described in 1911 by the physical chemist Frederick George Donnan (1870–1956), son of a Belfast merchant. The effect pertains not only to membrane equilibria but to any situation in which there is a tendency to produce a separation of ionic species. The asymmetric distribution of ions arises from the requirement of electroneutrality, and its magnitude decreases with increasing salt concentration and decreasing macromolecule concentration, as can be seen from Eqns. 5.19 and 5.20.

The Donnan effect is even more complicated for proteins than for DNA. This is because the net charge on a protein is highly dependent on pH. The greater the net charge, the greater the effect. The Donnan effect for proteins is minimized at the isoelectric point, where the net charge is zero. Though this causes us no conceptual difficulties, it may prove problematic in the laboratory. This is because protein solubility is usually relatively low at the isoelectric point. An example of where the Donnan effect is relevant is the erythrocyte, or red cell. The effect is caused mainly by the huge concentration of hemoglobin inside the cell and the inability of hemoglobin to penetrate the membrane. Other ions present, for instance sodium and potassium, do not contribute to the Donnan effect because they are impermeant and their effects counterbalance ($[K^+]_{plasma} \approx [Na^+]_{cell}$ and ($[K^+]_{cell} \approx [Na^+]_{plasma}$). Chloride, bicarbonate and hydroxyl, on the other hand, can cross the membrane, and they are relevant to the Donnan equilibrium. Experimental studies have shown that the cell–plasma ratios of these ions are 0.60, 0.685, and 0.63, respectively. The marked deviations from 0.5 arise from the confinement of hemoglobin to the cell. This has an impact on the pH of blood because both bicarbonate and hydroxyl are bases.

H. | Membrane transport

Glucose metabolism takes place within cells, and an individual cell in a higher eukaryote is separated from its surroundings by a plasma membrane. The membrane enclosing the cell is about 10 nm thick. It comprises two layers of phospholipids, with the charged groups on the outside (Chapter 4). The interior of the membrane is oily, making it highly impermeable to polar compounds. Charged substances are able to pass through membranes, but usually only by means of transport proteins embedded in the lipid bilayer.

Membrane transport is said to **passive** if a solute moves down its concentration gradient, and **active** if it moves against it. Active transport in cells is driven by the free energy of ATP hydrolysis. An example of active transport is the movement of Na^+ and K^+ across the cell membrane of red blood cells, nerves and muscle cells. The concentration of K^+ in muscle is about 124 mM, some 60 times greater than in serum. With Na^+ it's the other way around, the concentration being about 4 mM in muscle cells and about 140 mM in the serum. There are ways for these ions to flow down their concentration gradients, as one expects them to do on thermodynamic grounds. The gradients themselves are important to cell function, and they are maintained by a membrane-spanning enzyme called Na^+/K^+-transporting adenosine triphosphatase. In order to do this, however, the cell must pay the price of kicking out the unwanted ions that have come in and the total cost is that of recovering the wanted ones that have left. The ions must be moved from a region of low concentration to a region of high concentration. The Na^+/K^+-transporter acts a pump to move ions against their concentration gradient and is powered by ATP hydrolysis. Another example of active transport is the secretion of HCl into the gut of mammals by parietal cells. Unsurprisingly, perhaps, these cells have a relatively large number of mitochondria.

A numerical example will help to motivate the discussion. The change in chemical potential of a glucose molecule when it is transported down a 1000-fold glucose concentration gradient at 37 °C is given by Eqn. 4.32:

$$\Delta\mu = 8.314 \text{ J mol}^{-1}\text{K}^{-1} \times 310\text{ K} \times \ln(1/1000) = -17.8\text{ kJ mol}^{-1} \quad (5.21)$$

If the glucose concentration in the blood is high, as it is after a meal, and the concentration in the cell is low, there is a large force driving the sugar into the cell. Once the sugar enters, it gets 'tagged' with a highly charged phosphoryl group, making its escape through the oily membrane highly thermodynamically unfavourable. Another example is the following. The concentration of chloride in the blood is about 100 mM, while that in the urine is about 160 mM. Thus, work must be done to pump chloride out of the blood and into the urine. We can calculate the work done by the kidneys in this process: $\Delta\mu = 1.9872 \text{ cal mol}^{-1}\text{ K}^{-1} \times 310\text{ K} \times \ln(160/100) = 290 \text{ cal mol}^{-1}$. We can also estimate the number of chloride ions transported per ATP molecule hydrolyzed: free energy change of ATP hydrolysis in the cell / energy required to transport $Cl^- = 10\,000 \text{ cal mol}^{-1} / 290 \text{ cal mol}^{-1} \approx 34$. We have ignored charge effects in this second example.

We know from our study of the Donnan equilibrium, however, that if the solute particle is charged, as in the case of Cl^-, the situation is more complicated than for an electroneutral solute. Eqn. 4.32 still applies, but we also need to take into account the work done as the charged particle moves through the electrical potential of the membrane, ΔV. The magnitude of ΔV is 10–200 mV, depending on the cell type. In Chapter 4 we saw that the magnitude of the free energy change for electrical work is $\Delta\mu = nF\Delta V$ when the ionic charge is n. Adding this term to Eqn 4.32, gives

$$\Delta\mu = RT\ln[I]_i/[I]_o + nF\Delta V \tag{5.22}$$

where I represents an ionic solute and $\Delta V = V_i - V_o$. The reference state must be the same for both terms of the right hand side of this equation; in this case it is the extracellular matrix. The sign of the second term on the right-hand side is positive because $\Delta V < 0$. When there is no driving force to move an ion from one side of the membrane to the other, $\Delta G = 0$. Under this condition,

$$nF\Delta V = -RT\ln[I]_i/[I]_o \tag{5.23}$$

and

$$\Delta V = -\frac{RT}{nF}\ln\frac{[I]_i}{[I]_o} \tag{5.24}$$

We can use Eqn. 5.24 and the magnitude of the measured potential across the membrane to calculate the ratio of the concentrations of the ionic solute. Let's assume that we are working with a monovalent cation ($n = 1$) at 300 K and that $\Delta V = 120$ mV. Solving for $[I]_i/[I]_o$, we have

$$[I]_i/[I]_o = \exp(-nF\Delta V/RT)$$

$$= \exp[(-96.5\text{ kJ V}^{-1}\text{ mol}^{-1}\times 0.12\text{ V})/(8.314\text{ J mol}^{-1}\text{ K}^{-1}\times 300\text{ K})]$$

$$= 0.01 \tag{5.25}$$

So $[I]_o = 100[I]_i$.

At a number of points in this book we have mentioned nerve impulses (e.g. in the context of olfaction, perfume, and the Second Law), and now we wish to expand on the underlying mechanisms in mostly qualitative terms. We wish to see how the development of this section fits within the broader picture of how living organisms work. Neurons, like other cell types, have ion-specific 'pumps' situated in the plasma membrane. These protein machines use the energy of ATP hydrolysis to generate ionic gradients across the membrane in a manner that resembles how electron transport proteins use the energy of glucose metabolism to generate a proton gradient. When 'at rest', a neuron is not very permeable to Na^+ (concentrated outside the cell, dilute inside) or K^+ (concentrated inside, dilute outside). There is a voltage across the membrane on the order of 60 mV. Stimulation of a nerve cell results in a diminution or 'depolarization' of the membrane voltage. Voltage-sensitive channel proteins that are selective for specific ions are activated by the decrease in voltage, allowing Na^+ ions in and K^+ ions out. The combined effect of gain and loss of membrane permeability to these ions results in a millisecond-timescale spike in the membrane potential known as an action potential. Depolarization of one part of the membrane by an action potential triggers depolarization of the adjacent part of the membrane, and the impulse propagates down the axon of neuron. Nerve impulses travel in one direction only, because a certain amount of time is required for the recently depolarized part of the cell to regenerate its ion gradient.

The most famous of all transport proteins is ATP synthase (ATPase). As we have seen, this molecular motor plays a central role in bio-energetics. Eqn. 5.22 can be used to describe the energetics of the 'energy-transducing' membranes involved in ATP synthesis. In this case, the membrane of interest is the inner membrane of mitochondria and the ion is hydronium. The term $\ln[I]_i/[I]_o$ becomes $\ln[H^+]_i/[H^+]_o$, which as before can be rewritten as $2.3\Delta pH$ ($pH = -\log[H^+]$). Substituting this into Eqn. 5.22 gives the **proton motive force** of chemiosmotic theory:

$$\Delta\mu_{H+} = 2.3RT(pH_o - pH_i) + nF\Delta V \tag{5.26}$$

The magnitude of the potential across the inner membrane of a liver mitochondrion is about 170 mV (inside), and the pH of its matrix is about 0.75 units *higher* than that of its intermembrane space (outside). Thus,

$$\Delta\mu = [2.3 \times 8.314\,\text{J mol}^{-1}\,\text{K}^{-1} \times 298\,\text{K} \times (-0.75)]$$

$$+ [1 \times 96\,500\,\text{J V}^{-1}\,\text{mol}^{-1} \times (0.17\,\text{V})] \tag{5.27}$$

the result being about $+12\,\text{kJ mol}^{-1}$ for transport of a proton *out of* the matrix.

The basic principles discussed in this section apply not only to the synthesis of ATP from ADP and P_i and the separation of ions, but also to a broad range of transport processes occurring across plasma membranes and neuronal synaptic vesicles. Before closing this section, we want to take the opportunity to see how ATP synthesis is a matter of energy coupling on a grand scale. Doing this well help us to realize how things tie together, to see how marvelously integrated the various aspects of the living cell actually are.

As we have seen, glucose is oxidized by means of being coupled to the reduction of oxygen to water. Electron transport proteins play a key role in this. The overall redox reaction, which is energetically favorable, is used to pump protons *against* their concentration gradient to the opposite side of the membrane. In other words, the pH of solution on one side of the membrane is different from that on the other side. The voltage difference across the membrane, which is only about 10 nm thick, is about 200 mV. The electric field strength in the middle of the membrane is *huge*! Every living cell in your body is like that! As protons *diffuse* down their gradient through the membrane, they do so by means of a channel made of protein. They cannot move straight through a membrane made of lipids, because dielectric constant of hydrocarbon is very low, about 3. The channel is lined with polar chemical groups,

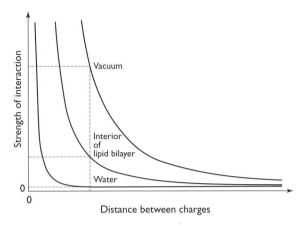

Fig. 5.13 Dependence of electrostatic energy on distance between charges and medium. The energy is inversely proportional to distance, so attraction or repulsion is greatest when the charges are close to each other. The energy also depends substantially on the stuff between the charges, varying inversely with the dielectric constant of the medium. The dielectric constant depends on temperature and pressure, and it must be determined empirically. The interaction between charges is greatest in vacuum. In water, where the dielectric constant is very large, charges must be very close for the interaction between them to be significant. Charge–charge interactions are relatively large in the hydrophobic core of a protein or in the plasma membrane, because the dielectric constant of hydrocarbon is much lower than that of water.

and these make passage energetically favorable (Fig. 5.13). The truly amazing thing about this channel is that the energy change of proton translocation is coupled to an energetically unfavorable process – ATP synthesis. This is not a trivial chemical coupling, because the free energy change on hydrolyzing ATP to ADP is about 10 kcal mol^{-1} at cellular concentrations. For all practical purposes, hydrolysis of ATP is *irreversible*! The cell (in fact, the mitochondria here, but the principles are exactly the same for entire cells) must do work to generate the proton gradient, but there's a purpose behind it. No wonder we need to eat so much.

I. | Enzyme–substrate interaction

In Chapter 2 we touched on the biological function of ribonuclease A (RNase A), a digestive enzyme that hydrolyzes RNA to its component nucleotides. We said that an inhibitor of the enzyme, 2'-cyclic monophosphate, can be used to study the enthalpy of nucleotide binding to RNase A. One aim of this chapter is to illustrate the general utility of Eqn. 4.2 by way of several examples. In this section we focus on how it applies to studies of the energetics of binding of small compounds to protein or DNA, taking RNase A as our example.

Figure 5.14 shows the reaction catalyzed by RNase A. The scheme is based on the isolation of 2',3'-cyclic nucleotides from RNase A digests of RNA. There are four types of 2',3'-cyclic nucleotides. RNase A hydrolysis of one of these, cytidine 2',3'-cyclic phosphate, has been studied extensively. The reaction is

cytidine 2',3'-cyclic phosphate(aq) + H$_2$O(l) →

    cytidine 3'-cyclic phosphate(aq)   (5.28)

How might one determine $\Delta G°$ for this reaction? Because the temperature is constant,

$$\Delta G = \Delta H - T\Delta S \qquad (5.29)$$

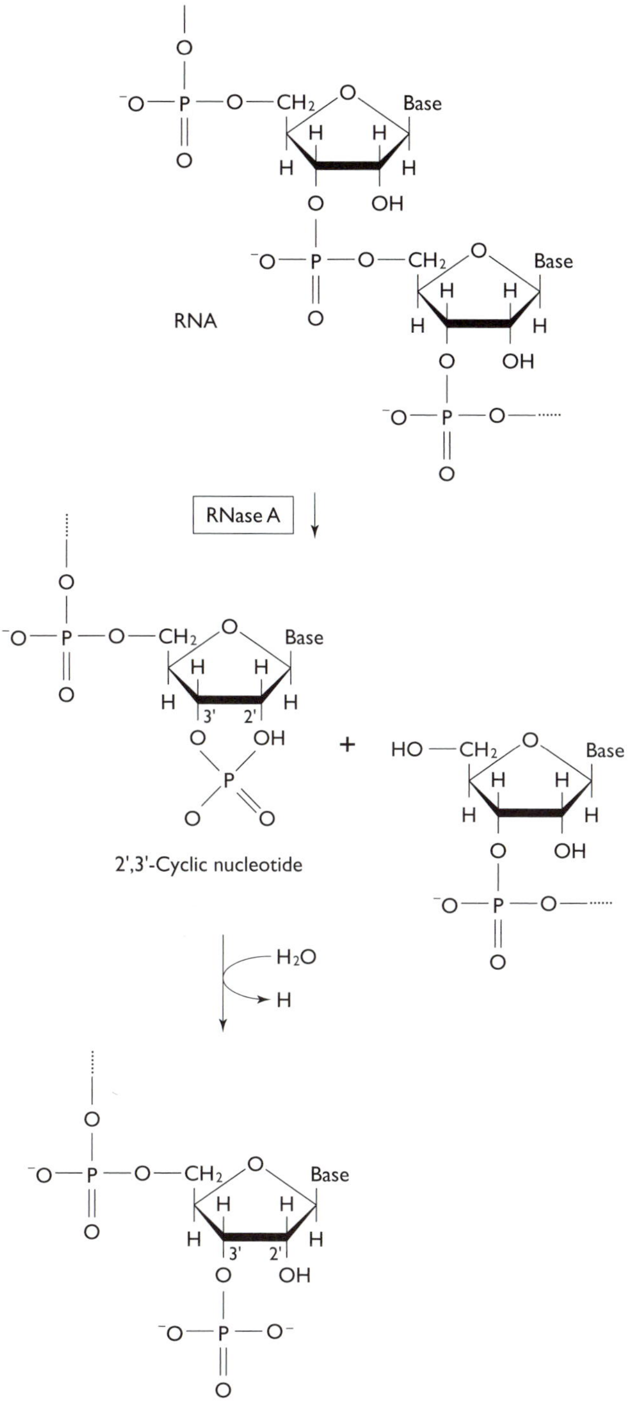

Fig. 5.14 Mechanism of RNase A activity. Bovine pancreatic ribonuclease A is an example of enzyme-mediated acid–base catalysis. The enzyme hydrolyzes RNA to its component nucleotides. The reaction scheme is based on the experimental finding that 2′,3′-cyclic nucleotides are present in RNase digests of RNA. RNase is inhibited by 2′-CMP. This binding interaction has been studied in considerable depth.

If the products and reactants are required to be in the standard state, Eqn. 5.29 becomes

$$\Delta G° = \Delta H° - T\Delta S° \tag{5.30}$$

To calculate $\Delta G°$, we need to know $\Delta H°$ and $\Delta S°$. The enthalpy change of the reaction, which can be estimated by calorimetry, is $-2.8\ kcal\ mol^{-1}$. What about ΔS? One approach would be to make use of the fact that S is a state function and combine measurements that, when summed, give $\Delta S°$ for Eqn. 5.28. The reaction scheme might look like this:

cytidine 2′,3′-cyclic phosphate(aq) →

    cytidine 2′,3′-cyclic phosphate(s)   (5.31)

cytidine 2′,3′-cyclic phosphate(s) + H_2O(l) →

    cytidine 3′-phosphate(s)   (5.32)

    cytidine 3′-phosphate(s) → cytidine 3′-phosphate(aq)   (5.33)

Eqn. 5.31 represents the dissolution of cytidine 2′,3′-cyclic phosphate, Eqn. 5.32 the conversion of cytidine 2′,3′-cyclic phosphate to cytidine 3′-cyclic phosphate in the solid state, and Eqn. 5.33 the dissolution of cytidine 3′-cyclic phosphate. The sum of these reactions is Eqn. 5.28, the conversion of cytidine 2′,3′-cyclic phosphate to cytidine 3′-cyclic phosphate in aqueous solution. If the entropy changes of these several reactions can be measured, $\Delta S°$ can be calculated for Eqn. 5.28. And that, when combined with $\Delta H°$ for the overall reaction, would give us $\Delta G°$ for the overall reaction.

    The entropy changes for Eqns. 5.31–5.33 have been determined at 25 °C. The values are: $+8.22\ cal\ mol^{-1}\ K^{-1}$, $-9.9\ cal\ mol^{-1}\ K^{-1}$, and $+8.28\ cal\ mol^{-1}\ K^{-1}$, respectively. The overall $\Delta S°$ for these reactions is just the sum of the individual contributions, $6.6\ cal\ mol^{-1}\ K^{-1}$. Combining this entropy change with enthalpy change measured by calorimetry, we can calculate the free energy change: $-2800\ cal\ mol^{-1} - 298\ K \times 6.6\ cal\ mol^{-1}\ K^{-1} = -4800\ cal\ mol^{-1}$. That $\Delta G°$ is negative tells us that cytidine 2′,3′-cyclic phosphate will hydrolyze *spontaneously* in aqueous solution, and this is confirmed by experiment. We'll study reaction rates in depth in Chapter 8.

J. | Molecular pharmacology

This topic is of great importance throughout the biological sciences. The equations we shall present are actually more general than the section title might suggest, as they can be applied not only to the interactions of drugs with membrane-bound receptor proteins, but also to proteins that bind DNA, small molecules or ions. Binding interactions play a role in regulation of enzyme activity and biosynthetic pathways, oxygen transport and regulation of blood pH, to name but a few topics in biochemistry where they are important. Here, though, we wish to think of binding in the context of pharmacology, to give some substance to the mathematics of this section. The equations are considerably less general than they could be, as they depend on a number of simplifying assumptions. This has been done to provide an introductory sense of how the concept of binding is related to Gibbs free energy. A more thorough treatment of binding is given in Chapter 7.

Equation 4.34 has been used to describe a chemical reaction in terms of reactants and products. The equation, however, could just as well represent the free energy difference between the 'bound' and 'free' states of a **ligand**, a small molecule or ion. Under suitable conditions, a ligand interacts with a macromolecule at a **binding site**. The specificity of binding depends on the interacting species and the conditions. In any case, $\Delta G°$ represents the driving force for binding under standard state conditions. Here we discuss binding in the context of a ligand interacting with a receptor.

Binding follows mass action and can be represented as

$$R + L \Leftrightarrow R \bullet L \tag{5.34}$$

where R is the receptor, L is the *free* (unbound) ligand, and R•L is the receptor–ligand complex. The **association constant** is *defined* as

$$K_a = [R \bullet L]/([R][L]) \tag{5.35}$$

and the **dissociation constant** as

$$K_d = K_a^{-1} = [R][L]/[R \bullet L] = ([R]_T - [R \bullet L])[L]/[R \bullet L] \tag{5.36}$$

where $[R]_T = [R \bullet L] + [R]$ is the total receptor concentration. The fractional occupancy of the ligand-binding sites, F_b, is

$$F_b = [R \bullet L]/[R]_T = [L]/(K_d + [L]) \tag{5.37}$$

A plot of F_b against [L] is shown in Fig. 5.15A. The shape of the curve is a **rectangular hyperbola**. We can see that K_d corresponds to the concentration of ligand at which the occupancy of binding sites is half-maximal. Most physiological dissociation constants are on the order of μM–nM. A nM binding constant is considered 'tight binding.' When Eqn. 5.37 is plotted as percentage response against dose (e.g. mg of drug per kg of body weight), it is called a dose–response curve. The dose is often plotted on a logarithmic scale, giving the curve a sigmoidal appearance (Fig. 5.15B), but the underlying relationship between dose and response is the same regardless of how the data are plotted.

We can rearrange Eqn. 5.36 to obtain

$$[R \bullet L]/[L] = ([R]_T - [R \bullet L])/K_d \tag{5.38}$$

This shows that $[R \bullet L]/[L]$, the concentration of bound ligand divided by the concentration of free ligand, is a *linear* function of $[R \bullet L]$. The slope of the curve is $-1/K_d$ (Fig. 5.16). The axis intercepts contain important information as well: the ordinate (*y*-axis) intercept is $[R_T]/K_d$; the abscissa (*x*-axis) intercept is the '**binding capacity**,' i.e. the concentration of ligand binding sites. A plot of bound/free *versus* bound is called a

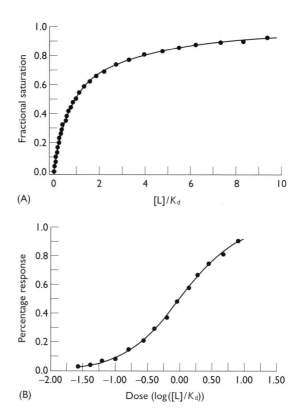

Fig. 5.15 Binding. (A) The circles are experimental data points, the solid line is a theoretical description of binding. There is one ligand-binding site per macromolecule. The shape of the curve is a rectangular hyperbola. Note that although half-saturation occurs when $[L] = K_d = 1/K_a$, $[L] = 9 \times K_d$ gives only 0.9 saturation and $[L] = 99 \times K_d$ but 0.99 saturation. In other words, most of the information about binding is in the free ligand concentration range $0–2 \times K_d$. Experiments should be designed accordingly. (B) A dose–response curve.

Scatchard plot, after the American physical chemist George Scatchard (1892–1973). Radioactive methods can be used to measure the amounts of bound and free ligand.

Competitive binding studies can be used to determine the dissociation constant of other ligands for the same ligand-binding site. For instance, suppose you wish to test the effectiveness of a number of candidate drugs to compete directly with L for a specific binding site on R. Let the candidate competitors be denoted $I_1, I_2, I_3 \ldots$. According to this model,

$$K_{d,I_i} = [R][I_i]/[R \bullet L] \tag{5.39}$$

for a general inhibitor, I_i. It can be shown that in the presence of an inhibitor, the receptor ligand complex, $[R \bullet L]$, is

$$[R \bullet L] = \frac{[R]_T[L]}{K_{d,L}\left(1 + \dfrac{[I_i]}{K_{d,I_i}}\right) + [L]} \tag{5.40}$$

The relative affinity of a ligand in the presence of an inhibitor can be found by dividing Eqn. 5.40 by Eqn. 5.38. This gives

$$\frac{[R \bullet L]_{I_i}}{[R \bullet L]_0} = \frac{K_{d,L} + [L]}{K_{d,L}\left(1 + \dfrac{[I_i]}{K_{d,I_i}}\right) + [L]} \tag{5.41}$$

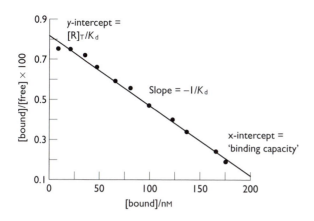

Fig. 5.16 Scatchard plot. The concentration of bound ligand divided by the concentration of free ligand is plotted against the concentration of bound ligand (nM). When binding data are presented in this way, the slope measures the negative inverse of the dissociation constant ($-1/K_d = -K_a$). The ordinate-intercept (y-intercept) is $[R]_T/K_d$, and the abscissa-intercept (x-intercept) is the binding capacity (the concentration of binding sites).

The left-hand side of Eqn. 5.41 is zero for all concentrations of I_i when there is no inhibition (compound I_i has no effect), and it is 1 at 100% inhibition. The concentration of competitor I_i that gives 50% inhibition is designated $[I_{i,50}]$. At this concentration,

$$K_{I_i} = \frac{[I_{i,50}]}{1 + \dfrac{[L]}{K_{d,L}}} \tag{5.42}$$

Fig. 5.17 shows the percentage inhibition for a number of different inhibitors. The shapes of the curves resemble that of Fig. 5.15.

The above mathematical relationships equations are very useful. They apply not only to natural ligands, e.g. epinephrine, a 'fight-or-flight' hormone, and to competitive inhibitors, e.g. the 'β-blocker' propranolol, which vies with epinephrine for binding sites on β-adrenergic receptors, but also to noxious chemical substances like botulinum toxin. (The effects of β-blockers were first described by Sir James W. Black (1924–), a Scot. Sir James was awarded the Nobel Laureate in Medicine or Physiology in 1988.) The equations also apply to DNA ligands, e.g. repressor proteins that physically block the enzymatic transcription of mRNA by binding to an operator site, as well as to the protein-protein interactions of signal transduction. An example of binding in a signal transduction cascade is the direct association of the SH2-domain (SH2, Src homology 2) of Grb2 to a specific phosphorylated tyrosine residue on a growth factor receptor (Fig. 5.18).

Phosphotyrosine-mediate binding is of particular interest for several reasons. One, it involves phosphorylated tyrosine, and the phosphoryl group is acquired *via* catalysis by a kinase from ATP, the energy molecule. Phosphorylation and dephosphorylation of tyrosine is a type of dynamic molecular switch that regulates cellular activity by controlling which proteins can interact with each other. Phosphorylation also places severe restrictions on the relative orientation of interacting proteins. An extremely important class of phosphotyrosine-mediated interactions is typified by phospholipase C-γ_1, an enzyme that interacts with phosphorylated growth factor receptors by means of its two SH2 (Src homology 2) domains. Grb2 interacts with phosphorylated

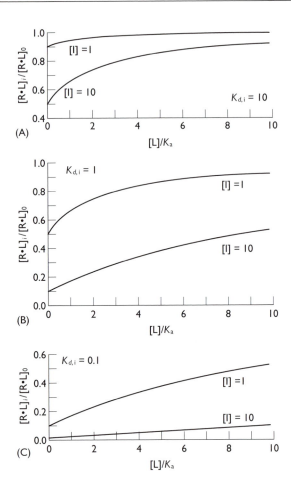

Fig. 5.17 Effect of inhibitor on ligand binding. When the concentration of inhibitor i is low and the inhibitor dissociation constant is high, as in panel (A), $[R \bullet L]_i / [R \bullet L]_0$ is nearly 1, even at low concentrations of ligand. Competition between ligand and inhibitor is more evident when the inhibitor concentration is increased by a factor of 10. In panel (B), the dissociation constant of the inhibitor is 10 times smaller than in panel (A). Note the marked impact this has on $[R \bullet L]_i / [R \bullet L]_0$. The effect of decreasing the dissociation by yet another factor of ten is shown in panel (C). A pharmaceutical company might use a study of this type to characterize the properties of inhibitors that could be used as drugs. Analysis of such compounds would include not only *in vitro* binding experiments (high-affinity specific binding) but also assessment of side effects (low-affinity non-specific or high-affinity unwanted binding).

tyrosines by means of its own SH2 domains, but Grb2 has no catalytic activity. Two, there are several different types of phosphotyrosine recognition module, and modules are found in many different proteins. The two best-known types are the SH2 domain and the PTB (phosphotyrosine binding) domain. In some cases, both module types are found in the same protein, e.g. Shc (Src homolog, collagen homolog) and tensin. Three, the range of possible interactions of a given type of module is greatly increased by subtle differences in structure. Typically, the amino acid side chains that interact directly with the phosphotyrosine ligand are conserved from module to module. Side chains nearby, however, are specific to the protein in which the module occurs. Such amino acid substitutions underlie the specificity of the interactions of an otherwise general modular protein structure. The sum total of the general and specific interactions gives the overall binding free energy. We can see here a close connection between biological information and energy.

As mentioned above, binding is a far more complex phenomenon that we have made it seem. For instance, if a macromolecule can interact with more than one type of ligand at different binding sites, there is the possibility that one kind of metabolite can 'sense' the concentration

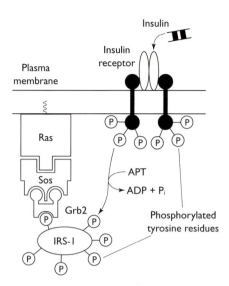

of another, despite the absence of a direct interaction between the metabolites. This aspect of the function of biological macromolecules, known as **allostery**, will be developed in Chapter 7.

K. | Hemoglobin

Most known organisms require oxygen for life; the only known exceptions are some types of bacteria. Reduction of molecular oxygen to water plays a key role in the generation of ATP. Oxygen is used as fuel in the combustion of glucose and production of ATP and carbon dioxide. Oxygen and glucose must be delivered to every cell in the body; carbon dioxide, a waste product, must be removed from every cell. Vertebrates carry out this food and gas transport by means of blood or a blood-like fluid that moves through a closed system of tubes called the vasculature (Fig. 1.10). The vasculature connects the lungs, gills or skin to the peripheral tissues.

Molecular oxygen and carbon dioxide are transported throughout the blood by an allosteric transport protein called hemoglobin. Hemoglobin has been called a molecular 'lung.' Vertebrate hemoglobin is a tetrameric protein, $\alpha_2\beta_2$ (it can also be thought of as a dimer of $\alpha\beta$ heterodimers); in invertebrates hemoglobins range in size from one to 144 subunits! Each subunit comprises a polypeptide chain called globin and a protoheme IX, a planar complex of iron and an organic molecule with rings called protoporphyrin IX (Fig. 5.1). The iron atom in the heme group coordinates the bound diatomic oxygen molecule. The ability of hemoglobin to bind oxygen depends on the partial pressure of oxygen. (The partial pressure of a gas is the contribution it makes to the overall gas pressure. By Dalton's law, named after the John Dalton of the atomic hypothesis, the total pressure is the just the sum of the partial pressures of the gases present. For example, if the air pressure is 1 atm, the partial pressures of nitrogen, oxygen, and carbon

Fig. 5.18 Protein–protein interactions and phosphoryl transfer in signal transduction. The extracellular concentration of the hormone insulin, a chemical signal, is communicated across the plasma membrane by means of dimeric insulin-specific transmembrane receptor molecules. The binding of insulin to its receptor results in receptor autophosphorylation, the catalysis by one receptor molecule of the transfer of a phosphoryl group from ATP to a tyrosine side chain of the other receptor molecule. Phosphorylation of tyrosine acts as a molecular switch in the recruitment of proteins that recognize specific phosphorylated tyrosine residues. One consequence of the chain of events elicited by insulin binding is the phosphorylation of insulin-receptor substrate-1 (IRS-1). A pharmaceutical company might be interested in testing and marketing compounds that inhibit interaction betwen IRS-1 and the insulin receptor. Once phosphorylated, IRS-1 can interact directly with the proteins Grb2, Sos, and Ras. The last of these plays a very important role in cellular signal transduction. Several of the protein–protein interactions involved in this and many other signaling cascades are mediated by phosphorylated tyrosine. Signal transduction is a form of biological communication and information processing; we shall return to this point in Chapter 9. Based on Fig. 20-48 of Lodish et al. (1995).

Table 5.3	Thermodynamics of hemoglobin dissociation	
Hemoglobin	Substitution in mutant	$\Delta G°$ of dissociation (kcal mol–1 of hemoglobin)
normal	—	8.2
'Kansas'	102β, Asn $\rightarrow$ Thr	5.1
'Georgia'	95α, Pro $\rightarrow$ Leu	3.6

Source: The data are from Chapter 4 of Klotz (1986).

dioxide sum to 1 atm.) It also depends on the amino acid chains near the protein-heme interface, as shown by amino acid replacement studies. Mutations in the region of the oxygen binding site can alter affinity for oxygen by up to about 30 000-fold! In this section we introduce a number of aspects of hemoglobin thermodynamics. The treatment will be brief. A more in-depth look at oxygen binding is reserved for Chapter 7.

The first thing we must realize is that if tetrameric hemoglobin is stable under normal physiological conditions, the tetramer must represent a minimum of free energy. That is, the tetrameric state must be a lower free energy state than the other possible combinations of subunits, for example $\alpha\beta$ dimers. A number of natural variants of human hemoglobin are known. One of these is the famous sickle-cell variant. As shown in Table 5.3, the free energy difference between the tetrameric and dimeric states of hemoglobin can depend substantially on the primary structure. Although the free energy difference between normal hemoglobin and hemoglobin Kansas is only 3.1 kcal mol^{-1}, the equilibrium constant differs by nearly 200 fold at 25 °C! Considerably less work must be done to dissociate tetrameric hemoglobin Kansas than wild-type hemoglobin into $\alpha\beta$ dimers. For comparison, it is well known that inhalation of too much carbon monoxide will normally be fatal, even if exposure lasts just a few minutes. The spectrum of pathological effects of CO poisoning includes damage to the peripheral nervous system, brain damage, cell death in the heart, cell death in other muscles, and pathological accumulation of fluid in the lungs. All this results from the binding of CO to hemoglobin with an affinity constant 'only' about 240 times greater than that of oxygen. Hemoglobin Georgia is even less stable than the Kansas variant; its behavior as an oxygen carrier is very noticeably altered relative to the normal protein.

The same reasoning that led us to think that the tetrameric state of hemoglobin represents a lower free energy state than the dimer explains why, under appropriate conditions, oxygen associates with hemoglobin. Binding occurs because the bound state is more thermodynamically favorable (has a lower Gibbs free energy) than the unbound state. Now, let us consider the oxygenation of hemoglobin in solution. For the moment, we'll take a rather simplistic view and assume that hemoglobin has just one binding site, even though it has one in each subunit. The reaction can be written as

$$Hb(aq) + O_2(g) \Leftrightarrow HbO_2(aq) \tag{5.43}$$

From experiments it has been determined that $K = 85.5$ atm^{-1} for the reaction as written. At 19 °C, $\Delta G° = -2580$ cal mol^{-1}. What is the free energy change when the partial pressure of oxygen is 0.2 atm and oxygen is dissolved in solution with an activity of 1 (as in the standard state)?

The free energy difference between $p = 1$ atm and $p = 0.2$ atm is found using Eqn. (4.5):

$$\Delta G = G(O_2, p = 0.2 \text{ atm}) - G°(O_2, p = 0.2 \text{ atm})$$
$$- \left(G(O_2, p = 1 \text{ atm}) - G°(O_2, p = 1 \text{ atm}) \right)$$
$$= RT \ln(O_2, 0.2 \text{ atm}) - RT \ln(O_2, 1 \text{ atm})$$
$$= RT \ln(0.2/1)$$
$$= -930 \text{ cal mol}^{-1} \tag{5.44}$$

where we have used G instead of μ, though technically μ is correct (remember than μ is molar Gibbs free energy). That ΔG is negative is just what we should expect, since a substance will move spontaneously from a region of higher concentration to one of lower concentration.

At equilibrium, $\Delta G = 0$ between the oxygen vapor and the dissolved oxygen. To calculate the free energy difference between the concentration of dissolved oxygen in the standard state $(a = 1)$ and the concentration at saturation, which is substantially lower, we need to account for the solubility in water. This is 0.000 23 molal (kg l^{-1}) at 19 °C. Thus, $\Delta G = RT \ln(1/0.000\,23) = 4860$ cal mol^{-1}. Because the Gibbs free energy is state function, the *net* free energy change on going from oxygen gas at 1 atm to dissolved oxygen at unit activity is just the sum of the individual contributions or -930 cal mol$^{-1} + 0 + 4860$ cal mol$^{-1} = 3930$ cal mol^{-1}. The free energy change of the reverse reaction is, of course, -3930 cal mol^{-1}.

Now, the two reactions we're interested in are:

$$Hb(aq) + O_2(g) \Leftrightarrow HbO_2(aq) \tag{5.45}$$

$$O_2(aq) \Leftrightarrow O_2(g) \tag{5.46}$$

which, when summed, give

$$Hb(aq) + O_2(aq) \Leftrightarrow HbO_2(aq) \tag{5.47}$$

By Eqn. 4.32 ΔG for this reaction is -2580 cal mol$^{-1} - 3930$ cal mol^{-1} $= -6510$ cal mol^{-1}. We can see from this that the driving force for oxygen association with hemoglobin is greater when the oxygen is solvated than when it is not solvated.

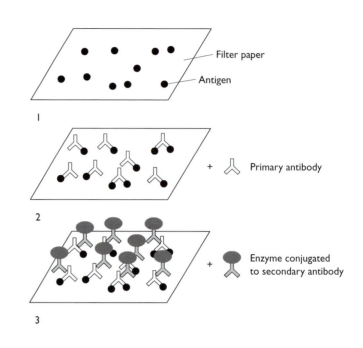

Fig. 5.19 ELISA. This very useful laboratory technique consists of three basic steps. First, the protein antigen is adhered to a solid support, usually a nitrocellulose filter. This partially denatures the antigen. Next, (primary) antibodies are allowed to bind to the antigen. Finally, (secondary) antibodies that recognize the first antibody bind are allowed to bind the primary antibodies. Attached to each secondary antibody is an enzyme that catalyzes a reaction that facilitates detection. One assumes that detection of enzyme activity implies detection of the antigen. This is usually the case because antibody recognition of antigen is highly specific, and in any case a control experiment in the absence of antigen can be done to assess the background signal. Nitrocellulose is very sticky, and the milk protein casein is often used to bind to sites not occupied by antigen in order to reduce the background signal arising from the non-specific adherence of primary antibodies. After step two, non-specifically adhered primary antibodies are rinsed off with buffer. Specifically bound antibodies are not lost in the rinsing procedure, because the rate of dissociation of the antibody from the antigen is very low. The equilibrium binding constant is inversely proportional to the dissociation rate (Chapter 4).

Labels on figure: Filter paper; Antigen; Primary antibody; Enzyme conjugated to secondary antibody

L. | Enzyme-linked immunosorbent assay (ELISA)

Antibody recognition of an antigen is mainly a matter of shape complementary and charge interactions in the antigen-binding site. The shape of the binding site must be a close match to a part of the surface of an antigen for specific binding to occur. Binding can be very tight indeed, with $K_a \sim 10^9 \, \text{M}^{-1}$ or greater. The following discussion, though it centers on ELISA, applies to a broad range of immuno-techniques, including for instance western blotting.

ELISA is a useful method for detecting small amounts of specific proteins and other biological substances in laboratory and clinical applications. For instance, it is used to detect the placental hormone chorionic gonadotropin in a commonly available pregnancy test. In the ELISA method, antibodies are used to detect specific antigens. The assay is very sensitive because antibody binding is very specific (K_{eq} is large) and the binding 'signal' can be amplified by means of an enzyme (often horseradish peroxidase, Fig. 5.19).

The ELISA protocol involves adsorbing a protein sample to an inert solid support (usually a type of filter paper). The binding of protein to the solid support can be very strong indeed, though it is relatively non-specific; most proteins bind well, in more or less the same way. This process partially denatures the protein. After adsorption, the sample is screened with an antibody preparation (usually a rabbit antiserum) and 'rinsed' to remove non-specifically bound antibody ($K_{eq} < 10^4$). The resulting protein–antibody complex on the solid support is mixed with an antibody-specific antibody to which the enzyme used for the detection assay is attached. This second antibody is usually from goat. What we wish to know here is why the rinse step does not ruin the experiment.

From the last chapter, we know that $K_{eq} = k_f/k_r$, where k represents the rate of reaction and 'f' and 'r' stand for 'forward' and 'reverse,' respectively. When binding is specific, $k_f \gg k_r$; the 'on rate' is much greater than the 'off rate.' So, even during rinsing the tightly bound antibodies will stay put, despite how the law of mass action requires antibodies to release the antigen and return to solution, where the antibody concentration is very low. To put things in perspective, for $K_a \sim 10 \text{M}^{-19}$, the free energy change on binding is about -50 kJ mol^{-1}. We can get a sense of how big this free energy change is by considering the energy required to raise a 100 g apple a distance of 1 m. It is easy to show that this energy is about 1 J. Now, 50 kJ could lift a 100 g mass about 50 km, over five times the height of Mt Everest! In other words, $K_a \sim 10^9$ M^{-1} is tight binding. This has been a very brief treatment of the thermodynamics of ELISA. We'll come back to binding and chemical kinetics in Chapter 7 and Chapter 8, respectively.

M. | DNA

Throughout this book we have put somewhat more emphasis on proteins than on DNA. In part that is a reflection of the expertise and interests of the author and not a conscious bias against nucleic acids or the people who study them! Besides, there is a great deal to be said about DNA in Chapter 9. To redress lingering impressions of imbalance, this section looks at the thermostability of DNA and the next one discusses the energetics of the polymerase chain reaction.

The structure of the DNA double helix is illustrated schematically in Fig. 5.20. The types of interactions that stabilize the structure are hydrogen bonds and 'stacking interactions.' Three hydrogen bonds are

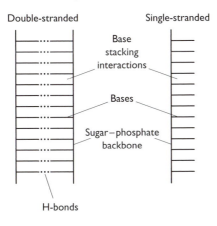

Fig. 5.20 Double-stranded and single-stranded DNA. DNA is composed of bases attached to a sugar–phosphate backbone. (See Fig. 5.14 for a higher resolution view of polynucleic acid.) There are two major types of interaction that stabilize double-stranded DNA: intermolecular hydrogen bonds (polar) and intramolecular base stacking interactions (non-polar). The number of hydrogen bonds depends on the bases involved: three are formed between cytosine (C) and guanine (G), and two between adenine (A) and thymine (T). Intermolecular hydrogen bonds are not present in single-stranded DNA. Based on Fig. 3.16 of van Holde (1985).

Table 5.4	Association constants for base pair formation
Base pair	$K(\text{M}^{-1})$
Self-association	
A•A	3.1
U•U	6.1
C•C	28
G•G	10^3–10^4
Watson–Crick base pairs	
A•U	100
G•C	10^4–10^5

Notes:
The measurements were made in deuterochloroform at 25 °C. The data are from Kyoguko *et al.* (1969). Similar values have been obtained for T in place of U. Non-Watson–Crick base pairs are relatively unstable. The entropic component of K is roughly the same for each Watson–Crick base pair. Thus, the enthalpic term must contribute substantially to the free energy.

formed between bases cytosine and guanine, two between adenine and thymine. (The adenine of DNA (and RNA) is exactly the same as the adenine of ATP, cAMP, NADH and $FADH_2$. As shown in Fig. 5.14, however, only one of the phosphate groups of ATP actually becomes part of the polynucleotide.) We might guess from this that the stability of double-stranded DNA relative to single-stranded DNA will depend on the proportion of C–G pairs, because this will influence the average number of hydrogen bonds per base pair, and in fact that is correct (Fig. 5.21, Table 5.4). (If K is so large for A•A and G•G, why are such base pairs not found in double-stranded DNA?) Analysis of structure of double-stranded DNA has revealed that the bases form extended stacks, interacting with adjacent residues in the same strand by van der Waals forces. Both hydrogen bonds and van der Waals interactions contribute to the overall stability of the double helix.

The equilibrium between double- and single-stranded DNA can be symbolized as

$$D \Leftrightarrow S \tag{5.48}$$

The equilibrium constant for this reaction is

$$K = [S]/[D] \tag{5.49}$$

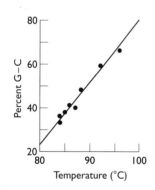

Fig. 5.21 Influence of G–C content on DNA melting temperature. As the percentage of G–C pairs increases, the number of intermolecular hydrogen bonds per base pair increases. The stabilizing effect this has on double-stranded DNA is reflected in the relationship between G–C content and melting temperature.

This equilibrium constant is the product of the Ks for the individual base pairings, as each base pair contributes to the overall stability of the double-stranded molecule.

Thermal denaturation of double-stranded DNA has been studied extensively. As mentioned above, C–G composition is an important determinant of the stability of duplex DNA and therefore of the conditions under which $K = 1$. One means of promoting the dissociation of double-stranded DNA is to add heat. Just as with proteins, heat absorbed by DNA increases its thermal energy, fluctuations of structure increase

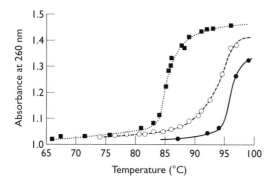

Fig. 5.22 DNA melting curves. The melting temperature varies not only with G-C content but also with size. In other words, a 100 base-pair-long double-stranded DNA molecule will have a higher melting temperature than a 50 base-pair-long double-stranded DNA, if the percentage of G–C pairs and their distribution is the same in both cases. Differences in G–C content and distribution and molecular size lead to differences in melting temperature.

in magnitude, and the disordered state becomes more favorable than the ordered one. Measurement of the temperature at which double-stranded DNA is 50% 'melted,' the **melting temperature**, is one way of comparing the genetic material of one genome to another (Fig. 5.22). Research in this area has been used to work out empirical rules for the melting temperature of DNA as a function of C–G content, length and concentration of ions, principally Mg^{2+}. Magnesium 'counterions' play an important role in neutralizing the electrostatic repulsion between the negatively charged phosphate groups in the sugar–phosphate backbone by decreasing the range and strength of the repulsive Coulombic (electrostatic) interactions between the phosphate groups on opposite strands of the double helix. (Coulomb's law is an empirically derived mathematical description of the electrical interaction between stationary charged particles. It is named after the French physicist and military engineer Charles Augustin de Coulomb (1736–1806). The effect of counterions is explained by the Debye–Hückel theory of strong electrolytes. See Fig. 5.13) Decreases in the concentration of 'counterions' reduce the melting temperature of double-stranded DNA.

Figure 5.23 shows the percentage of double-helix as a function of temperature for the forward and reverse reactions in Eqn. 5.48. It is clear that there is some hysteresis (Greek, a coming late) in this process. This arises because the rate of the forward process differs considerably from the rate of the reverse process throughout the temperature range, and the sample is not at equilibrium throughout the experiment. From a structural point of view, the dissociation of strands is a much simpler reaction

Fig. 5.23 Melting and cooling profile for double-stranded DNA. Lower solid line: the melting of double-stranded DNA is cooperative. Relatively few base pairs are broken below the melting temperature. Once melting has begun, however, relatively small increases in temperature result in the rupture of a relatively large number of hydrogen bonds. The melting profile differs greatly from the cooling profile (upper solid line). Unless cooling is carried out very slowly, the system will not be in a near-equilibrium state at every point on the reaction pathway (every temperature in the range studied). Rapid cooling of melted DNA will not yield perfectly formed double-stranded DNA. Single-stranded DNA has a higher absorbance than double-stranded DNA. Broken line: melting temperature is influenced not only by G–C content and distribution and molecular size but also by ion concentration, particularly divalent cations. These ions interact favorably with the phosphate groups of the DNA backbone. The consequent reduction of electrostatic repulsion results in increased stability of the DNA duplex. Based on Fig. 3.15 of van Holde (1985) and Fig. 5.15 of Bergethon (1998).

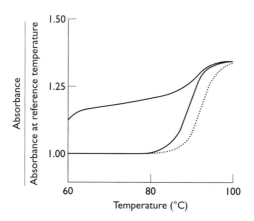

than the formation of perfectly matched double-stranded DNA. And as we have said, reversibility of a process depends on the system being taken through a series of equilibrium or near-equilibrium states. During the reverse reaction, if the system is not given sufficient time to come to equilibrium, some mismatching of bases is likely to occur, preventing or at best strongly inhibiting return to the initial conditions.

Up to now we have treated DNA in rather general terms. There are different types of DNA, however, and these have different thermodynamic properties. Genomic DNA of higher eukaryotes, for example, is linear; there is a definite 3'-end and a definite 5'-end. In contrast, plasmid DNA, which is of great utility as an *E. coli* expression vector in recombinant DNA technology, is circular. Circular DNA can exhibit a variety of conformations ranging from no **supercoiling**, or no twisting, to tight supercoiling; DNA is not a static structure in aqueous solution at 300 K. The topology of circular DNA suggests that energetics of plasmid DNA melting will differ from that of linear DNA, even if the basic principles we have discussed thus far apply to both types, and there is no change in length or base composition.

A double-helical DNA molecule with covalently attached ends will have a certain number of coils, or twists. A DNA coil is analogous to a twist one can introduce in a belt *before* the buckle is fastened. It is easy to show that the number of coils cannot be changed without cutting the belt if fastening is irreversible. In the same way, coils in circular DNA cannot be undone without cutting the polynucleotide strand. From a mathematical point of view, supercoiling can be expressed in terms of three variables as

$$L = T + W \tag{5.50}$$

L, the **linking number**, is the integral number of times that one DNA strand winds around the other; it is the number of coils in our belt analogy. The **twist**, T, is the number of complete revolutions that one polynucleotide strand makes about the duplex axis (usually the number of base pairs divided by 10.6, the approximate number of base pairs per turn of DNA). T can be positive or negative, depending on the direction of the helix, and it can vary from one part of a molecule to another. W, the **writhe**, is the number of turns that the duplex axis makes about the superhelix axis. Like T, W can be positive or negative.

If the duplex axis of the DNA is constrained to lie in a single plane, $W = 0$: there is coiling but no supercoiling, $L = T$, and the twist must be an integral number. From Eqn. 5.50 it is clear that different combinations of W and T are possible for a circular DNA molecule with L, which is a property of the molecule that is constant in the absence of a break in a polynucleotide strand. At equilibrium, one expects a given circular DNA molecule to fluctuate between a variety of conformations, each of which must have linking number L.

As a specific example of DNA supercoiling, consider the circular DNA molecule of the SV40 virus. This molecule is about 5300 base pairs long is therefore expected to have $L = T \approx 500$ in the absence of supercoiling. The prediction is based on the most energetically favorable number of bases per turn. DNA isolated from SV40, however, is supercoiled, at least under normal salt conditions (the predominant conformation depends on salt concentration and temperature). This supercoiling probably arises from an untwisted region being present at the end of DNA replication. Such 'underwinding' is energetically unfavorable, because the average number of bases per turn is low. The conformation of the molecule changes until the lowest free energy state is reached, but regardless of the conformation adopted L is constant (the DNA backbone is not severed). It has been found experimentally that $|W| \approx 25$, so by Eqn. 5.50 $W \approx -25$. The sign of W tells us that the supercoils are negative, and they form to compensate for the effects of helix underwinding. Because T is mainly a property of chain length, $T \approx 500$, and $L \approx 475$. Supercoiling increases the elastic strain in circular DNA, just as it does in any other circularized object, for instance a rubber band. . ..

N. | Polymerase chain reaction (PCR)

PCR is an extremely useful laboratory process in which double-stranded DNA is replicated rapidly. Under favorable circumstances, a very small amount of a DNA can yield a large, readily analyzed sample of DNA. The technique was developed in the mid-1980s by Kary R. Mullis (1944–) and colleagues at the Cetus Corporation. Mullis, an American, was awarded the Nobel Prize in Chemistry for this work in 1993.

The procedure works as follows. DNA is dissolved in aqueous solution containing a DNA polymerase from a thermophilic bacterium (e.g. *Bacillus stearothermophilus*), polymerase buffer, free nucleotides (dATP, dCTP, dGTP and dTTP, where the 'd' means 'deoxy'), and oligonucleotide primers. The primers are short sequences of single-stranded DNA that one designs to bind to either end of the DNA segment of interest. One primer binds one end of the desired segment on one of the two paired DNA strands, and the other binds the other end of the segment on the complementary strand (Fig. 5.24).

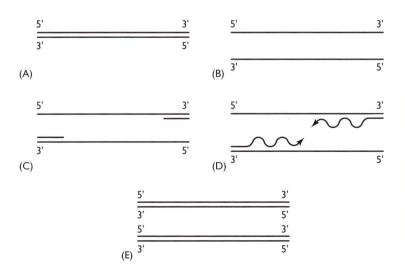

Fig. 5.24 Schematic diagram of PCR. When heated, a double-stranded template DNA (A) melts into two single-stranded molecules (B). If cooling is rapid, double-stranded template will not be able to reform. In the excess of primer, however, a short piece of DNA, binding will occur at temperatures below the primer T_m (C). Increasing the temperature to the optimal value for polymerase activity results in extension of the primer (D). This completes the reaction cycle, and the yield is two double-stranded DNA molecules from one (E). In the ideal case, there is an exponential increase in the amount of template DNA with each cycle.

In a typical PCR experiment, the solution described above is cycled repeatedly through three different temperatures. The first temperature is usually 95 °C. Thermal energy is used to break the hydrogen bonds and base stacking interactions that stabilize double-stranded DNA. The solution is then cooled to about 55 °C, at which temperature the primers bind to sites on the separated strands of the template DNA. The separated strands do not reanneal in the region of the primer binding site because the primers, which are present in great excess, out-compete them. The third temperature of each cycle is usually about 72 °C. At this temperature, DNA polymerase activity is high, and it builds a new strand by joining free nucleotide bases to the 3′-end of the primers at a rate of several hundred bases per minute. Each time the thermal cycle is repeated, a strand that was formed with one primer is available to bind the complementary primer, the result being a new two-stranded molecule. In this way, the region of DNA between the primers can be selectively replicated. Further repetitions of the process can produce billions of identical copies of a small piece of DNA in 2–3 hours. A well-designed PCR experiment yields the desired product with better than 99% purity.

A question of considerable practical importance to the molecular biologist is: 'How long should the oligonucleotide primer be?' There are two chief considerations. One is cost: why spend more than necessary? The other is that although specificity of an oligo increases with length, length is not necessarily an advantage. In order to make headway here, let's think about how double-stranded DNA is held together.

Above we said that hydrogen bonds contribute to double-helix stability. How do they do that? After all, the bases in single-stranded DNA are able to form hydrogen bonds with water! In the double helix, the interstrand hydrogen bonds are not being made and broken constantly, as they are in single-stranded DNA when it interacts with itself non-specifically. Assuming that the enthalpy change of hydrogen bond formation is of roughly the same magnitude in both cases, the hydrogen bond donor or acceptor is, on the average, bonded more of the time in duplex DNA than in single-stranded DNA. We should therefore expect there to be a difference in enthalpy between double-stranded DNA and its separated strands. If heat must be added to 'melt' the double helix, then, according to Eqn. 2.3 the single strands represent a higher enthalpy state than the double helix at a temperature favoring the double helix. This means that the longer the oligonucleotide primer, the larger ΔH_d, where the subscript 'd' means 'denaturation of the double helix formed by the primer binding to the template.' Below T_m, base pair formation is energetically favorable, from an enthalpic point of view.

So much for the enthalpy. What about entropy? In terms of the strands of DNA alone, the double-stranded state has a much lower entropy than the single-stranded one. This is because two particles sticking together in a very specific way is a more orderly situation than two particles floating freely in solution. Formation of a base pair entails a *decrease* in entropy and is, therefore, energetically unfavorable. The unfavorability comes not only from requiring the two strands of DNA to be in the same place in the sample volume, but also from the restric-

tions on the shape of both strands that are compatible with helix formation and on the orientation in space of individual bases.

As is often the case in science, thinking will get you only so far. At
some stage it becomes necessary to do experiments to find out whether
or not the world really is how you imagine it to be! There are two key
experimental findings that will help us here. One is that G–C pairs contribute more to duplex-DNA stability than A–T pairs. This cannot be
rationalized in terms of base stacking interactions, as the surface area
of an A–T pair is not appreciably different from that of a G–C pair. The
extra stability must come from the G–C pair's extra hydrogen bond. The
other empirical finding is that oligonucleotide primers must be about
20 bases long in order for PCR to work well, the exact length depending
on the G–C content of the oligo and the temperature at which anneal-
ing occurs (usually 55 °C). What this tells us is that we need to form
about 50 hydrogen bonds for the favorable enthalpic contribution to
the free energy change of double helix formation to exceed the unfavorable entropic contribution.

This reasoning helps us to see why we would not want to make our oligonucleotide primers too short. We also know, from the financial point
of view, that we would not want to make them too long. But there is
another, thermodynamic reason why oligos should not be too long.
Assuming a random base sequence in the template DNA strand, the absolute specificity of an oligonucleotide can only increase with length.
However, if the oligo is long, there will be many, many sites at which
partial binding could occur with about the same affinity on the template
DNA. Moreover, a long oligonucleotide molecule might be able to bind
not only to more than one place on the same template at the same time,
but also to more than one template molecule! Such a situation, which
can be energetically favorable from an entropic point of view, will
promote a huge number of side reactions and yield a very messy product.

O. Free energy of transfer of amino acids

The free energy of transfer is the free energy change on transfer of a compound from one medium to another. Suppose we have two perfectly
immiscible solvents in the same beaker. There is an interface, with one
solvent on one side and the other on the other side. There is a large free
energy barrier to mixing; this is what it means for solvents to be immiscible. Now, if a solute is dissolved in one of the solvents, when the three-
component system comes to equilibrium, solute will be found in the
second solvent as well, if the solute can be dissolved in it. Perhaps you
recognize this as the basis for the extraction of chemical compounds
using organic solvents. Strictly speaking this phenomenon, which is
often grouped together with freezing point depression, boiling point
elevation and osmosis, is *not* a colligative property. We are interested in
it here because it will help us to have a better understanding of the solubility and thermodynamic stability of biochemically functional conformations of biological macromolecules, topics we shall study below.

Let's look at a specific example. The solubility of phenylalanine (Phe) in water at 25 °C on the molality scale is 0.170 mol (kg solvent)$^{-1}$; in 6 M urea it substantially higher: 0.263 mol (kg solvent)$^{-1}$. Urea improves the solubility of hydrophobic side chains in aqueous solution; this is what makes urea such a good chemical denaturant of proteins. Using the given information, we can calculate the standard state free energy of transfer of Phe from water to aqueous urea solution. We shall assume that the activity coefficient of Phe is approximately the same in both media, though in principle it varies with the solvent.

The situation can be pictured as follows. In one process, Phe dissolved in water is in equilibrium with crystalline Phe; the solution is saturated. In another, Phe dissolved in urea solution is in equilibrium with crystalline Phe. Both of these processes can be studied experimentally and the solubilities can be measured. In a third process, which is a sort of thought experiment, Phe in one solution is in equilibrium with Phe in the other solution. We construct a notional boundary between the solutions, and require it to be permeable to Phe but *not* to water or urea. This is a notional boundary! There will be a net flow of Phe across the boundary until equilibrium is reached. The fourth 'process', which completes the thermodynamic cycle (Fig. 2.3), is just solid Phe in equilibrium with solid Phe.

In mathematical terms,

$$\mu_{water,\,sat.} - \mu_{solid} = 0 \tag{5.51}$$

$$\mu_{water,\,a=1} - \mu_{water,\,sat.} = -RT\ln a_{water,\,sat.} = +1050 \text{ cal mol}^{-1} \tag{5.52}$$

$$\mu_{urea,\,sat.} - \mu_{solid} = 0 \tag{5.53}$$

$$\mu_{urea,\,a=1} - \mu_{urea,\,sat.} = -RT\ln a_{water,\,sat.} = +790 \text{ cal mol}^{-1} \tag{5.54}$$

$$\mu_{solid} - \mu_{solid} = 0 \tag{5.55}$$

The energy barrier between saturation and unit activity is greater in water than in urea because the solubility of phenylalanine is lower in water than in urea. The difference between Eqns. 5.52 and 5.54, which is what we set out to find, is $\mu_{urea,\,a=1} - \mu_{urea,\,sat.} - (\mu_{water,\,a=1} - \mu_{water,\,sat.}) = \mu_{urea,\,a=1} - \mu_{solid} - (\mu_{water,\,a=1} - \mu_{solid}) = \mu_{urea,\,a=1} - \mu_{water,\,a=1} = 790 \text{ cal mol}^{-1} - 1050 \text{ cal mol}^{-1} = -260 \text{ cal mol}^{-1}$. This is the standard state driving force for phenylalanine to transfer from saturated water to saturated 6 M urea. We can see by the sign of the reaction that the transfer is spontaneous. This is exactly what we should expect on the basis of the solubility data.

Though this example has involved a complete amino acid, there is in principle no reason why the experiments could not be done with small organic molecules. Comparison of the thermodynamic data with structural information would then provide clues to the thermodynamics of transfer of individual chemical groups. Table 5.5 gives thermodynamic values for the transfer of various chemical groups from non-polar organic solvent to water.

There are some practical lessons we can draw from the above analysis. One is that the hydrophobic surface of phenylalanine (or indeed of

Table 5.5 Thermodynamics of transfer at 25 °C from non-polar solvent to water of various chemical groups

Chemical group	ΔG_{tr} (cal mol^{-1} Å^{-2})	ΔH_{tr} (cal mol^{-1} Å^{-2})	ΔC_p (cal K^{-1} mol^{-1} Å^{-2})
Aliphatic: $-CH_3$, $-CH_2-$, CH	+8	−26	0.370
Aromatic	−8	−38	0.296
Hydroxyl	−172	−238	0.008
Amide & amino: $-NH-$, NH_2	−132	−192	−0.012
Carbonyl C: C=	+427	+413	0.613
Carbonyl O: =O	−38	−32	−0.228
Thiol and sulfur: $-SH$, $-S-$	−21	−31	−0.001

Source: The data are from T. Ooi & M. Oobataka, *J. Biochem.*,**103**, 114–120 (1988).

any amino acid side chain), forms more favorable interactions with the solvent when urea is present than when it is absent (urea is a co-solvent). Looked at another way, urea could be said to weaken hydrophobic interactions. This leads to point number two. Experimental studies have shown that urea is a good chemical denaturant. We mentioned something about this in the context of dialysis but did not elaborate. The example of this section helps to rationalize the empirical finding. We know from X-ray analysis of the folded states of proteins that, although hydrophobic side chains appear on the protein surface, protein cores are mainly hydrophobic. In the presence of urea, where the solubility of the hydrophobic side chains is considerably increased relative to the absence of urea, the unfolded state of the protein is correspondingly more thermodynamically favorable. This phenomenon can be used to investigate protein stability, as we shall see below and in Chapter 6.

P. | Protein solubility

Here we are interested not so much in solubility of a substance *per se* but in solubility of proteins and proteins–nucleic acid complexes; the discussion is qualitative and practical rather than quantitative and theoretical.

A protein molecule is of course a very complex polyion; there are numerous ionizable groups and a variety of pK_as. The solubility of a protein in aqueous solution depends strongly on ionic strength and pH (Fig. 5.25). This is of the greatest practical significance for the choice of techniques that one might use to study a protein molecule. For instance, nuclear magnetic resonance spectroscopy (NMR) is a very high-resolution structural technique, making it valuable for protein structure determination and other aspects of biochemical research. But NMR is also an extremely insensitive technique, meaning that very large

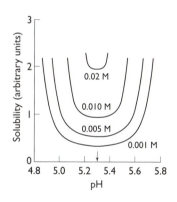

Fig. 5.25 Protein solubility. This depends not only on the net charge of the protein, which varies with pH, but also on ionic strength and temperature. The data shown are for the milk protein called β-lactoglobulin at different concentrations of NaCl. Solubility is very low at pH 5.3, the approximate isoelectric point of the protein. At this pH, protein solubility increases exponentially with increasing ionic strength (solubility $\approx 0.255e^{10\mathrm{[pH]}}$). The data are from Fox & Foster (1957).

concentrations of protein are needed for detection, on the order of 1 mM. At the same time, NMR structural studies require a relatively low rate of exchange of labile protons in tryptophan and tyrosine side chains, a pH-dependent phenomenon (see Chapter 8). In other words, the protein must not only be highly soluble, it must soluble in a suitable pH range (near neutral pH). These requirements (and others!) severely restrict which protein structures can be studied effectively by NMR.

Here's an example of the effect of pH on protein solubility. The PTB domain of chicken tensin is highly soluble at pH 2, where its net charge is about +30. The net charge on each PTB molecule at acidic pH is so great that electrostatic repulsion inhibits the formation of protein aggregates, as long as the ionic strength is low. In contrast, at pH 7, the net charge is zero and the domain is not very soluble at all. At neutral pH, there is nothing to prevent recombinant PTB domains from aggregating. Many proteins exhibit such behavior.

Isoelectric point precipitation of a protein can be used as a step in its purification. For instance, recombinant PTB domain can be separated from bacterial proteins by adjusting the bacterial cell lysate to pH 7. This pH, at which a protein has a net charge of zero, is known as the **isoelectric point**, or p*I*. At pH7, the recombinant protein precipitates, allowing it to be separated from soluble proteins by centrifugation. The isoelectric point of a protein depends primarily on **amino acid composition**; more specifically, the number of amino acids with ionizable side chains and the proportion of acidic and basic ones. If the number of basic side chains is relatively large and the number of acidic side chains relatively small, as with hen egg white lysozyme, the isoelectric point of the protein will be high, and the net charge will be positive through most of the usual pH range (2–12).

In general, the situation is more complex than we have made it seem thus far. At a given pH, a typical protein will have both positive and negative charges. Depending on the location of the charges on the protein surface, if the ionic strength is low, the proteins can interact with each other by electrostatic attraction. Because of this, it is often found that **the solubility of a protein at low ionic strength increases with salt concentration**. This phenomenon is known as '**salting in**.' This depends on the protein being charged, so the effect is least pronounced at the isoelectric point. In contrast, at high ionic strength the protein charges are strongly shielded. Electrostatic repulsion is negligible. Solubility is reduced. This effect is known as '**salting out**.' The shape of a solubility curve with ionic strength is thus, roughly speaking, an upside down U. Salting out is thought to arise not only from the screening of charges but also from the numerous interactions between the salt ions and water, resulting in a *decrease* in the water molecules available to solvate the protein.

Salting out is also an important method of purifying proteins. For instance, ammonium sulfate is often used to purify antibodies. Below a certain ionic strength, antibodies and some other proteins are soluble, but many other proteins are insoluble. The insoluble proteins can be removed from solution by centrifugation. Above a certain ion strength, the antibodies themselves precipitate. They can be separated from the rest of the solution and subjected to further purification. A similar procedure can be used to purify many different proteins. Once a protein is sufficiently pure, it is sometimes possible to crystallize it by dissolving it in a salt solution near the solubility limit of the protein. From a thermodynamic point of view, crystallization occurs because the crystalline state of the protein has a lower Gibbs free energy than the dissolved state.

Q. Protein stability

This section is on cooperative and reversible order–disorder transitions. It builds on the several previous sections, including the ones on DNA and PCR. A major difference between protein stability and duplex DNA stability is the size of ΔC_p between the ordered and disordered states: in proteins it is relatively large, in DNA relatively small. As we shall see, the magnitude of ΔC_p has a marked effect on the thermostability of a protein.

When protein unfolding is cooperative, effectively only two states are populated at equilibrium: the folded (native) state and the unfolded (denatured) state. The transition occurs over a relatively narrow range of the independent variable, be it temperature, pH or chemical denaturant concentration. In such cases, the equilibrium can be represented as

$$F \Leftrightarrow U \tag{5.56}$$

The equilibrium constant (Eq. 4.34) is then

$$K_{eq} = [U]/[F] \tag{5.57}$$

Note that Eqns. 5.56 and 5.57 are consistent with the possibility that all the information required for a protein molecule to fold into its native form might be present in the amino acid sequence. That's because the free energy difference between the folded state of a protein and its unfolded state is independent of the path! In other words, regardless of the process by which a protein folds, the free energy change difference between states is the same (given the same temperature, ion concentrations, pH, etc.). Unfolding followed by refolding would be sufficient evidence that the polypeptide chain contained all the necessary information for folding, if not for the presence of disulfide bonds in proteins.

The earliest attempts to give what could be called a thermodynamic description of reversible protein denaturation and coagulation appeared in the 1920s and 1930s. This was the work of American physical biochemists Alfred Ezra Mirsky (1900–1974), Mortimer Louis Anson (1901–1968), and Linus Carl Pauling.[1] Later, in the 1950s, Rufus Lumry (1920–) and Henry Eyring (1901–1981), also both Americans, provided a

---

[1] L. Pauling (1901–1994) was awarded the Nobel Prize in Chemistry in 1954 for his work on protein structure. His model of DNA structure, which had the bases pointing outwards, was no longer tenable after publication of the famous work of James Watson and Francis Crick. Something worth remembering: Nobel laureates are unusually accomplished rational animals, but they are capable of error. But, it is also fair to say that no one's discovered anything who hasn't also made a mistake.

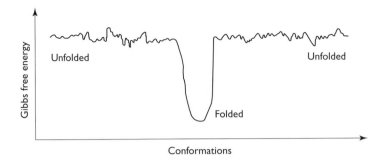

Fig. 5.26 Free energy profile of a small protein. There are only two stable states: the folded state and the unfolded state. The number of unfolded conformations is vastly greater than the number of folded conformations. Although the unfolded conformations differ in energy, these differences are relatively small. Therefore the collection of unfolded conformations can be thought of as a single state. The energy difference between the unfolded state and the folded state is comparatively very large.

more substantial mathematical analysis of reversible protein denaturation. This experimental and theoretical work implied that the folded state of a protein is in a (local) minimum of free energy, also known as an **energy well**, when conditions favor the folded state (Fig. 5.26). An important lingering difficulty, however, was the generality of the applicability of the thermodynamic description. Did it work for some proteins and not others? Did it apply to proteins with disulfide bonds? It was Christian Boehmer Anfinsen's investigations of the reversible denaturation of proteins that showed conclusively that the native state of a protein with disulfide bonds could be recovered spontaneously, even when the disulfides were not formed in the starting state. This led to the general acceptance of the 'thermodynamic hypothesis' for the folding of proteins, according to which all the information required for attainment of the native structure is present in the amino acid sequence. Anfinsen (1916–1995), an American, was awarded the Nobel Prize in Chemistry for this work in 1972. Since then, and particularly since the late 1980s, the goal of working out the physical basis of protein folding and thermostability has been pursued with considerable intensity throughout the world.

We have already discussed protein denaturation (in Chapters 2 and 3) in the context of illustrating the physical meaning of H and S and showing the utility of a van't Hoff analysis of equilibrium constant data. Now we wish to think about protein denaturation in terms of free energy. At constant temperature and under standard state conditions, Eqn. 4.2 becomes

$$\Delta G_d = \Delta H_d - T\Delta S_d \qquad (5.58)$$

where as before the subscript 'd' signifies 'denaturation.' ΔG_d° is the difference in Gibbs free energy between the unfolded state and the folded state. In most cases, the energy of the unfolded state is measured relative to the energy of the folded state; i.e. the folded state is the reference state. There are two main reasons for this: the folded state has the least ambiguous conformation, and more often than not *equilibrium* studies investigate transitions *from* the folded state *to* the unfolded state; the folded state is often the starting state (this is obviously not true of kinetic refolding experiments!). ΔG_d tells us nothing about the relative magnitudes of ΔH_d or ΔS_d; an *infinite* number of combinations of these thermodynamic functions would be consistent with a given value of ΔG_d. Of course, many of these combinations of ΔH_d and ΔS_d will have little or no physical meaning for the system under study, and only one combination will actually describe the system. In order to determine the values, at least one more experiment must be done. ΔH_d and ΔS_d for proteins can be very large in comparison with ΔG_d. For instance, it is common for the maximum value of ΔG_d for a protein in solution to be about 15 kcal mol^{-1}, and for ΔH_d at the denaturation temperature to be more than an order of magnitude greater. ΔG_d for proteins is thus a delicate balance of ΔH_d and ΔS_d.

At the melting temperature, also called the **heat-denaturation temperature**, the fraction of molecules in the folded state equals that in the unfolded state; the free energy difference between them, ΔG_d, is zero. This leads to Eqn. 3.24 and enables one to calculate the entropy of unfolding from a measurement of ΔH_d°. Being explicit about the temperature dependence of ΔH and ΔS, Eqn. 5.58 becomes

$$\Delta G_d(T) = \Delta H_d(T_r) + \Delta C_p(T - T_r) - T[\Delta S_d(T_r) + \Delta C_p \ln(T/T_r)] \qquad (5.59)$$

where the subscript 'r' means 'reference.' $\Delta G_d(T)$ is not $(\Delta G_d \times T)$ but ΔG_d evaluated at temperature T. As an example, suppose that our reference temperature is 25 °C and that both ΔH_d and ΔS_d are known at this temperature. What is ΔG at 35 °C? If $\Delta H(25\,°C) = 51$ kcal mol^{-1}, $\Delta S(25\,°C) = 100$ cal mol^{-1} K^{-1}, and $\Delta C_p = 1500$ cal mol^{-1} K^{-1}, then $\Delta G(35\,°C) = 51$ kcal mol^{-1} + 1500 cal mol^{-1} K$^{-1} \times (308\ K - 298\ K) - 308\ K \times$ [100 cal mol^{-1} K^{-1} + 1500 cal mol^{-1} K$^{-1} \times \ln(308\ K/298\ K)] = 20$ kcal mol^{-1}. $\Delta G_d(T)$ is known as the **stability** of a protein. It tells us how much energy must be added (more specifically, the minimum amount of work that must be done) to unfold the protein at a given temperature. A plot of $\Delta G_d(T)$ versus temperature (or any other independent variable, e.g. pH or concentration of chemical denaturant) is called a **stability curve** (Fig. 5.27).

Note that the stability curve as a function of temperature looks like a parabola and has a peak at which ΔG_d is a maximum. It can be shown (using Eqn. 5.58 and some calculus) that at this temperature, known as the **temperature of maximum stability**, $\Delta S_d = 0$ (compare Fig. 4.4). That is, the stability of the folded state of a protein is a maximum when the entropy of the folded state is equal to the entropy of the unfolded state. At this temperature, which is often 40 or 50 degrees below the heat-denaturation temperature, enthalpic interactions alone hold the folded state together. Somewhat *below* T_m (for heat denaturation), ΔG_d is positive (if the folded state is the reference state). On the average, unfolding will not occur spontaneously, because $\Delta G_d > 0$. (This statement needs some qualification. In fact, unfolding *can* and *does* occur spontaneously when $\Delta G > 0$, but with lower probability than spontaneous refolding of unfolded protein (see Chapter 6). The more positive ΔG, the less probable spontaneous unfolding. The situation is exactly the opposite when $\Delta G < 0$.) To bring about unfolding by a further temperature increase, we expect ΔH_d to be positive; this is roughly speaking the energy required to disrupt non-covalent interactions in the folded state. We also expect ΔS_d to be positive, as the polypeptide chain will be much more disordered in the unfolded state than in the folded one. But *on the balance*, $\Delta G_d > 0$ below T_m. Above T_m, the balance is shifted towards the entropy, $|T\Delta S_d| > |\Delta H_d|$, and there is a preponderance of unfolded protein.

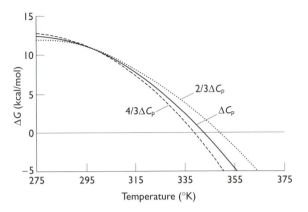

Fig. 5.27 Protein stability curves. A stability curve is a plot of ΔG versus T (or some other independent variable). This gives the free energy difference between the unfolded state and the folded state (the minimum amount of work that must be done to induce a transition from the folded state to the unfolded state). The curvature in ΔG v. T arises from the positive and relatively large ΔC_p of protein unfolding. The stability curve crosses the T-axis at two points, the heat- and cold-denaturation temperatures. In the figure only the heat denaturation temperatures are visible.

ΔC_p plays a key role in protein stability. Because both the enthalpy change and the entropy change of denaturation depend on ΔC_p, the free energy change must depend on ΔC_p. Figure 5.27 shows how the Gibbs free energy difference between the unfolded and folded states changes as the magnitude of ΔC_p changes. If $H_d(T_r)$ and $\Delta S_d(T_r)$ are held constant, *decreasing ΔC_p increases* the breadth of the stability curve, and *increasing ΔC_p decreases* the breadth; all the curves intersect at the reference temperature (298 K in this case). This tells us that if a protein has a small heat capacity change on unfolding, it is likely to have a relatively high transition temperature, and this is exactly what is observed experimentally. In contrast, when ΔC_p is relatively large, the stability curve is more sharply peaked and can cross the temperature axis in more than one place in the experimentally accessible range (solvent in the liquid state).

The second intersection of the stability curve with the temperature axis, which occurs well below the heat-denaturation temperature, is known as the **cold-denaturation temperature**. The 'mere' mathematical form of the stability curve, which is based on solid experimental evidence of *heat*-denaturation, suggests that protein unfolding can be induced either by heating or by cooling. This prediction has been confirmed by experimental studies in a number of cases, greatly underscoring the value of good mathematical modeling of experimental results for prediction of the behavior of biochemical systems. (Things do not always work out this way: in some cases modelling yields solutions that are not physically meaningful.) Cold denaturation seems rather counterintuitive. For in order to melt a crystal, one expects to have to *add* heat, in which case $\Delta H > 0$. The entropy change on protein unfolding, $\Delta S_d = \Delta H_d/T_m$, is therefore positive, in accord with intuition. In contrast, in cold denaturation $\Delta H_d < 0$! It follows that $\Delta S_d < 0$, strange as that might seem. ΔG_d can pass through zero more than once because both the enthalpy change and the entropy change depend on temperature.

So far we have been discussing ΔG as a function of temperature. There are, however, other independent variables we could consider. These include chemical denaturant concentration and pH. As chemical denaturant is added to a protein solution, the folded state becomes destabilized *relative to* the unfolded state, and the protein unfolds. At the so-called midpoint concentration, $\Delta G = 0$, and the fraction of molecules

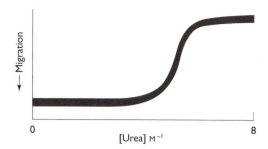

Fig. 5.28 Urea denaturation gel electrophoresis. Structural transitions in protein can be studied by polyacrylamide gel electrophoresis. The abscissa is a linear gradient in urea concentration. When the concentration is sufficiently high, the protein unfolds. This is detected as a change in the mobility of the protein in the gel matrix. Compact folded proteins generally migrate faster in the gel than extended unfolded proteins.

in the folded state equals the fraction in the unfolded state. Note how the midpoint concentration of chemical denaturation resembles the transition temperature of thermal denaturation. In pH denaturation, either acid or base is added to the protein solution to induce unfolding. At the midpoint pH, half of the molecules are in one state and half are in the other, and $\Delta G = 0$.

Chemical denaturation and pH denaturation are such common forms of studying protein stability that further details can be provided here. The stability of the folded state of a protein in the presence of a chemical denaturant is often modelled as

$$\Delta G_d = \Delta G_d^\circ - mc \tag{5.60}$$

where ΔG_d° is the free energy change in the absence of denaturant, c, the concentration of denaturant (usually in molar units), is the only independent variable, and m is a parameter that depends on temperature, pH, buffer, and of course the protein. It is usually assumed that the energetics of protein denaturation are independent of concentration in the absence of oligomerization and aggregation. Note that at the concentration midpoint, which can be determined experimentally, $\Delta G_d^\circ = mc$. So if ΔG_d° is known independently, m can be measured. Though the form of Eqn. 5.60 describes the behavior of many proteins, the physical meaning of m is not entirely clear. What one can say is that it represents the increase in surface area of the protein in contact with the solvent upon unfolding. Figure 5.28 illustrates the effect on protein stability of changing the denaturant concentration. The denaturant is urea, and a change in the conformation of the protein is detected as a change in electrophoretic mobility. The more compact the protein, the higher the mobility. An unfolded protein migrates more slowly than a folded one. This model of protein stability is plausible for a number of reasons. At the same time, however, it says nothing at all about what the denaturant does to make the protein unfold. Moreover, experiments show that in some cases the dependence of ΔG_d on denaturant concentration is distinctly non-linear. The origin of such non-linear behavior is often unclear.

As we have said, changes in pH can also cause protein unfolding. This happens primarily for two reasons. One is that a charged side chain can be partly buried, giving it an anomalous pK_a. Ionization of the side chain can greatly destabilize the folded state. Another is that at extremes of pH, the net charge on the protein can be very large, resulting in very large destabilizing contribution to the overall thermostability. The change in stability of the folded state, $\Delta(\Delta G_d)$, varies with a change in pH as

$$\Delta(\Delta G_d)/\Delta(\text{pH}) = 2.3RT\Delta Q \tag{5.61}$$

where ΔQ is the difference in number of bound protons between the unfolded state and the folded state. When $\Delta Q = 1$ for $\Delta(\text{pH}) = 1$, ΔG_d increases by about 1.4 kcal mol^{-1}.

It has frequently been observed that point mutations in proteins lead to relatively large changes in ΔS and ΔH but a small change in ΔG. The changes in ΔS and ΔH are often difficult to rationalize in terms of changes in protein structure as assessed by NMR spectroscopy or X-ray crystallography. This 'cancellation' of changes in the enthalpy and entropy terms is known as **enthalpy–entropy compensation**. Subtle differences in structure and solvation can apparently have significant thermodynamic consequences. This points up both the remarkable plasticity of the folded state of a protein and the relatively shallow understanding we have of the microscopic origins of macroscopic properties. We'll return to this point in Chapter 9.

Now let's bring a bit of biology into the picture. We wish to cover two subtopics: the engineering of enzymes to enhance their thermostability without altering specificity and the role of stability in protein degradation.

Protein engineering. Enzymes are biological catalysts (Chapter 8). The ability to produce massive quantities of an enzyme using recombinant DNA technology has dramatically increased the feasibility of using enzymes in biomedical, chemical, and industrial applications. Often, though, the physical environment of the enzyme in a practical application will be different from the environment in which it is found in nature. Enzymes can be engineered using standard molecular biological techniques to tailor their properties to specific needs. One way in which the stability of a protein can be increased is to decrease the disorder of its unfolded state. In some cases this leads to a substantial decrease in ΔS_d but effectively no change in ΔH_d. This can be done by replacing suitably located Gly residues with any other residue type. Gly residues make the polypeptide backbone very flexible, while the amino acids with side chains restrict bond rotations in the backbone. Pro residues allow practically no backbone flexibility, and these can be introduced into regions of the amino acid sequence that form turns in the native protein. Yet another approach is to add disulfide bonds. These link different parts of the polypeptide chain and reduce its mobility in the unfolded protein. (See Chapter 6, Section C.) A complementary approach to stabilization of a protein is to increase the enthalpy of the folded state. This often leads to a substantial increase in ΔH_d but effectively no change in ΔS_d. Unfavorable electrostatic interactions can be replaced by favorable ones, and negatively charge side chains (particularly that of Asp) can be placed at the beginning of an a helix to interact favorably with the helix dipole. Amino acid substitutions can be made within helices to increase their strength, and within the protein core to increase the hydrophobic surface area. All such changes can lead to a protein of increased stability.

Protein degradation. Because K_{eq} measures the ratio of the forward and reverse rate constants (discussed in detail in Chapter 8), when the folded state of a protein is stable (i.e. when ΔG_d, if the folded state is the reference state), the rate of folding is greater than the rate of unfolding. When the folded state is the more stable one, however, there will still be some molecules in the unfolded state, even though that proportion will be small except in the region of the transition (see Chapter 6). When K_{eq} = 1, not only is the free energy difference between states zero, there is an equimolar mixture of folded state and unfolded state. In other words, the bottom of the energy well is at the same level as the ground surrounding the well! The proportion of molecules in one state or the other changes as the conditions are adjusted. A question we'd like to be able to answer is, because the turnover rate of endogenous protein in an organism is very high (the body is constantly recycling its protein, see Chapter 9), does the body clear protein by having a specialized degradation protein bind to and digest unfolded proteins? Does the degradation of an unfolded protein then diminish the population of folded proteins by mass action and thereby stimulate the synthesis of replacement protein? Or, does the body make degradation proteins that actively unfold proteins at random? If the former, it is thermal energy and the specific solution conditions of the body (salt concentration, pH, etc.), that play an important role in clearing proteins. If the latter, then energy must be supplied to clear proteins, since work must be done to denature a stable folded protein. In fact, there appear to be proteins that can unfold and degrade stable, native proteins in the presence of ATP. One such protein in bacteria, ClpA, is a member of the Clp/Hsp100 'chaperone' family. Another question we'd like to be able to answer is: if the body continually recycles protein, it must continually make it; and because proteins are synthesized on ribosomes from mRNA templates, genes must continually be transcribed; and if mutations in genes can lead to pathological proteins, and if mutations accumulate as the body ages, does the body somehow 'program' its own death?

R. | Protein dynamics

In a previous chapter we described the folded state of a protein as an organic crystal. If this were not so, it probably would not be possible to crystallize proteins! More importantly, without a relatively fixed geometry in the catalytic site, how could an enzyme carry out a specific function? These considerations, however, might give the false impression that the folded state of a protein has a rigidly fixed structure. Instead, folded states of proteins, though sturdy and crystal-like, are nevertheless flexible, and they exhibit many very rapid small-scale fluctuations. Evidence for dynamical behavior which arises from thermal motion tells us that the native structure of a protein is a large ensemble of similar and rapidly inter-converting conformations that have the same or nearly the same free energy. As we shall see, structural mobility in the native state has crucial functional significance.

We wish to see how this section links to the previous one, on protein stability. The melting temperature of a protein showing two-state behavior depends on the balance of the enthalpy change and entropy change. For a given enthalpy change and entropy of the unfolded state, if the folded state of a protein were extremely rigid, then the change in entropy on unfolding would be large, and the protein would never fold; T_m would simply be too low (Eqn. 3.24). On the other hand, if the entropy of the unfolded state of a protein were not very different from the entropy of the folded state, ΔS_d would be small, and $\Delta H_d/\Delta S_d$ would be so large that the protein would never unfold. This could be a disastrous for the cell if a particular protein inadvertently became modified in such a way as to be pathogenic.

Interestingly, protein flexibility is demonstrated very clearly from protein crystals! This comes about in two major ways. One is that high resolution X-ray diffraction data provide valuable information on the motions of atoms having more electrons than hydrogen. Modeling of the protein structure is a matter of fitting a molecule with the known covalent constraints to an electron density map. The map does not reveal precisely where the center of mass of an atom will be, but only a volume of space where an atom is likely to be found. Analysis of such data shows that some atoms in a protein move very little, while others move a great deal. There is another way in which protein crystals reveal that motions have physiological relevance. X-ray studies of the oxygen transport and storage proteins hemoglobin and myoglobin show that there is no obvious route for O_2 to move from the solvent to the binding site. One concludes that O_2 (and CO_2) binding and release depend on fluctuations in structure known as '**breathing motions**.'

The length- and time-scales of such motions depend on the height of free energy barriers between stable conformations. There are three basic types of dynamical motion: atomic fluctuations, collection motions and triggered conformational changes. Atomic fluctuations occur on a time scale on the order of picoseconds and are relatively small in magnitude, while conformational changes are typically much slower and larger in magnitude. X-ray analysis of the active site cleft of hen lysozyme shows that some of its atoms move by ~1 Å on binding the substrate.

Other techniques that reveal the motional behavior of folded proteins are NMR spectroscopy and hydrogen exchange. NMR can be used to measure the rate of 180°-flipping of the ring of a Phe or Tyr side chain about the C_β–C_α bond. Rates vary, but they generally fall in the microsecond to second range. New developments in NMR technology permit more general analysis of polypeptide backbone and amino acid side chain dynamics. Most of the rapidly interconverting conformations have about the same free energy. Proton NMR can also be coupled with the exchange of labile protons in the polypeptide backbone of a protein. Exchange is temperature-dependent for two reasons: the stability of the native state of a protein varies with temperature, as we saw above, and the exchange rate depends on temperature. These experiments involve D_2O, and deuterium is exchanged for hydrogen. This is particularly attractive from the point of view of experimental design and analysis of results, because hydrogen and deuterium have very different NMR characteristics. As we shall see in Chapter 8, hydrogen exchange and NMR can be used to study protein stability.

S. | Non-equilibrium thermodynamics and life

At some point in your study of thermodynamics you may have wondered: if the First Law requires the total energy of the universe to be constant, and the Second Law requires that every process be accompanied by an increase in the entropy of the universe, then how is life possible at all? Do the tremendously complex forms of matter we call living organisms violate the laws of thermodynamics? Clearly, the answer must be no, if the laws of thermodynamics as formulated actually do describe our universe.

In Chapters 4 and 5 we have used a combination of the First and Second Laws to analyze a number of biological processes at equilibrium. We have seen how useful the Gibbs free energy function is for helping to describe these processes. Although aspects of biochemistry can be described in this way, we should always remember that no living organism is at equilibrium! This holds not only for the organism as a whole but also for each of its cells. Moreover, it applies to every last bacterium living anywhere. Important for us, a non-equilibrium process is by definition irreversible (though possibly not completely irreversible).

Let's think about this somewhat more carefully. An *in vitro* biochemical reaction is a closed system (Chapter 1). As such, it will change until equilibrium is reached. A living organism, on the other hand, be it an amoeba, a bombardier beetle, or a wildebeest, is an open system (Chapter 1). It is therefore *never* at equilibrium. An organism takes in high-enthalpy and low-entropy compounds from its surroundings, transforms them into a more useful form of chemical energy, and returns low-enthalpy and high-entropy compounds to its surroundings. By means of such energy flow, living organisms degrade the quality of the energy of the universe. **Non-equilibrium systems 'dissipate' the useful energy of the universe**.

Energy flow through an organism is like water flow through a channel. (But this does not mean that energy is a material particle or even a collection of particles!) The rate of flow through an organism in adulthood is approximately constant, a situation known as **steady state**. A steady state system changes continually, but there is no *net change* in the system. Steady state in an open system is the analog of equilibrium in a closed system. A steady inward flow of energy is the most stable state an open system can achieve. As depicted in Fig. 2.5B, if the inward flow of energy differs from the rate at which energy is consumed, a change in weight occurs.

Must a living organism be a non-equilibrium system? If it were not, it could not do useful work. An equilibrium system cannot do useful work. This is because at equilibrium, there is no free energy difference between reactants and products. An equilibrium process cannot be directed. It is 'rudderless.' The schematic diagrams of earlier chapters highlighted the machine-like qualities of living organisms, and indeed there are many similarities. For instance, both organisms and machines are made of matter, and the processes carried out by both are, at least to some extent, irreversible. Organisms and machines can do work. Because body temperature is not very different from and often greater than the temperature of the surroundings, an organism cannot do work by means of heat transfer. Instead, its main free energy source is food. Like machines, organisms 'wear out' with use. At the same time, though, machines are basically static structures. For example, the plastic of which a computer keyboard is made is not changing dramatically from moment to moment. The keyboard molecules are not simultaneously being degraded by enzymes and replaced by newly synthesized ones. In contrast, living organisms need free energy because they must renew themselves to live. Proteins are constantly being destroyed and new ones being made to take their place, and DNA is constantly being replicated. Organisms certainly are like machines, but they are also qualitatively different from machines.

What is the source of the irreversibility in a living organism? If many individual biochemical reactions are reversible, at what length scale does irreversibility set in? What is the microscopic origin of irreversibility in biological organisms? There are, undoubtedly, numerous contributions to the overall effect, but a particularly important one is non-productive hydrolysis of ATP. As we have seen, ATP is hydrolyzed spontaneously in water. If hydrolysis is not coupled to a metabolic reaction, the energy released will be given off as heat. This is important in thermogenesis. There are as well three irreversible steps in the metabolism of glucose to pyruvate. These occur between glucose and G6P, F6P and FDP and PEP and pyruvate (Fig. 5.4). This turns out to be extremely important for cellular function, for it is really only at the irreversible steps of a process that control can be exerted; it permits regulation of the speed of the reaction. Such regulation is of considerable importance to any reactions that occur downstream.

Non-equilibrium systems present a number of problems for the quantification of thermodynamic quantities. Though the First Law has been verified for living organisms, the same cannot be done so easily for the Second Law. Entropy, free energy and chemical potential cannot be measured for non-equilibrium systems. There is nevertheless a way of connecting a non-equilibrium system with something more amenable to study and analysis. That is the internal energy, U. Suppose we wish to measure a change in the internal energy of a non-equilibrium system. This can be done by isolating the system and waiting for it to come to equilibrium. Because the system is isolated, the internal energy will be the same at equilibrium as in any non-equilibrium state. If U of the equilibrium state is then measured with respect to some reference value, U of the non-equilibrium state will be known.

T. | References and further reading

Adair, G. (1925). The osmotic pressure of hæmoglobin in the absence of salts. *Proceedings of the Royal Society of London A*, **109**, 292–300.

Anfinsen, C. B. (1973). Principles that govern the folding of protein chains. *Science*, **181**, 223–230.

Arakawa, T. & Timasheff, S. N. (1985). Theory of protein solubility. *Methods in Enzymology*, **114**, 49–77.

Atkins, P. W. (1994). *The Second Law: Energy, Chaos, and Form*, ch. 8. New York: Scientific American.

Baker, T. A. (1999). Trapped in the act. *Nature*, **401**, 29–30.

Barth, R. H. (1992). Dialysis. In *Encyclopedia of Applied Physics*, ed. G. L. Trigg, vol. 4, pp. 533–535. New York: VCH.

Bergethon, P. R. (1998). *The Physical Basis of Biochemistry: the Foundations of Molecular Biophysics*, ch. 13.1. New York: Springer-Verlag.

Brandts, J. F. (1964). The thermodynamics of protein denaturation. I. The denaturation of chymotrypsinogen. *Journal of the American Chemical Society*, **86**, 4291–4301.

Brandts, J. F. (1964). The thermodynamics of protein denaturation. II. A model of reversible denaturation and interpretations regarding the stability of chymotrypsinogen. *Journal of the American Chemical Society*, **86**, 4302–4314.

Bridger, W. A. & Henderson, J. F. (1983). *Cell ATP*. New York: John Wiley.

Brønsted, J. N. (1923). *Recueil des Travaux Chimiques des Pays-Bas*, **42**, 718–728.

Chothia, C. (1984). Principles that determine the structure of proteins, *Annual Review of Biochemistry*, **53**, 537–572.

Christensen, H. N. & Cellarius, R. A. (1972). *Introduction to Bioenergetics: Thermodynamics for the Biologist: A Learning Program for Students of the Biological and Medical Sciences*. Philadelphia: W.B. Saunders.

Cramer, W. A. & Knaff, D. B. (1991). *Energy Transduction in Biological Membranes. A Textbook of Bioenergetics*. New York: Springer-Verlag.

Creighton, T. E. (1991). Stability of folded proteins. *Current Opinion in Structural Biology*, **1**, 5–16.

Dawes, E. A. (1962). *Quantitative Problems in Biochemistry*, 2nd edn, ch. 1. Edinburgh: E. & S. Livingstone.

Donnan, F. G. (1911). Title. *Zeitschrift für Elektrochemie und Angewandte Physikalische Chemie*, **17**, 572.

Encyclopædia Britannica CD 98, 'Colligative Property,' 'Dialysis,' 'Metabolism,' 'Photosynthesis,' 'Saturation,' and 'Vapour Pressure.'

Epstein, I. R. (1989). The role of flow in systems far-from-equilibrium. *Journal of Chemical Education*, **66**, 191–195.

Fersht, A. R. (1999). *Structure and Mechanism in Protein Science: a Guide to Enzyme Catalysis and Protein Folding*. New York: W. H. Freeman.

Fox, S. & Foster, J. S. (1957). *Introduction to Protein Chemistry*, p. 242. New York: John Wiley.

Frauenfelder, H., Parak, F. & Young, R.D. (1988). Conformational substates in proteins. *Annual Review of Biophysics and Biophysical Chemistry*, **17**, 451–479.

Freeman, B. (1995). Osmosis. In *Encyclopedia of Applied Physics*, ed. G. L. Trigg, vol. 13, pp. 59–71. New York: VCH.

Fröhlich, H. (1969). Quantum mechanical concepts in biology. In *Theoretical Physics and Biology*, ed. M. Marios. Amsterdam: North-Holland.

Franks, F. (1995). Protein destabilization at low temperatures. *Advances in Protein Chemistry*, **46**, 105–139.

Fruton, J. S. (1999). *Proteins, Enzymes, Genes: the Interplay of Chemistry and Biology*. New Haven: Yale University Press.

Garrett, J. (1990). Thermodynamics in sheep. *Education in Chemistry*, **27**, 127.

George, P. & Rutman, R. J. (1960). The 'high energy phosphate bond' concept. *Progress in Biophysics and Biophysical Chemistry*, **10**, 1–53.

Gillispie, Charles C. (ed.) (1970). *Dictionary of Scientific Biography*. New York: Charles Scribner.

Girandier, L. & Stock, M. J. (eds) (1983). *Mammalian Thermogenesis*. London: Chapman & Hall.

Gutfreund, H. (1949). In *Hæmoglobin: a Symposium Based on a Conference Held at Cambridge in June 1948 in Memory of Sir John Bancroft*, ed. F. J. W. Roughton and J. C. Kendrew, p. 197. London: Butterworths.

Haase, R. (1969). *Thermodynamics of Irreversible Processes*. New York: Dover.

Harold, F. M. (1986). *The Vital Force: a Study of Bioenergetics*. New York: W. H. Freeman.

Harris, D. A. (1995). *Bioenergetics at a Glance*, ch. 1. Oxford: Blackwell Science.

Hatefi, Y. (1985). The mitochondrial electron transport and oxidative phosphorylation system. *Annual Review of Biochemistry*, **54**, 1015–1069.

Haynie, D. T. (1993). *The Structural Thermodynamics of Protein Folding*, ch. 4. Ph.D. thesis, The Johns Hopkins University.

Hinckle, P. C. & McCarty, R. E. (1978). How cells make ATP. *Scientific American*, **238**, no. 3, 104–123.

Karplus, M. & McCammon, J. A. (1986). Protein dynamics. *Scientific American*, **254**, no. 4, 30–39.

Katchalsky, A. & Curran, P. F. (1967). *Nonequilibrium Thermodynamics in Biophysics*. Cambridge, Massachusetts: Harvard University Press.

Kauzmann, W. (1958). Some factors in the interpretation of protein denaturation. *Advances in Protein Chemistry*, **14**, 1–63.

Klotz, I. M. (1986). *Introduction to Biomolecular Energetics*, cc. 3–7. Orlando: Academic Press.

Kondepudi, D. & Prigogine, I. (1998). *Modern Thermodynamics: from Heat Engines to Dissipative Structures*, ch. 8.2. Chichester: John Wiley.

Kyoguko, Y., Lord, R. C. & Rich, A. (1969). An infrared study of the hydrogen-bonding specificity of hypoxanthine and other nucleic acid derivatives. *Biochimica et Biophysica Acta*, **179**, 10–17.

Lodish, H., Baltimore, D., Berk, A., Zipursky, S. L., Matsudaira, P. & Darnell, J. (1995). *Molecular Cell Biology*, 3rd edn, cc. 2, 4 & 21. New York: W. H. Freeman.

Makhatadze, G. I. & Privalov, P. L. (1995). Energetics of protein structure. *Advances in Protein Chemistry*, **47**, 307–425.

Matthews, B. W. (1995). Studies on protein stability with T4 lysozyme. *Advances in Protein Chemistry*, **46**, 249–278.

McCammon, J. A. & Harvey, S. C. (1987) *Dynamics of Proteins and Nucleic Acids*. Cambridge: Cambridge University Press.

Millar, D., Millar, I., Millar, J. & Millar, M. (1989). *Chambers Concise Dictionary of Scientists*. Cambridge: Chambers.

Mitchell, P. (1976). Vectorial chemistry and the molecular mechanisms of chemiosmotic coupling: power transmission by proticity, *Biochemical Society Transactions*, **4**, 398–430.

Morowitz, H. J. (1978). *Foundations of Bioenergetics*, ch. 3E. New York: Academic Press.

Nicholls, D. G. & Ferguson, S. J. (1992). *Bioenergetics 2*, ch 3. London: Academic Press.

Noyes, R. M. (1996). Application of the Gibbs function to chemical systems and subsystems. *Journal of Chemical Education*, **73**, 404–408.

Pardee, G. S. & Ingraham, L. L. (1960). Free energy and entropy in metabolism. In *Metabolic Pathways*, ed. D.M. Greenberg, vol. I. New York: Academic Press.

Pauling, C. L. (1970). Structure of high energy molecules. *Chemistry in Britain*, **6**, 468–472.

Pepys, M. B., Hawkins, P. N., Booth, D. R., Vigushin, D. M., Tennent, G. A., Soutar, A. K., Totty, N., Nguyent, O., Blake, C. C. F., Terry, C. J., Feest, T. G., Zalin, A. M. & Hsuan, J. J. (1993). Human lysozyme gene mutations cause hereditary systemic amyloidosis, *Nature*, **362**, 553–557.

Peusner, L. (1974). *Concepts in Bioenergetics*, cc. 3, 5, 6, 7 & 10–8. Englewood Cliffs: Prentice-Hall.

Plum, G. E. & Breslauer, K. J. (1995). Calorimetry of proteins and nucleic acids. *Current Opinion Structural Biology*, **5**, 682–690.

Prigogine, I. (1967). *Introduction to Thermodynamics of Irreversible Processes*. New York: John Wiley.

Prigogine, I. (1969). Structure, Dissipation and Life. In *Theoretical Physics and Biology*, ed. M. Marios. Amsterdam: North-Holland.

Prigogine, I., Nicolis, G. & Babloyants, A. (1972) Thermodynamics of evolution. *Physics Today*, **25**, no. 11, 23–28.

Prigogine, I., Nicolis, G. & Babloyants, A. (1972) Thermodynamics of evolution. *Physics Today*, **25**, no. 12, 38–44.

Record, M. T., Jr, Zhang, W. & Anderson, C.F. (1998). Analysis of effects of salts and uncharged solutes on protein and nucleic acid equilibria and processes: a practical guide to recognizing and interpreting polyelectrolyte effects, Hofmeister effects, and osmotic effects. *Advances in Protein Chemistry*, **51**, 281–353.

Roberts, T. J., Marsh, R. L., Weyland, P. G., & Taylor, C. R. (1997). Muscular force in running turkeys: the economy of minimizing work, *Science*, **275**, 1113–1115.

Schellman, J. A. (1987). The thermodynamic stability of proteins, *Annual Review of Biophysics and Biophysical Chemistry*, **16**, 115–137.

Secrest, D. (1996). Osmotic pressure and the effects of gravity on solutions. *Journal of Chemical Education*, **73**, 998–1000.

Segal, I. H. (1976). *Biochemical Calculations: How to Solve Mathematical Problems in General Biochemistry*, 2nd edn, ch. 3. New York: John Wiley.

Shavit, N. (1980). Energy transduction in chloroplasts. *Annual Review of Biochemistry*, **49**, 111–139.

Shaw, A. & Bott, R. (1996). Engineering enzymes for stability. *Current Opinion in Structural Biology*, **6**, 546–550.

Shortle, D. (1996). The denatured state (the other half of the folding equation) and its role in protein stability, *Federation of the American Societies for Experimental Biology Journal*, **10**, 27–34.

Smith, C. A. & Wood, E. J. (1991). *Energy in Biological Systems*, cc. 1.3 & 1.4. London: Chapman & Hall.

Spolar, R., Livingstone, J. & Record, M. T., Jr (1992). Use of liquid hydrocarbon and amide transfer data to estimate contributions to thermodynamic functions of protein folding from the removal of nonpolar and polar surface from water. *Biochemistry*, **31**, 3947–3955.

Snell, F. M., Shulman, S., Spencer, R. P. & Moos, C. (1965). *Biophysical Principles of Structure and Function*, ch. 8. Reading, Massachusetts: Addison-Wesley.

Tanford, C. (1968). Protein denaturation (parts A and B). *Advances in Protein Chemistry*, **23**, 121–282.

Timasheff, S. N. (1993). The control of protein stability and association by weak interactions with water: how do solvents affect these processes? *Annual Review of Biophysics and Biomolecular Structure*, **22**, 67–97.

Timasheff, S. N. (1998). Control of protein stability and reactions by weakly interacting cosolvents: the simplicity and the complicated. *Advances in Protein Chemistry*, **51**, 355–432.

Timbrell, J. A. (1991). *Principles of Biochemical Toxicology*, 2nd edn, ch. 7. London: Taylor & Francis.

van Holde, K. E. (1985). *Physical Biochemistry*, 2nd edn, cc. 2.1, 2.3, 2.4, 3.4 & 3.5. Englewood Cliffs: Prentice-Hall.

Voet, D. & Voet, J. G. (1995). *Biochemistry*, 2nd edn, cc. 3, 4, 15-4–15-6, 16, 18-1, 19-1, 20, 22, 28-3, 28-5A & 34-4B. New York: John Wiley.

Weber-Ban, E. U., Reid, B. G., Miranker, A. D. & Horwich, A. L. (1999). Global unfolding of a substrate protein by the Hsp100 chaperone ClpA. *Nature*, **401**, 90–93.

Williams, S. (1999). Life as a part-time plant. *Wellcome News*, **20**, 38.

Williams, T. I. (ed.) (1969). *A Biographical Dictionary of Scientists*. London: Adam & Charles Black.

Woodcock, A. & Davis, M. (1978). *Catastrophe Theory*. Harmondsworth: Penguin.

Wrigglesworth, J. (1997). *Energy and Life*, cc. 3, 5.7.2, 7.1, 7.3 & 7.5.1. London: Taylor & Francis.

U. | Exercises

1. Speculate in broad terms on the effect on Earth of the cessation of photosynthesis.

2. The energy conversion process by which sunlight is converted into biomass is not completely efficient. What happens to the energy that does not become biomass? Rationalize your answer in terms of the First and Second Laws of Thermodynamics.

3. Animal life as part-time plant? Sue Williams of the Department of Botany, University of Western Australia, says that the green-tinged sea slugs she studies 'enslave' chloroplasts from the seaweed they ingest, and use them as a means of capturing up to 25% of their energy. Explain how this might work.

4. Use the following information to determine the standard free energy change of ATP hydrolysis:

   Glucose + ATP $\Leftrightarrow$ glucose 6-phosphate + ADP $\Delta G^{\circ\prime} = -16.7 \text{ kJ mol}^{-1}$

   Glucose 6-phosphate $\Leftrightarrow$ glucose + P$_i$ $\Delta G^{\circ\prime} = -13.8 \text{ kJ mol}^{-1}$

   Show all work.

5. Buffers containing ATP are ordinarily made up fresh and not stored as a stock solution. When a stock solution is made, it must usually be kept at 4 °C (short term storage) or at -20 °C (long term storage). Rationalize these practices. What bearing does this have on the necessary molecular machinery of a cell?

6. ATP is the energy currency of the cell. ATP is essential for life as we know it. Comment on the stability of ATP in aqueous solution and the constraints this may place on theories of the origin of life.

7. The free energy status of a cell can be described in various ways. One of these, called the **adenylate energy charge** (AEC), was first proposed by Daniel Edward Atkinson (1921–). The AEC is defined as

$$AEC = ([ATP] 1 0.5[ADP])/([ATP] + [ADP] + [AMP])$$

and it varies between 1.0, when all the adenine nucleotide is ATP, and 0, when all the β- and γ-phosphoanydride bonds have been hydrolyzed. The relative amounts of ATP, ADP, AMP can be determined by comparing the sizes of the respective peaks in a high-performance liquid chromatography (HPLC) profile. The AEC of a healthy cell is about 0.90–0.95. Malignant hypothermia is an inherited muscle disease in humans and pigs. Patients suffer rapid rises in body temperature, spasms in skeletal muscle, and increases in the rate of metabolism, which can be fatal if not treated with a suitable muscle relaxant. The following data were obtained before the onset of symptoms and just prior to the death of a pig afflicted with the disease:

	[ATP]	[ADP]	[AMP]
	μmol g^{-1} tissue		
Before symptoms	4.2	0.37	0.029
Before death	2.1	0.66	0.19

Calculate the AEC before the symptoms began to occur and just before death. Comment on the magnitude of the values and what they indicate.

8. A 1 M solution of a glucose gives an osmotic pressure more than 25 times greater than that of the atmosphere. A 1 M solution of a salt gives an even larger osmotic pressures. Explain.

9. Suppose we have an osmometer that is constructed from a capillary tube with a membrane in the middle, and that the tube is oriented *horizontally* (why?). Now let some osmotic particles suddenly appear on one side of the tube only. Explain what happens.

10. You have a U-tube osmometer with cells of equal shape and volume. On one side, you place a sphere of volume V, and at the same time but on the other side you place a cube of volume V. Neither particle is membrane-permeant. Suppose that these particles are able to interact with the solvent. Explain what will happen in the following situations: (a) the particles are so dense that they sit on the bottom of the cells of the osmometer; (b) the density of the particles is such that they are able to diffuse throughout the volume of their respective cells.

11. What causes the membrane of a red blood cell to burst when the cell is placed in hypotonic solution? Be as specific as possible.

12. Suppose you have an osmometer in which the solute particles are confined to a fixed volume, for instance an indestructible membrane of fixed volume. What happens? Why?

13. Suppose you have an osmometer with a membrane that is permeable to water but not to larger molecules. Add glucose to one side to a final concentration of 0.001 M and hemoglobin to the other side to

a concentration of 0.001 M. Will a pressure head develop? If yes, on which side will the water level be higher? If no, why not?

14. Suppose you are involved in preparing recombinant protein for a physical biochemistry experiment. The approach involves 8 M urea, formation of which from urea crystals and water is highly endothermic. The bacteria are lysed in 8 M urea, a chemical denaturant, and the recombinant protein is separated from the bacterial proteins by column chromatography. Separation of the recombinant protein from urea is done by dialysis in two stages. In each, c. 100 ml of lysate is dialyzed against 5 l of water. The dialysis membrane allows the passage of water and urea but not protein. Will the volume of the protein preparation change in this procedure, and if so, how? Assuming that the volume of the protein solution at the end of dialysis is 100 ml, what is the final concentration of urea? Explain, in enthalpic and entropic terms, the driving force for the reduction in urea concentration in the first step of dialysis. Explain from a thermodynamic point of view what drives the further reduction in urea concentration in the second step of dialysis.

15. Recall what happens to a red blood cell when it's placed in a hypotonic solution. What must be done to ensure that dialysis tubing doesn't burst?

16. Prove that Eqns. 5.19 and 5.20 follow from the preceding equations.

17. Show that Eqn. 5.38 follows from Eqn. 5.36.

18. Derive Eqn. 5.40. (Hint: Start with $R_T = R + R \cdot I + R \cdot L$, express R and $R \cdot I$ in terms of $R \cdot L$, and solve for $R \cdot L$.)

19. Eqn. 5.41 is 0 for all concentrations of I_i when there is no inhibition (compound I_i has no effect), and it is 1 at 100% inhibition. Explain.

20. Analysis of gene regulation involves study of structural and thermodynamic aspects of how proteins bind nucleic acid. One area of such research is the recognition of DNA operator sites by repressor molecules. Suppose protein P binds a single specific sequence on a molecule of DNA D. This is a common mechanism for the regulation of gene expression. At equilibrium, $P + D \Leftrightarrow P \cdot D$. A bacterial cell contains one molecule of DNA. Assume that cell is cylindrical, and that its diameter and length are 1 μM and 2 μM, respectively. Calculate the total concentration of D. Assume that $K_{eq} = 10^{-10}$ M. Calculate $[P \cdot D]$, assuming that $[P] = [D]$. The concentration of *bound* D is just $[P \cdot D]$. Calculate the concentration of *unbound* D. Calculate $[P \cdot D]/[P]$. Give an interpretation of this quantity. The subject of binding will be discussed in detail in Chapter 7.

21. The previous problem involved the association and dissociation of two types of macromolecule, proteins and DNA. A basic feature of such situations is the dependence of the equilibrium on the total concentrations of the interacting species. The concept can be illustrated by means of the monomer–dimer equilibrium. Consider the equilibrium

$$2M \Leftrightarrow D \tag{5.62}$$

The total concentration of monomer, $[M]_T$, is $[M] + 2[D]$, where the factor 2 accounts for there being two monomers in each dimer. This equation can be solved for [D]. Write down an expression for the equilibrium constant for the reaction in Eqn. 5.62. Combine this with your equation for [D] and solve the resulting quadratic equation for [M]. Show that $[M]/[M]_T \rightarrow 1$ as $[M]_T \rightarrow 0$, and that $[M]/[M]_T \rightarrow 0$ as $[M]_T \rightarrow \infty$. How does one interpret these limiting conditions?

22. What might be the structural basis for the low stability of Georgia hemoglobin relative to normal hemoglobin?

23. Hemocyanin is a Cu-containing oxygen-binding protein that is found in some invertebrates. In squid hemocyanin, when the partial pressure of oxygen gas is 0.13 atm at 25 °C, the oxygen binding sites are 33% saturated. Assuming that each hemocyanin molecule binds one molecule of oxygen gas, calculate the equilibrium constant. What are the units of the equilibrium constant? Calculate the standard state free energy change when hemocyanin interacts with $O_2(aq)$. The solubility of pure oxygen in water at 1 atm and 25 °C is 0.00117 mol $(kg\ H_2O)^{-1}$.

24. In ELISA, what type of interactions are likely to be most important for protein adsorption to the solid support? Why are antibodies able to bind to partially denatured protein?

25. Explain in thermodynamic terms how a single 30-cycle PCR experiment can yield billions of copies of double-stranded DNA.

26. Under normal conditions, complementary strands of DNA form a double helix. In the section on PCR we provided a way of rationalizing the stability of DNA. Compare and contrast our view with that put forward by Voet & Voet, authors of a popular biochemistry textbook (see pp. 866–870 of the second edition (1995)). Can the data in Table 28-4 be trusted? Why or why not?

27. Eqn. 5.50 for DNA supercoiling resembles the First Law of Thermodynamics. List and explain the similarities and differences.

28. A certain machine of a biotechnology company provides a controlled environment for the automation of sequence-specific DNA analysis and performs all the reaction steps required for capture and detection of nucleic acids. A main feature of the product is its capture specificity. For instance, suppose a 300 bp PCR fragment derived from the filamentous bacteriophage M13 was specifically captured using a series of complementary oligonucleotide probes 24 residues in length, and that the capture probes incorporated 0–6 mismatches with the target. Explain how optimizing the hybridization conditions (i.e. by adjusting the temperature) could distinguish sequences differing by a single base.

29. 'Hot start.' When plasmid DNA is used as the template in a PCR reaction, the enzyme buffer, plasmid, and oligonucleotide primers are often incubated at 95 °C for several minutes before starting thermal cycling. Why?

30. The release of insulin from pancreatic β cells on uptake of glucose is a complex process. The steps of the process in rough outline are as follows. The resting membrane potential of a β cell is determined by open ATP-sensitive K^+ channels in the plasma membrane. After a meal, glucose is taken into the cell and phosphorylated. Eventually,

there is an increase in [ATP]/[ADP] ratio in the cell, and this closes the K^+ channels. The membrane depolarizes, stimulating the opening of Ca^{2+} channels. Calcium enters the cell, stimulating the release of insulin through exocytosis of secretory granules. Describe each step of this process in moderately detailed thermodynamic terms.

31. Isothermal titration calorimetry. The key condition underlying this technique is thermodynamic equilibrium. When an aliquot of titrant is injected, the Gibbs free energy of the system increases. A spontaneous chemical reaction occurs until G reaches a new minimum and equilibrium is established once again. An ITC study of a ligand binding to a macromolecule was carried out at three temperatures, T_1, T_2 and T_3, where $T_1 < T_2 < T_3$. At T_1, $\Delta H_b > 0$; at T_2, $\Delta H_b = 0$, and at T_3, $\Delta H_b > 0$. The ligand is known to bind the macromolecule all three temperatures by means of independent experiments. Explain what is happening in the reaction cell at each stage of a general ITC experiment, viz. before an injection and during an injection. Rationalize the results obtained.

32. Speculate on the possibility of observing the cold-denaturation of DNA. What about tRNA?

33. The folded and unfolded states of a protein are in equilibrium as shown in Eqn. 5.56. Suppose that you are working with a solution of RNase A at a concentration of 2.0×10^{-3} M, and that fraction of protein in the *unfolded* state are as follows: 50 °C: 0.00255; 100 °C: 0.14. In the thermal denaturation of this protein, there are essentially just two states, the folded one and the unfolded one, so the fraction of protein in the folded state is just one minus the fraction in the unfolded state. Calculate ΔH and ΔS for *unfolding* of RNase A. What key assumption must be made about temperature-dependence? Calculated ΔG for *unfolding* of RNase A at 37 °C. Is this process spontaneous at this temperature? Determine the melting temperature of RNase A under standard state conditions (for a two-state reaction, at T_m half of the proteins are folded and half are unfolded).

34. The role of ΔC_p in protein stability and its molecular origin was discussed in publications by the American biochemist John Brandts as early as 1964. Use Eqn. 4.3 to investigate the role of ΔC_p in the thermostability of a protein. One relatively easy way to do this is to assume values for ΔH and ΔS at some reference temperature, say 298 K, and then to use spreadsheet to calculate ΔG throughout a temperature range that includes 0–100 °C. Plot ΔG v. T for several different values of ΔC_p. Note that the curve crosses the T-axis at two points. What are the names of these intercepts? What if $\Delta C_p < 0$? Is this physically meaningful? Is it relevant to biological macromolecules?

35. Suppose you have designed a four-helix bundle. A four-helix bundle is just a polypeptide that folds into four helices of approximately equal length, and whose helices are bundled together giving a compact structure. The helices interact with each other in the core of the protein. Various structural techniques show that at room temperature the structure is highly dynamic and not very much like

an organic crystal, though all four helices are intact. Thermal denaturation studies, however, indicate that the unfolding temperature
of your designed protein is over 100 °C! Explain. How could the
design be modified to reduce the melting temperature and increase
the specificity of interactions in the protein core?

36. Living organisms have been described as 'relatively stable' systems
that 'show an organized collective behavior which cannot be
described in terms of an obvious (static) spatial order' and are 'not
near thermal equilibrium.' Explain.

37. The synthesis of ATP under standard conditions requires
7.7 kcal mol^{-1}, and this is coupled to the movement of 2H$^+$.
Calculate the pH difference across the inner mitochondrial membrane needed to drive ATP synthesis at 25 °C.

38. Oxidation–reduction reactions in *E. coli* generate a pH gradient of
$+1$ (outside to inside) and a voltage gradient of -120 mV (outside to
inside). Which type(s) of free energy is (are) made available by this
proton motive force? β-galactosides are transported along with H$+$
ions. Calculated the maximum concentration ratio of β-galactoside
that can result from the coupling of its transport to the proton
motive force.

39. An empirical expression for the melting temperature of double-
stranded DNA in the presence of NaCl is

$$T_m = 41.1X_{G+C} + 16.6\log[Na^+] + 81.5 \qquad (5.63)$$

Where X_{G+C} is the mole fraction of G–C pairs. Given a 1000 base pair
gene with 293 Gs and 321 Cs, calculate the sodium ion concentration at which it will have a melting temperature of 65 °C.

40. Use the following osmotic pressure data for horse hemoglobin in
0.2 M phosphate and at 3 °C to determine the molecular mass of the
protein.

Concentration of hemoglobin (g/100 ml)	Osmotic pressure (cm H_2O)
0.65	3.84
0.81	3.82
1.11	3.51
1.24	3.79
1.65	3.46
1.78	3.82
2.17	3.82
2.54	3.40
2.98	3.76
3.52	3.80
3.90	3.74
4.89	4.00
6.06	3.94
8.01	4.27
8.89	4.36

41. The effect of pH on the osmotic pressure of sheep hemoglobin has been investigated by Gilbert Adair (see Chapter 7). The following data were obtained.

pH	Osmotic pressure (mmHg/1g protein/100 ml)*
5.0	21.5
5.4	13.4
6.5	3.2
6.7	2.4
6.8	2.4
6.8	3.5
6.8	4.5
7.2	5.0
9.6	15.6
10.2	21.4

*1 mmHg = 133.322 ... Pa.

Graph the data and use them to deduce the isoelectric point of sheep hemoglobin.

42. Why would it not be a good idea to water your houseplants with boiling water?

43. Suggest a biochemical means by which one might test the origin of the heat produced by *Arum maculatum* (see Chapter 3). (Hint: use tissue extracts of the spadix and appendix of the plant and consider poisons that block either electron transport or oxidative phosphorylation.)

44. It is sometimes said that the two terminal phosphoanhydride bonds of ATP are 'high-energy' bonds. This implies that the energy released as free energy when the bond is cleaved is stored within the bond itself. Why is the term *high-energy bond* misleading?

45. Mg^{2+} ions interact with ATP under physiological conditions. What is the likely effect of this on the free energy of hydrolysis of ATP? Why?

46. If the association constants for formation of A•A and G•G are so large (Table 5.4), why is such base pairing disfavored in polynucleotides?

For solutions, see http://chem.nich.edu/homework

Chapter 6

Statistical thermodynamics

A. | Introduction

Classical thermodynamics is a phenomenological description of nature. The mathematical relationships of thermodynamics are precise, but they do not tell us the molecular origin of the properties of matter. This chapter discusses a means of gaining a *molecular interpretation* of thermodynamic quantities. If you've spotted a trend we set from page one of this book, you will have guessed that mathematics will play an important role here. The required mathematical background certainly is greater than earlier on, but not substantially so. And, as before, all the main ideas can be expressed relatively well in figures or words. Though it is of course important to be able to use the mathematics, what is far more important is to have a good sense of *what* the mathematics says! Indeed, this is what distinguishes the physical biochemist from the mathematician. One should always bear in mind that although *mathematics* applies to *everything*, it is the physical biochemist who hires the mathematician, not the mathematician who hires the physical biochemist.

The need for a rethinking of the foundations of thermodynamics was first realized when the work of British chemist and physicist John Dalton (1766–1844) and Russian chemist Dmitri Ivanovich Mendele'ev (1834–1907) on the atomic theory of matter began to be widely accepted after the middle of the nineteenth century.[1] Classical thermodynamics

---

[1] The belief that the world is made of atoms is in fact much older than this; it simply was not widely accepted until the nineteenth century. Its earliest known exponent is the Greek philosopher Democritus (*c.* 460–*c.* 370 BC), who held that the atoms of the heavy elements combined to form the Earth, while the atoms of the light ones formed the heavenly bodies (planets and stars). If you have ever doubted the significance of the idea that world is made of atoms, consider the following quotation. In his *Lectures on Physics*, American Nobel Laureate in Physics Richard Phillips Feynman (1918–1988) says, 'If, in some cataclysm, all of scientific knowledge were to be destroyed, and only one sentence passed on to the next generation of creatures, what statement would contain the most information in the fewest words? I believe it is the *atomic hypothesis* (or the atomic *fact*, or whatever you wish to call it) that *all things are made of atoms – little particles that move around in perpetual motion, attracting each other when they are a little distance apart, but repelling upon being squeezed into one another*. In that one sentence, you will see, there is an *enormous* amount of information about the world, if just a little imagination and thinking are applied.' (Emphasis in original.)

is built on the tacit assumption that many particles are present, and it deals with **macroscopic properties** of such collections of particles. Thermodynamics itself does not give a molecular view of what it describes; it does not explain thermodynamic quantities in terms of the **microscopic properties** of individual particles.

An example will help to illustrate the difference between macroscopic properties and microscopic ones. As discussed in Chapter 2 (and Appendix B), DSC can be used to make quantitative measurements of the heat absorbed by a solution of macromolecules as the molecules undergo an order–disorder transition. For instance, DSC has been employed in the study of the thermodynamics of tRNA folding/unfolding. The measured value of the heat capacity, however, says practically nothing about *why* a particular value should be observed. What kinds of bonds are broken as the temperature increases? Which bonds? How many bonds? How does the heat absorbed in an order–disorder transition correspond to the structure that is melted? The result of a DSC experiment does not answer these questions in the absence of additional information. Moreover, the mathematical relations of thermodynamics do not provide the answers.

In contrast to thermodynamics, statistical mechanics provides a molecular theory or interpretation of thermodynamic properties of macroscopic systems. It does this by linking the behavior of individual particles or parts of macromolecules to classical thermodynamic quantities like work, heat and entropy. Statistical mechanics can be thought of as a bridge between the macroscopic and the microscopic properties of systems. Using statistical mechanics, one can begin to rationalize, often in structural terms, how changes in a system connect up to the results of thermodynamic experiments.

One can easily show that the capture of light energy by plants is important to their growth. The light energy provided and the change in mass of the plants can quantified and analyzed in terms of classical thermodynamics. It is something else to describe energy capture in terms of the wavelength of photons involved and electronic bound states (Chapter 1). As discussed in Chapter 2, the inflation of a bicycle tire is a matter of stuffing a tube with air. The amount of air inside the tube can be measured as air pressure, but this says nothing at all about whether is air is a continuous compressible fluid or a gaseous collection of particles separated by vacuum. The physical properties of the different phases of water are of biological importance, and one can use classical thermodynamics to describe transitions between these phases. Such descriptions, however, say nothing at about the structure of an individual water molecule or about how water molecules interact to give rise to the bulk properties of the solid phase, the liquid phase, or the gas phase (Fig. 6.1). Here we are interested in a molecular interpretation of macroscopic thermodynamic properties, and when this can be given, a connection is formed between the world made of atoms and measurable thermodynamic quantities.

Fine. But *how* does one go about providing a detailed description of molecular behavior? After all, a typical macroscopic system might have on

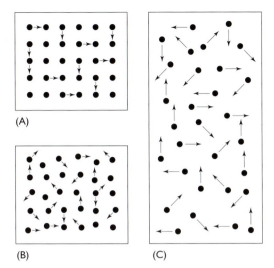

(A)

(B)   (C)

Fig. 6.1 Molecular interpretation of the three states of matter. In the solid phase (panel A), the molecules are in a regular array. Interactions between molecules occur, but overall there is practically no translational motion; each molecule stays put. Panel B depicts the liquid state. Molecules are free to translate in any direction. The volume, however, is not much different from in the solid state. In the gas phase (panel C), the volume occupied by the molecules is much larger than in the liquid phase or the solid phase. The average molecular speed is relatively high. After Figs. 4-1, 4-2, and 4-3 of Peusner (1974).

the order of 10^{23} particles, and on a practical level the *complete* description of each particle and each particle's motion seems to be *impossible*. There simply isn't a way to keep track of what each particle is doing! This is not simply a matter of the number of particles involved; for our current understanding of physics provides an analytical expression for two interacting bodies but not three or more! A middle way has to be chosen between an explicit quantum mechanical description of every last particle and a thermodynamic description of the system as a whole. The approach is to use *statistical* methods and to assume that the average values of the *mechanical* variables of the molecules in a thermodynamic system (e.g. pressure and volume) are the same as the measurable quantities of classical thermodynamics, at least in the limit that the number of particles is very large ($>10^{10}$). The statistical treatment of the mechanical properties of molecules as they relate to thermodynamics is called statistical mechanics or statistical thermodynamics.

In Chapter 2 we described the Zeroth Law of Thermodynamics, showing how it justifies the concept of temperature. Then, and throughout this book, heat has been treated as a sort of fluid that can pervade matter. In fact, historically people thought of heat transfer as the flow of a fluid from one body to another.[2] When viewed from the statistical perspective, the concept takes on additional meaning: **temperature is a measure of the average kinetic energy of the molecules of a system**. Increases in temperature reflect increases in translational motion – the motion of movement from one place to another. On this view, when two systems are in contact, energy is transferred between molecules as a result of *collisions*, not flow of a fluid-like heat from one system to the other. The transfer continues until a statistical uniformity is reached, and this corresponds to thermal equilibrium.

---

[2] Count Rumford (Benjamin Thompson), founder of the Royal Institution and husband of the widow of the great French chemist Lavoisier, was born in America. (Lavoisier was guillotined during the French Revolution.) Rumford showed that heat is not a fluid and proposed that it is a form of motion, or energy. He lived 1753–1814.

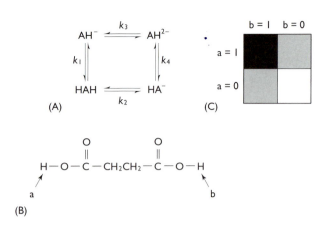

(A)

(B)

(C)

(D)

Fig. 6.2 Molecular interpretation of acid–base equilibria. Panel (A) shows a thermodynamic cycle for a diprotic acid. The microscopic equilibrium constants contain information on the energetics of each step of the process. Panel (B) shows the molecular structure of succinic acid, an intermediate in the citric acid cycle. This molecule is symmetrical. The two dissociable protons are labeled 'a' and 'b.' A schematic diagram of the possible states of a succinic acid molecule is given in panel (C). I = on, 0 = off. There are two possible ways in which the single-proton state can be formed from the two-proton state. Panel (D) shows the molecular structures of the various states of the molecule. Note that the two single-proton states are *indistinguishable*.

Two chapters ago we discussed thermodynamic properties of acids and bases. We introduced the concept of pK_a, a macroscopic property, and learned something of its significance for biochemistry. That discussion required only a very qualitative idea of the structures of the acid and base involved. The acid was HA, the base A$^-$. How do we explain the measured macroscopic properties in terms of microscopic properties? Fig. 6.2A shows a reaction scheme for a diprotic acid. The ks are **microscopic** equilibrium constants, each corresponding to the loss of a specific proton from a specific structure. In principle, either proton can be lost first in going from the fully protonated state to the fully deprotonated state. If a diprotic acid is symmetrical, the two singly deprotonated states can be indistinguishable. Succinic acid, for example, an intermediate in the citric acid cycle, is a symmetrical molecule (Fig. 6.2B). The various states of the molecule can be represented diagrammatically as in Fig. 6.2C. The structures of the various species and the relative proportion each possible type is given in panel (D). Thinking in such terms is part and parcel of giving a molecular interpretation of a thermodynamic phenomenon.

B. | Diffusion

We have already encountered diffusion on more than one occasion. In our discussion of gas pressure in Chapter 2, the gas particles moved around the system in no preferred direction, i.e. at random. Then, in Chapter 3, we thought about perfume becoming distributed throughout a room. Convection currents play a role in that, of course, at least in real life. But if a room is at thermal equilibrium, the only means by which the aromatic molecules can spread out on their own is by diffusion (Brownian motion). We also learned about diffusion in the sections on osmosis and dialysis in Chapter 5. So, we should have a good idea of what diffusion is by now. This section is devoted to developing a more quantitative and molecular understanding of this phenomenon. We are interested in diffusion because it is important in photosynthesis, respiration, molecular transport in cells, the absorption of digested food from the gut into the bloodstream, and many other topics in biology.

Unlike the thermodynamic concepts of energy, enthalpy, entropy, and all other strictly thermodynamic quantities for that matter, diffusion involves *time* explicitly. For this reason one might prefer to see a discussion of it in a different place, perhaps in the chapter on kinetics. It is included here, however, because it helps to illustrate a very important aspect of statistical thermodynamics: that macroscopic properties of a system can be 'built up' from the average behavior of its individual components. Moreover, it is an essential aspect of the *dynamics* of equilibrium.

Suppose we have a system at equilibrium, for instance a small volume of an aqueous solution of glucose. Some of the molecules will be moving relatively rapidly, others slowly, but the average speed will correspond to the kinetic energy that is identical to the thermal energy (Chapter 2). Any glucose molecule in the system will experience a sequence of collisions with other glucose molecules. An important point is that these collisions will occur *at random*; there is no apparent preferred direction of change on the microscopic level of a single molecule. On the macroscopic level, however, there will be no net change at all, because the system is at equilibrium.

We can describe the collisions made by our glucose molecule using simple mathematics. We say 'simple' because the equations *are* simple. Some students, though, find the thinking involved here more like trying to run a five-minute mile than walking from one class to another. So don't be surprised if you find that you want to read this section more than once.

In a unit of time **t**, our molecule will make an *average* of N collisions with other glucose molecules. If we increase the length of the observation time, say to 5t, the molecule will make and *average* of 5N such collisions. The *average* time between collisions, τ, is just

$$\tau = \text{t}/N \qquad (6.1)$$

An analogy might be advertisements on television. Suppose there are 9 breaks during a two-hour program. Then the *average* time between breaks is 120 min / 9 breaks = 13.3 minutes. (Because there are so many interruptions, it might be a better use of time to turn the tube off and find something else to do!)

Now, we want to know the *odds* that one glucose molecule will collide with another glucose molecule during a small unit of time, Δt. This is just $\Delta t/\tau$. The larger our small unit of time, the larger the ratio, and the greater the *probability* that a collision will occur. As $\Delta t \rightarrow \tau$, this ratio approaches unity, so there is a 100% chance (on the average) of a glucose–glucose collision (in the average time between collisions). Let's think about this in another way. Instead of one molecule making N collisions, let's have N molecules make as many collisions as they will in a short period of time Δt. The number of collisions is then $N \times \Delta t/\tau$. Switching back to the TV analogy, instead of calculating the *odds* from the observed number of breaks on a single channel during a block of time that a single break will occur in the next minute, we assume that what was observed on one channel applies to all of them and use the single-channel data to calculate the *odds* that a break will occur in the next minute on any one of N channels. So, just as we would expect, for a fixed value of τ, the *odds* that an advertisement will occur within the next minute on any channel increases proportionately with the number of channels.

Can we calculate the *probability* that a molecule will *not* make a collision during an interval of time? Yes! Suppose there are N molecules in total, and suppose we watch the molecules for a time t. Suppose also that we find that the number of them that have *not* made a collision during this time is $N(t)$. As in previous chapters, this notation means 'the variable N changes value with time,' not '$N \times t$.' After a short period of time Δt, the number of molecules that will *not* have made collision is $N(t) + \Delta N(t)$. Despite the plus sign, this number must be *smaller* than $N(t)$; the number of molecules that have not had a collision cannot increase! In other words, $\Delta N(t) < 0$. From above, the odds that N molecules will make collisions in a short period of time is $N\Delta t/\tau$, so

$$N(t) + \Delta N(t) = N(t) - N(t)\Delta t/\tau \qquad (6.2)$$

Rearrangement gives

$$\Delta N(t)/\Delta t = -N(t)/\tau \qquad (6.3)$$

This tells us that the change per unit time in the number of molecules that have *not* collided is proportional to the ratio of the number of molecules that have *not* collided divided by the time between collisions. In other words, the decrease per unit time in the number of television channels on which an advertisement has *not* appeared after the start of observation would be proportional to the number of channels divided by the average time between breaks. When Eqn. 6.3 is solved for $N(t)$, one obtains (using a little calculus)

$$N(t) = N_o \exp(-t/\tau) \qquad (6.4)$$

where N_o is the number of molecules being observed. A plot of $N(t)$ against time is shown in Fig. 6.3. $N(t)$ decreases exponentially with time, and the amount of time required for a decrease by a factor of e^{-1} is τ.

We can also think about collisions in terms of the distance a molecule travels between them. The **mean free path** of a particle, l, is defined as the average time between collisions, τ, times its *average* speed, v:

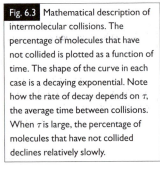

Fig. 6.3 Mathematical description of intermolecular collisions. The percentage of molecules that have not collided is plotted as a function of time. The shape of the curve in each case is a decaying exponential. Note how the rate of decay depends on τ, the average time between collisions. When τ is large, the percentage of molecules that have not collided declines relatively slowly.

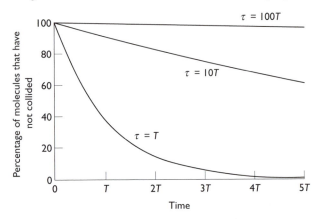

$$l = \tau v \qquad (6.5)$$

The greater the average speed of a molecule, the longer the mean free path for a given value of τ. If the concentration of a type of particle increases but the temperature is held constant, the average speed of the particles does not change but τ decreases. Correspondingly, l decreases. Here's a numerical example. If the average time between collisions involving a bicyclist and flying insects is 6 min, and the average speed of the cyclist on her way to university is 15 m.p.h., the mean free path would be just 1/40 of a mile. In other words, it might be a good idea to wear goggles and nose and mouth protection! The *odds* of a collision along a short stretch of cycle path would depend on the number of bugs per unit volume (distributed at random and flying in all directions), N; the size of the cyclist (actually, the frontal 'cross-sectional' area of the cyclist on the bike – it would clearly be a good idea to be fit and trim!), A; and the length of the short stretch of cycle path, Δx:

$$\text{Odds of biker–bug collision in } \Delta x = AN\Delta x \qquad (6.6)$$

During certain weeks of the year when N is large, it might be a good idea to walk or use public transportation to get to university!

The television and biker analogies we have used here are, admittedly, somewhat artificial. Why? Someone must *decide* what the programming of a TV channel will be! Diffusion is qualitatively different, for as far as anyone can tell gas particles move around at *random*. This is the basic assumption for Einstein's description of diffusion; it is based on the Brownian motion, the microscopic and rapid motions that bits of particulate matter can be observed to exhibit when suspended in a solvent (Fig. 6.4).

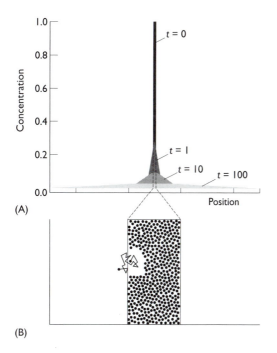

(A)

(B)

Fig. 6.4 Diffusion. Panel (A) shows how a concentrated substance spreads out in time. The area below the curve is the same in each case (the number of solute particles is constant). A magnified view of the concentrated solute is shown in panel (B). At this resolution it is possible to see individual solute particles. The trajectory of one particle is shown to the left, with the position indicated at the end of each time increment. The macroscopic result of all solute particles showing such microscopic behavior is shown in panel (A). The 'random walk' of an individual particle is known as Brownian motion. Based on Figs. 4.7 and 4.8 of van Holde (1985).

Now we have a deeper sense of the nature of collisions between molecules in solution. Why is this important in the biological sciences? Suppose we introduce a small amount of concentrated glucose into a volume of water. What happens? The glucose distributes itself throughout the volume according to the Second Law of Thermodynamics! Diffusion tells us about particle behavior as the system moves towards equilibrium. It also tells us about particle behavior after the system has reached equilibrium. In fact, from a qualitative point of view, there is no *essential* difference between the Brownian motion of a particle in a volume in a non-equilibrium state and in an equilibrium state. The average length of time between glucose–glucose collisions, however, is a maximum at equilibrium and a minimum before the concentrated solution is diluted. When the glucose is still relatively concentrated, collisions between glucose molecules are much more probable than when the glucose is randomly distributed throughout the volume.

Why else should we know about diffusion? It's the means by which fatty acids enter the small intestine! Sucrose, lactose, maltose, and glucose chains, on the other hand, are broken down into monosaccharides by enzymes on hair-like protrusions of the small intestine called microvilli. The sugars then cross the epithelium and enter the bloodstream by active transport or **facilitated diffusion**; they are too big for simple diffusion across the membrane. Active transport (for example of Na^+ and K^+) requires a membrane protein and additional energy; facilitate diffusion involves a channel through which sugar can move down its concentration gradient.

Diffusion also helps to understand such important concepts in biology as enzyme function. Suppose we have a soluble substrate molecule in solution and an enzyme that recognizes it. Suppose also that we can measure rate of catalysis. What limits the rate? Is it the speed at which catalysis occurs? If so, that would be a matter of the molecular mechanism of the reaction, or biochemistry. But if the fundamental limitation on rate is the time required for the substrate to make a collision with the enzyme, that is a matter of how quickly the molecules are moving in solution (the temperature of the system) and concentration. There is an important qualitative difference between these two ways of thinking about what limits the rate of an enzymatic reaction.

C. | Boltzmann distribution

Let's return to the subject of osmosis (Chapter 5). As we have seen, the underlying conditions for the phenomenon to occur under isothermal and isobaric conditions are solute particles that can move throughout a volume of solvent, absence of a constraint on the volume of solution, and separation of the solution from solvent by a semi-permeable membrane. We stated with confidence that osmosis is an entropy-driven process, providing perhaps less evidence than you might have liked. We wish to take this opportunity to complete the discussion.

In Chapter 3 we derived the formula $\Delta S = nR\ln(V_f/V_i)$ for the rever-

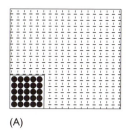 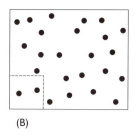

(A)   (B)

Fig. 6.5 Expansion of perfume throughout a room. In panel (A), the molecules are close together, as in a solid or a liquid. Only 25 of the 288 volume elements are occupied. In panel (B), the 25 molecules are spread throughout the volume. In other words, the volume accessible to the molecules is much greater in panel (B) than in panel (A), by a factor of 288/25. The entropy of the molecules is much greater in panel (B) than in panel (A). Based on Fig. 1.2 of Wrigglesworth (1997).

sible expansion of an ideal gas. Let's now divide up the volume of the gas into notional volume elements of identical size. When the volume doubles, the number of elements doubles. To simplify matters, let's work in two dimensions, so that each volume element can be represented as an area. Our gas container can now be drawn on a piece of paper, as shown in Fig. 6.5. If expansion of the gas corresponds to a volume change of a factor of 288/25, the number of volume elements increases by a factor of 288/25.

Why all this discussion about volume elements? They are a helpful way of considering the number of ways of arranging particles. In Fig. 6.5A, for example, where all 25 gas molecules are stuffed into a tiny space, 25 volume elements are occupied (1 molecule per volume element), and 288 elements are empty. The arrangement of molecules in panel (A) resembles the close packing of a crystal, but we wish to keep things simple by ignoring phase changes. The gas pressure is high! In contrast, in panel (B), 25 gas molecules are spread throughout the entire volume (288 volume elements). There are many more ways of placing 25 molecules in 288 elements than in 25 elements! As we shall see below, combining the division of space into volume elements with the $\Delta S = nR\ln(V_f/V_i)$ relationship from classical thermodynamics leads to the famous Boltzmann equation of statistical mechanics.

In Chapter 1 we discussed how energy flows from one part of the world to another, how energy is transformed from one to form to another, and how in the midst of change the total amount of energy stays the same. We saw how there are many ways in which a given quantity of energy can be distributed. Here we are concerned not so much with the possibility of more than one distribution as with the relative *probability* of each possible distribution. For instance, given a closed *macroscopic* system, we should always expect the equilibrium distribution to be the *most probable* distribution. Fluctuations from equilibrium will occur, of course, but unless the system is particularly small, all probable fluctuations will be of negligible magnitude. This suggests that **if the number of molecules in a system is large, the behavior of the system will coincide with that predicted from a statistical consideration of the behavior of individual molecules of the system**.

A way of describing the basic principles involved is the following. As we saw in Chapter 1, photosystems absorb light within a certain band of wavelengths, not photons of just any energy. This is precisely why plants look the way do: they absorb red and blue light but reflect green light. The absorption of a photon of a certain wavelength requires a suitable electronic bound state. This is how photons interact with matter. When

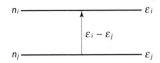

Fig. 6.6 Energy level diagram. There are n_i particles in energy level ε_i and n_j particles in energy level ε_j. The difference in energy between these states is $\varepsilon_i - \varepsilon_j$. When a photon is absorbed by chlorophyll or some other pigment molecule, the energy of the photon (hc/λ) is $\varepsilon_i - \varepsilon_j$, where i and j correspond respectively to the excited and ground states of the electron.

the light-absorbing molecule absorbs a photon of the right wavelength, the electron is elevated from its ground state to an excited state, and the difference in energy between the two states is effectively *identical* to the energy of the photon. Energy is conserved! The important point here is that there are electronic energy *levels* in light-absorbing molecules: the electron cannot be found at an intermediate energy; it is either in the ground (low-energy) state or in some excited state, not in between (Fig. 6.6). Energy levels arise from a quantum mechanical analysis of the bound state of a particle.

Let's expand on this. We choose as our system an ensemble of *indistinguishable* particles. The particles are indistinguishable because there is no way of telling them apart; they're something like identical siblings, only *impossible* to distinguish unless labels are attached. We assume that the system is such that its energy spectrum is discrete, as in Fig. 6.6. In other words, the particles are bound. Note that this situation is qualitatively different from Eqn. 1.1, in which the energy of a photon varies smoothly with wavelength. A photon travelling through space can have 'any' energy, at least in principle; it is a free particle.

Note that an energy state of a single particle is *not* the same thing as a thermodynamic state of an entire system of particles. How can we distinguish between these 'states'? Suppose a house is drawing $3x$ amperes of current, and that this current corresponds to three identical lights being on. We could say that the state of the system, the house, is $3x$, a value that can be measured by putting a meter on the electric line connecting the house to the main electrical supply. This measured value, however, tells us nothing at all about whether a light is on in a particular room of the house. Are the three rooms the kitchen, sitting room, and bathroom, or the dining room, foyer, and a bedroom? There are clearly different combinations of lights that correspond to a current of $3x$, and if we are monitoring the current from outside the house, we could know the state of the system but not the state of each light, at least in the absence of additional information. Each room of the house corresponds to a particular energy state or level: 0 if the light is off, and x if the light is on. A light cannot be 'half-way' on (this house was built before the advent of dimmer switches and has not been rewired!), so the state of the house must be an integral multiple of x.

Now suppose we have a definite number of particles, say seven, and let the measured total energy of the system be 15ε, where ε represents a unit of energy. We wish to know *how many* different ways there are of arranging the particles so that two criteria are met: all the particles are accounted for ($7 = \Sigma n_i$; in words, the sum over all energy levels (Σ) of the number of particles in energy level i, n_i, is 7), and the total energy is 15ε ($15\varepsilon = \Sigma n_i \varepsilon_i$; in words, the sum over all energy levels of the number of particles in energy level i times the energy of that level, ε_i, is 15ε). These two constraints make the system look the same from the outside, regardless of the specific arrangement of particles on the inside. There are only seven ways of arranging seven *indistinguishable* particles if 6 have energy ε; the seventh one *must* have energy 9ε. Fig. 6.7 shows three other ways in which the particle and energy con-

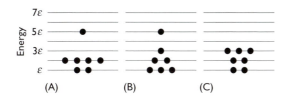

Fig. 6.7 Different ways of arranging seven indistinguishable particles under the constraints of constant total energy and constant number of particles. Arrangement B appears to be the most stable one of the three. Based on Fig. 1.2 of van Holde (1985).

straints can be met simultaneously. But are these configurations equally probable?

To answer the question, let's *assume* that **the *most probable* distribution is the one that corresponds to the *largest number of ways* of arranging particles in a given configuration**. It is easy to show that **configuration** B is more probable than A or C. (This does not necessarily mean that configuration B is the *most probable* distribution of all possible distributions satisfying both constraints). Suppose we have three identical peas, and let the peas be indistinguishable. The seeds can be arranged into two 'piles' of configuration {1,2}; there are two seeds in one pile and one seed on its own. How many different ways can the seeds be arranged in two piles? Three! The lone seed can be any of three of them, as shown in Fig. 6.8, and the other two constitute the second pile. If there are six seeds distributed into three piles with configuration {1,2,3}, the number of arrangements is 60 (prove this).

In general, the **number of ways**, Ω, of arranging identical N particles of configuration $\{n_1, n_2, \ldots\}$, i.e. n_1 in one group, n_2 in another, n_3 in another, and so on, is

$$\Omega = N!/(n_1! n_2! n_3! \ldots n_i! \ldots) \qquad \text{Statistical weight} \qquad (6.7)$$

where $x!$ ('x factorial') denotes $x(x-1)(x-2)\ldots 1$. An implicit assumption here is that for any particle of the system, any state of the system is equally probable. For by the **postulate of equal *a priori* probability**, nothing predisposes a particle to occupy a particular state. The two constraints, however, of particle number and total energy, must still be satisfied. Ω is called the **statistical weight** of the configuration. An example of how to use Eqn. 6.7 is as follows. Suppose we have 20 identical objects ($N = 20$) with configuration $\{1, 0, 3, 5, 10, 1\}$. Then $\Omega = 20!/(1! \times 0! \times 3! \times 5! \times 10! \times 1!) = 9.31 \times 10^8$, a large number! ($0! = 1$.) Note that **the statistical weight depends only on the distribution of particles**, and not on whether there is one particle in energy state 1, three particles in energy state 3, 5 particles in energy state 4, and so on.

On the assumption that the most probable distribution is the one has the largest number of arrangements of particles, identifying the most probable distribution is the same as maximizing Ω under the constraint that the number of particles and the total energy are constant. *Maximizing* the number of ways of arranging particles? That sounds

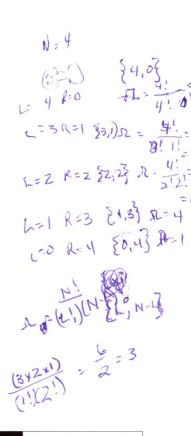

Fig. 6.8 Three possible ways of arranging three indistinguishable particles of configuration {1,2}. Labels have been attached to the particles to distinguish one from another.

A ●	C ●	A ●	B ●	C ●	A ●
B ●		C ●		B ●	
Pile 1	Pile 2	Pile 1	Pile 2	Pile 1	Pile 2

Arrangement 1	Arrangement 2	Arrangement 3

curiously like maximizing the entropy, a necessary condition of equilibrium. Ω does indeed resemble the entropy function discussed in Chapter 3, particularly in the context of $S_f - S_i = \Delta S = nR\ln(V_f/V_i) = nR\ln V_f - nR\ln V_i$. In fact, the connection between S and Ω, known as the **Boltzmann equation**, is

$$S = N_A k_B \ln\Omega = R\ln\Omega \qquad (6.8)$$

where k_B, the Boltzmann constant (1.381×10^{-23} J K^{-1}), is a fundamental constant of physics, and N_A is Avogadro's number (named after the Italian physicist Amadeo conte di Quaregna Avogadro (1776–1856). If the system changes from state 1 to state 2, the molar entropy change is $S_2 - S_1 = N_A k_B(\ln\Omega_2 - \ln\Omega_1) = R\ln(\Omega_2/\Omega_1)$.

Note that by Eqn. 6.8 the entropy is an *extensive* thermodynamic quantity; if our system comprises two parts, A and B, the total entropy is $S_A + S_B$. If Ω_A is the number of ways of arranging particles in part A, and Ω_B is the number of ways of arranging particles in part B, and S is the total entropy, then

$$S = S_A + S_B = N_A k_B \ln\Omega_A + N_A k_B \ln\Omega_B =$$

$$N_A k_B (\ln\Omega_A + \ln\Omega_B) = N_A k_B \ln(\Omega_A \Omega_B) = N_A k_B \ln\Omega \qquad (6.9)$$

where $\Omega = \Omega_A \Omega_B$. Relationships 6.8 and 6.9 were put forward in the late nineteenth century by Ludwig Boltzmann (1844–1906), an Austrian theoretical physicist. They are derived from $U = \Sigma n_i \varepsilon_i$ (Chapter 3). This tells us that Boltzmann's discovery was not incredibly fortuitous; he did not just 'happen' upon a *mathematical* relationship (Eqn. 6.8) that works so well. The fact of the matter is that he had a good rough idea of what he was looking for. He started from the *physical* point view and made a number of clever guesses and approximations. For instance, he assumed not only that all gas particles of an ideal gas move in all possible directions, but also that all the particles move with the same speed. But as we have seen in Chapter 1, there is a *distribution* of particle speeds. Boltzmann's particle-speed approximation works because thermal energy is proportional to the average kinetic energy, and the average energy is consistent with many possible distributions, including the one where all particles move at the same speed! Boltzmann also built on the work of Carnot and Clausius and went far beyond what they had done. This is not to diminish the work of Boltzmann, for there is no doubt that he made an extremely important contribution to physics! It is, rather, to put Boltzmann's work in perspective, and to provide clues as to why his mathematical results have proved so valuable.

Boltzmann did not stop with Eqn. 6.9. He was able to show by means of a few mathematical tricks, that when $N = \Sigma n_i$ is very large, say on the order of N_A,

$$n_i = n_1 \exp\left[-\alpha(\varepsilon_i - \varepsilon_1)\right] \qquad (6.10)$$

where n_i is the number of particles with energy ε_i, n_1 is the number of particles with the lowest energy, ε_1, and α is a constant. This equation is called **the Boltzmann energy distribution** or ordering principle. **The**

Boltzmann distribution is the most probable distribution for a large system at or very near equilibrium.

If energy level i of a single particle corresponds to ω_i arrangements, then

$$n_i = n_1 \exp\left[-\alpha(\varepsilon_i - \varepsilon_1)\right](\omega_i/\omega_1) \tag{6.11}$$

State 1 is the **reference state**, and measurement of the energy of state i is made relative to it. The constant term in the argument of the exponential, α, is $1/k_BT$, where T is the absolute temperature. ω_i is known as the **degeneracy** of state i. Note that we are now using the lower case of Ω. This is because we are now interested in the possible ways of ordering the atoms of a single molecule, not the number of ways of arranging a given number of molecules in space. In other words, the entropy of a system encompasses not only to the different ways of arranging particles in space, but also to the ways in which the atoms of individual molecules can be arranged. This way, the entropy includes the various possible arrangements of a chemical group owing to bond rotation and such like.

To illustrate the usefulness of Eqn. 6.11 by way of example, we turn to the subject of disulfide bonds. These covalent links form between cysteine residues, usually of the same polypeptide. One of our favorite proteins, hen egg white lysozyme, has four intramolecular disulfide bonds. One of these is between residues 6 and 127. There are only 129 amino acids in hen lysozyme, so disulfide bond 6,127 connects the N- and C-termini of the polypeptide chain (Fig. 6.9). It so happens that 6,127 is mostly solvent-exposed (this can be seen by examination of the crystallographic structure), while the other disulfides are solvent-inaccessible. This makes 6,127 susceptible to selective reduction and chemical modification. One can carry out a certain chemical reaction and obtain a three-disulfide derivative of hen lysozyme in which 6,127 alone is modified (3SS-HEWL).

To come to the point. The thermodynamic properties of 3SS-HEWL have been investigated by scanning microcalorimetry (Cooper *et al.*, 1991). The results showed not only that the modified lysozyme exhibited 'two-state' behavior under all conditions studied, but also that ΔH_d was essentially the same for 3SS-HEWL as for the wild-type (WT) enzyme at the same temperature. Thus, the main thermodynamic consequence of removal of disulfide 6,127 is to increase the *entropy* change of unfolding. Because the enthalpy of unfolding was the same for both WT-HEWL and 3SS-HEWL, one would suspect that the native fold forms of both types have approximately the same structure, the same number of hydrogen bonds, the same number and type of van der Waals interactions, and so on. This guess has been confirmed by determination of the structure of 3SS-HEWL at atomic resolution.

The difference in the change in entropy between the modified and unmodified proteins, $\Delta\Delta S$, is interpreted as an increase in **conformational entropy**. This term is defined as the number of different ways that the covalent structure of a protein can be arranged in space at a

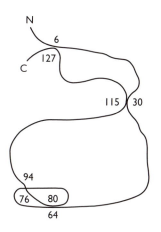

Fig. 6.9 Schematic diagram of the topology of hen egg white lysozyme (and α-lactalbumin) on the level of the amino acid sequence. The N-terminus and C-terminus are marked, as are the residue numbers of the cysteine residues, all of which are involved in disulfide bonds. Note that disulfide 6,127 joins the chain termini together, forming a loop with disulfide bridge 30,115. Breaking disulfide 6,127 opens this loop, increasing the mobility at equilibrium of all the encompassed amino acid residues but especially those at the chain termini.

given energy. The experiments showed that $\Delta\Delta S \approx 25$ cal mol^{-1} K^{-1} at 25 °C; the entropy of unfolded 3SS-HEWL is much greater than the entropy of the unfolded intact protein, in good agreement with intuition.

Equation 6.8 can be used to calculate the increase in the number of conformations of the unfolded state (U) as follows.

$$\Delta\Delta S_{conf,4SS\rightarrow3SS} = \Delta S_{d,3SS} - \Delta S_{d,WT}$$

$$= N_A k_B[\ln(\omega_{U,3SS}/\omega_{F,3SS}) - \ln(\omega_{U,WT}/\omega_{F,WT})] \qquad (6.12)$$

Assuming that the entropy of the folded state is the same in both forms of the enzyme, the ω_F terms cancel out, leaving

$$\Delta\Delta S_{conf,4SS\rightarrow3SS} = N_A k_B(\ln\omega_{U,3SS} - \ln\omega_{U,WT}) = N_A k_B\ln(\omega_{U,3SS}/\omega_{U,WT}) \qquad (6.13)$$

Substituting in what is known,

$$25 \text{ cal mol}^{-1} \text{ K}^{-1} = (6.02 \times 10^{23} \text{ mol}^{-1}) \times (1.381 \times 10^{-23} \text{ J K}^{-1})$$

$$\times 0.239 \text{ cal J}^{-1} \times \ln(\omega_{U,3SS}/\omega_{U,WT}) \qquad (6.14)$$

The entropy change per residue is 25/129 cal mol^{-1} K^{-1}, from which

$$\omega_{3SS}/\omega_{WT} \approx 1.1 \qquad (6.15)$$

That is, breaking disulfide bond 6,127 of hen lysozyme results in a *c.* 10% increase in the number of conformations per residue in the unfolded state, assuming that the entire magnitude of $\Delta\Delta S$ can be attributed to a redistribution of the preferred conformations of the unfolded state. A more realistic interpretation would be that if the folded state is uniformly rigid, there is a larger increase in the number of conformations per amino residue at the N- and C-termini and a smaller increase elsewhere. An even more realistic interpretation would be that the folded state is not uniformly rigid and that the entropy of folded 3SS-HEWL is somewhat greater than the entropy of WT-HEWL!

Let's return to our simple assumption that the folded-state entropies are the same. According to Eqn. 6.11,

$$\omega_{U,WT}/\omega_{U,3SS} \propto n_{U,WT}/n_{U,3SS} \qquad (6.16)$$

This tells us that the distribution of conformations in the unfolded state of the three-disulfide derivative is more probable than the distribution of conformations in the unfolded state of the wild-type protein when disulfide bond 6,127 is severed. In short, **disulfide bonds play a key role in stabilizing proteins.**

D. | Partition function

We can see from Eqn. 6.11 that as $T \rightarrow 0$, the argument of the exponential becomes very large and negative, and $n_i \rightarrow 0$ for all $i > 0$. As $T \rightarrow \infty$, the denominator in the exponential term becomes so large that the magnitude of the numerator becomes irrelevant, and all energy levels are equally populated. Between these extremes, the probability that a particle is in state j is

$$P_j = \frac{n_j}{N} = \frac{n_j}{\Sigma n_i} = \frac{n_j}{n_1} \frac{n_i}{n_i \sum_i \exp\left(\dfrac{-(\varepsilon_i - \varepsilon_1)}{k_B T}\right) \dfrac{\omega_i}{\omega_1}} \qquad (6.17)$$

$$= \frac{\omega_j \exp\left(\dfrac{-(\varepsilon_j - \varepsilon_1)}{k_B T}\right)}{\sum_i \omega_i \exp\left(\dfrac{-(\varepsilon_i - \varepsilon_1)}{k_B T}\right)} \qquad (6.18)$$

$\Sigma P_j = 1$. It is clear from Eqn. 6.17 that P_j increases with increasing n_j and decreasing $\Delta\varepsilon$. The closer energy state j is to the lowest energy state (the ground state), the more likely it is to be occupied. If we now divide the numerator and denominator of Eqn. 6.18 by ω_1 and rearrange, we obtain

$$P_j = \frac{\dfrac{\omega_j}{\omega_1} \exp\left(\dfrac{-(\varepsilon_j - \varepsilon_1)}{k_B T}\right)}{\sum_i \dfrac{\omega_i}{\omega_1} \exp\left(\dfrac{-(\varepsilon_i - \varepsilon_1)}{k_B T}\right)} \qquad (6.19)$$

$$= \frac{\exp[\ln\omega_j - \ln\omega_1]\exp\left(\dfrac{-(\varepsilon_j - \varepsilon_1)}{k_B T}\right)}{\sum_i \exp[\ln\omega_i - \ln\omega_1]\exp\left(\dfrac{-(\varepsilon_i - \varepsilon_1)}{k_B T}\right)} \qquad (6.20)$$

$$= \frac{\exp\left(\dfrac{-[(\varepsilon_j - \varepsilon_1) - k_B(\ln\omega_j - \ln\omega_1)T]}{k_B T}\right)}{\sum_i \exp\left(\dfrac{-[(\varepsilon_i - \varepsilon_1) - k_B(\ln\omega_i - \ln\omega_1)T]}{k_B T}\right)} \qquad (6.21)$$

Putting the arguments of the exponential terms on a molar basis (multiplying numerator and denominator by N_A), we have

$$P_j = \frac{\exp\left(\dfrac{-[(H_j - H_1) - T(S_j - S_1)]}{RT}\right)}{\sum_i \exp\left(\dfrac{-[(H_i - H_1) - T(S_i - S_1)]}{RT}\right)} = \frac{K_j}{\sum_i K_j} = \frac{K_j}{Q} \qquad (6.22)$$

where $N_A\varepsilon_i$ and $N_A k\ln\omega_i$ have been *interpreted*, respectively, as the thermodynamic functions H_i (cf. Eqn. 2.10) and S_i (cf. Eqn. 6.8 and Chapter 3). Note that $\exp(-\Delta G/RT)$ has been written as K, as though K were an equilibrium constant. That's because K is an equilibrium constant! Each exponential term of the form $\exp(-\Delta G/RT)$ is called a **Boltzmann factor**, and there is one of these for each accessible state of the system except the reference state (for which $K = 1$). Note also that if all the probabilities are summed, the result is one, as necessary. The sum in the denominator of Eqn. 6.22, the sum of all the Boltzmann factors, is called the (canonical) **partition function**. It is often symbolized as Q. Eqn. 6.22 tells us that, given a large collection of molecules, the fraction of them that will be in state j at any given time is given by the ratio of the Boltzmann factor for state j divided by the sum of all the Boltzmann factors, Q.

The partition function contains *all* the thermodynamic information available on the system; it is a key mathematical concept in statistical thermodynamics. As useful as the partition function is, however, one's being able to it write down for a given situation and manipulate it flawlessly does not necessarily imply a good understanding of the physics of the situation; a point one does well to remember. One might be tempted to think that because statistical mechanics is so useful learning classical thermodynamics is a waste of time. Mistake! For it often turns out that a classical description of a situation is more intuitive and just plain simpler than a full-blown statistical mechanical treatment. Life is difficult enough with making things harder than they need to be! There are times, however, when statistical mechanics does not only what classical thermodynamics cannot do but also just what is needed. This is why it's a good idea to prefer both classical thermodynamics and statistical mechanics, and not just one or the other.

E. | Analysis of thermodynamic data

We now wish to make a definite step towards connecting Eqn. 6.22 to measurements that can be made in the laboratory. P_j stands for the population of state j, or the fraction of molecules in state j. As we shall see below, this is of particular importance for rationalizing the measured value of an observable property of a system. It does not matter whether the observable is the ellipticity of a sample at a given wavelength, the intrinsic fluorescence emission, partial heat capacity, or intrinsic viscosity of sample at a given pH. Moreover, Eqn. 6.22 applies whether the sample is protein, DNA, lipids, carbohydrates, or different combinations of these types of macromolecule. It is amazing that the relatively simple mathematical theory outlined above can be used to analyze the results of such a wide variety of methods and samples. That is the beauty of physics.

We have seen in Chapter 2 that according to DSC the thermal denaturation of hen lysozyme in aqueous solvent is a two-state, first-order phase transition. That is, the observed folding/unfolding properties of the protein can be modeled as arising from a folded state and an unfolded state only. This is not to say that intermediates are *not* present! But it does mean that the fraction of molecules in a partly folded state must be so small that a deviation from two-state behavior is within the error of other aspects of the experiment (e.g. determination of protein concentration). Two-state behavior holds not only for protein folding/ unfolding, where the free energy difference between states corresponds to a relatively intuitive change in structure, but for *any* system in which there are two main states. Another example would be the binding of a ligand to a macromolecule to a single binding site.

The two states in this case are the unbound state and the bound state, and the free energy difference between states is the free energy of binding. Binding energy is an all-encompassing thermodynamics quantity that can and usually does include contributions from structural rearrangement. One can easily imagine, however, a situation where the structure of the macromolecule does not change substantially on binding, e.g. when binding is mainly electrostatic in origin and the geometry of the binding site does not change on binding. All these situations can be rationalized in terms of the so-called **two-state approximation**.

In carrying out biochemical experiments, one reduces the importance of thermal fluctuations and increases the signal-to-noise ratio by increasing the number of molecules present. There are often very practical reasons why experiments are done at a certain concentration. For instance, in a ^{1}H NMR experiment at 600 MHz one needs a minimum of about 1 mM sample in order to get a good signal. Unfortunately, though, many proteins are not soluble at such a high concentration (Chapter 5)! And before the advent of recombinant DNA technology and production of recombinant proteins, very few proteins could be obtained in such large quantities. In contrast, fluorescence emission is so sensitive that orders of magnitude less protein is needed for an experiment. In fact, it is can be a liability to have too high a sample concentration when using a fluorimeter, even if there is no difficulty with solubility of the sample. Regardless of the technique, though, in such experiments one invariably makes use of the **ergodic hypothesis**. According to this principle, **the average relatively short-duration behavior of a large collection of identical particles is *assumed* to be identical to the average long-duration behavior of a single particle under given conditions**. In other words, it is not just the protein molecule or DNA molecule alone that is the system in a reaction carried out in solution; rather, the system is the entire collection of solvated macromolecules and any other difference between the solution and the buffer in which the macromolecules are dissolved. Observed values are normalized for solute concentration and *assumed* to represent the average properties of a single macromolecule.

Bearing all this in mind, we assign the variable O to the **observable quantity** of interest, be it heat capacity, fluorescence, or what-have-you. The measured value is the sum of the fractional contributions of each state:

$$<O> = \Sigma_i P_i o_i \tag{6.23}$$

where $<O>$ is the average value of observable O and o_i is the contribution to O of state i, i.e. the characteristic value of O for this state under the conditions of the experiment. Regardless of the number of states, the set of numbers $\{P_i\}$ defines the *distribution* of states. The distribution tells us the fraction of particles in state 1, the fraction in state 2, the fraction in state 3, and so on.

For the simple case of a two-state system (A ⇔ B), Eqn. 6.23 becomes

$$\langle O \rangle = P_A o_A + P_B o_B = \frac{1}{1+K} o_A + \frac{K}{1+K} o_B \qquad (6.24)$$

where $0 < P_i < 1$, $i = A, B$. For example, suppose you are using a fluorimeter to monitor the binding the reversible binding of a protein to DNA. Assume that the protein and protein–DNA complex have the necessary fluorescence emission properties to distinguish the bound state from the unbound state of the protein. Assume also that the ligand exhibits a negligible contribution to the fluorescence emission in the unbound state. To analyze the resulting experimental data, and assuming that there have been no problems in collecting them, it is not necessary to know anything about how a fluorimeter works or the physical basis of fluorescence emission. In fact, it is not even necessary to know that there are just two states! For the sake of simplicity, however, we assume here that there are two and only two states and that there is a difference in the observed value of fluorescence when binding occurs.

Now, suppose our experiment shows that the fluorescence intensity of our folded but unbound protein is 225 (arbitrary units), and that the intensity of the same concentration of protein at the same temperature but in the presence of a sufficient amount of DNA ligand to saturate the binding sites is 735 (arbitrary units). Then $o_A = 225$, $o_B = 735$. In the absence of ligand, $P_A = 1$ and $O = 225$; in the presence of a large quantity of ligand, $P_B = 1$ and $O = 735$. Intermediate concentrations of DNA ligand give a fluorescence intensity between these extremes. The measured value O is an average of o_A and o_B that is weighted by P_A and P_B, the proportion of molecules in the bound state and unbound state, respectively.

It is possible to calculate the population, say, of state A. To do this, we rewrite P_B in Eqn 6.24 as $1 - P_A$ and solve for P_A. The result is

$$P_A = \frac{\langle O \rangle - o_B}{o_A - o_B} \qquad (6.25)$$

In other words, the population of state A (or state B) can be found from the baseline measurements (225 and 735 in the example above), and the measured value of some intermediate degree of saturation.

From a qualitative point of view, the analysis would proceed in exactly the same way if NMR chemical shift were the observable instead of fluorescence intensity. Suppose you were using NMR to measure the pK_a of a titratable side chain. Changes in the protonation state of a side chain can have a marked effect on chemical shift. In principle chemical shift changes can be detected not only in one dimension, as in the fluorescence experiment described above (fluorescence intensity is the sole dependent variable), but in two dimensions, using one of numerous NMR methods.

In the typical situation, things will not be as simple as they have been made to seem above. That's because the baseline value of fluorescence, resonance frequency, radius of gyration, or whatever need not be constant (Fig. 6.10). Moreover, it need not be linear! Often, though, each baseline can be modeled as a straight line. In general, the slope and

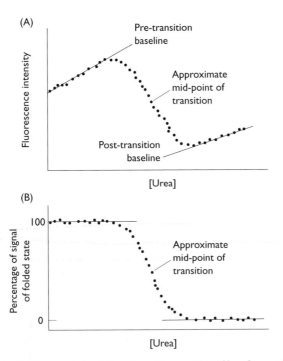

(A)

Fluorescence intensity

Pre-transition baseline

Approximate mid-point of transition

Post-transition baseline

[Urea]

(B)

Percentage of signal of folded state

100

0

Approximate mid-point of transition

[Urea]

Fig. 6.10 Cooperative unfolding of a protein at equilibrium. The protein contains at least one partially buried tryptophan side chain. Tryptophan flourescence depends on the polarity of the environment of the side chain. Panel (A) shows the fluorescence intensity as a function of urea concentration. Note that although the baselines are approximately linear, neither is constant. Panel (B) shows the result of a mathematical transformation of the data of panel (A). The baselines have been accounted for, the fluorescence signal has been converted to a percentage change.

y-intercept of the pre-transition baseline will differ from the slope and intercept of the post-transition baseline. The procedure one uses for calculating the population of states is the same as before, except that a linear function is used in place of a constant value for the baseline.

There are different ways of determining the slope and intercept of the baselines. One method is as follows. A guess is made on the basis of experience as to which data points represent a baseline, and a linear least-squares fitting procedure is used to find the slope and intercept of the best fit line to these points. The slope and intercept define the baseline. The same procedure is carried out to find the other baseline. One then uses automated non-linear least-squares regression to fit an assumed model to the remaining data points. The fitting procedure adjusts the values of the parameters to optimize the fit of the model to the data, and this determines the energetics of the transition. For instance, if the object of study is protein stability and the independent variable is urea concentration, as in Fig. 6.10, one might use $\Delta G = \Delta G° - m[\text{urea}]$ to model the relationship between denaturant concentration and free energy difference between the folded and unfolded states of the protein. One then plugs the free energy formula into the expression for K in the expression for P in Eqn. 6.23 and fits the model to the data by varying the values of the adjustable parameters, in this case $\Delta G°$ and m. (An **adjustable parameter** is not the same as a variable! A variable represents a condition of the experiment, e.g. T or [urea]. It is called a variable because its value can be set a desired value. A parameter is an entity whose value is fixed by fitting a model to experimental data that were obtained at known values of the relevant variables.)

Once the non-linear least-squares procedure has yielded sensible values of $\Delta G°$ and m, one might consider making the baseline slopes and intercepts adjustable parameters as well. Doing this, however, increases the number of adjustable parameters to be determined *simultaneously*; they are now $\Delta G°$, m, *slope1*, *intercept1*, *slope2* and *intercept2*. Of course, this is the same number of adjustable parameters we've had to determine from the beginning, but now we're making no assumption about which data points represent the baseline, though we are still assuming that the baselines are sufficiently linear to be modelled in this way. This approach to analysis can be taken even further. For instance, if one of the baselines is not very straight, one could substitute a second-order polynomial function for the baseline in place of the straight line (first-order polynomial). What if temperature is an independent variable? Because ΔC_p of protein unfolding is large, ΔH_d and ΔS_d are sensitive functions of temperature, and there are not two but three adjustable parameters, unless there are grounds on which parameters can be fixed or ignored. And if a two-state model seems inadequate, because there are good reasons to believe that more than two states are populated, even more fitting parameters must be included. And so on.

But hold on! It is rightly said that given enough adjustable parameters it would be possible fit an elephant. For increasing the number of parameters will *always* improve the appearance of the fit and certain quantitative gauges of its quality. One should therefore never forget that *received convention* might not support the determination of so many adjustable parameters, even if there are reasons why they some of them might be included. One should be particularly cautious if the number of data points is not very large or the data are especially noisy. Such conventions are ignored at the risk of one's credibility. Expertise in data analysis, however, is no substitute for creativity or native ability in science. Entire books have been written on data analysis, so enough about that here!

F. | Multistate equilibria

Systems of three or more states do and do not differ in kind from two-state systems. A two-state transition is a first-order phase transition. If three states are present, the order-disorder transition as a whole is not all-or-none, but it might still be possible to model the multistate transition as a sum of two-state transitions (Fig. 6.11). For instance, suppose one has a protein molecule in which several α-helices but no β-sheets are present in the folded state. In the first approximation (see below), because the helices in proteins tend to be short and very unstable in the absence of the rest of the protein, the folding/unfolding transition of each helix can be modeled as a two state transition: each helix is either completely folded or completely unfolded. In a four-helix bundle there are four helices, and each can be in one of two states, so the total number of states is $2 \times 2 \times 2 \times 2 = 2^4 = 16$. Some (often all) of the partly folded forms of the protein, however, might be so unstable at equilibrium as to be but negligibly population and below the level of detection. On the

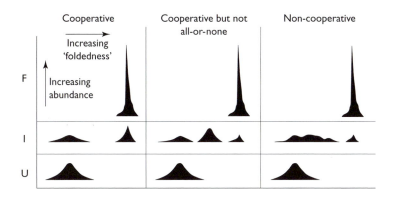

| Cooperative | Cooperative but not all-or-none | Non-cooperative |

Increasing 'foldedness'

F Increasing abundance

I

U

Fig. 6.11 Cooperative *versus* non-cooperative folding/unfolding. F, folded; I, intermediate; U, unfolded. When folding/unfolding is highly cooperative, only two states are present at equilibrium. Each of these states is a distribution of conformations, but the distributions are relatively sharply peaked. Moreover, there is a large gap between the two distributions. Cooperative denaturation need not involve just two states, but the distributions of the various states must be distinct. In a non-cooperative transition there is no clear distinction between states. The folded state is likely to be unambiguous, but there may be a continuum of denatured states.

other hand, a multistate transition might involve a continuum of partly folded conformations (Fig. 6.11), and in practice it could be very difficult to distinguish between these two types of multistate equilibrium.

Even when there are more than two states present at equilibrium, it can be difficult to provide sufficient evidence of this. It is much harder, however, to say just how many states are actually present. In part this is because thermodynamic data do not unambiguously establish a reaction mechanism. The first and most important step in dealing with a multistate transition is to show that at least one property of system simply cannot be explained on a two-state model. For instance in optical spectroscopy, a necessary but not a sufficient condition for a two-state behavior would be an **isosbestic point**, a wavelength at which a family of curves has the same value of the observable quantity. Each curve of the family might correspond to a different concentration of denaturant at a fixed temperature.

Thermodynamic techniques are very useful for assessing cooperativity. In a previous chapter, we said that one criterion of two-state behavior in the absence of oligomerization and aggregation was the equivalence of the van't Hoff enthalpy and the calorimetric enthalpy. The van't Hoff enthalpy change for a process can be determined from the data of any technique that allows one to follow the relative concentrations of reactant and product; the van't Hoff enthalpy change is a measure of the rate of change of the ratio of the concentration of production to the concentration of reactant. The van't Hoff enthalpy is calculated on the assumption that two and only two states are substantially populated throughout the process. In contrast the calorimetric enthalpy change is a measure of the heat absorbed during a reaction. This quantity does not depend at all on the model one might try to use to rationalize the data (excluding baseline analysis). Thus, under normal circumstances, comparison of the van't Hoff enthalpy and the calorimetric enthalpy provides a criterion for assessing the number of states populated during a process.

There are many specific examples one could present in a discussion of multi-state behavior. To simplify matters, we shall focus on a very well-known and well-studied case: α-lactalbumin and its relationship to hen lysozyme. We have already said a good deal about lysozyme, and this should make the present discussion seem less like *terra incognita*.

Fig. 6.12 Populations of states.
Depending on the experimental
approach, it might be possible to
make independent measurements of
the populations of the folded state
and the unfolded state under the
same conditions. This can be done,
for example, using circular dichroism
spectroscopy. Ellipticity is measured
at two wavelengths, one in the 'near
UV' (3° structure) and one in the 'far
UV' (2° structure). If the curves are
not coincident, as shown here, at
least one partly folded state must be
present. Note that the population of
the partly folded state first increases
and then decreases with increasing
denaturant concentration. At low
concentrations of denaturant, the
partly folded state is less stable than
the folded state, and at high
concentrations of denaturant, the
partly folded state is less stable than
the unfolded state. Somewhere
between these extremes, the
population of partly folded state will
be maximal. (Compare Fig. 6.15.)

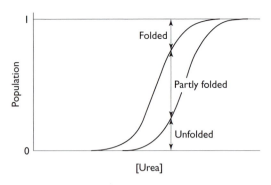

More importantly, lysozyme and α-lactalbumin have *very similar* three-dimensional structures in the folded state. Both proteins have a relatively large helical 'domain' comprising the same number of α-helices, and a relatively small 'domain' where several β-strands are found. There are four disulfide bonds in both proteins at the same places. It is very likely that the lysozyme gene and the α-lactalbumin gene descend from the same proto-gene. So many similarities make the differences that much more interesting. As discussed in Chapter 2, hen lysozyme shows cooperative folding/unfolding under wide range of conditions. In contrast, α-lactalbumin displays marked deviations from two-state behavior. This was first studied systematically in the 1970s.

If the two proteins are so similar, *why* do they display such different properties? The amino acid sequences are less than 50% identical. The percentage identity of amino acids, however, is not likely to be the main cause of the differences in folding characteristics. Instead, α-lactalbumin but not lysozyme binds a divalent metal ion with high affinity. Because selective binding to the folded state stabilizes the folded conformation relative to all other accessible states (Chapter 7), removal of the cation by EDTA reduces the stability of the folded state, probably because the aspartate side chains in the binding pocket repel each by electrostatic interactions. When the folded state is destabilized, the relative stability of partly folded states of α-lactalbumin is increased to a detectable level. An equilibrium partly folded state of a protein is called a **molten globule** when fixed tertiary structure (i.e. specific side chain interactions) is absent but elements of secondary structure are present and the protein compact. (The four major levels of protein structure – primary, secondary, tertiary, and quaternary – were first described by Linderstrøm-Lang in 1952, several years before the first protein structure was visualized at atomic resolution.) The molten globule state might correspond to a general partly ordered conformation on the protein folding pathway.

Supposing that three states are present at equilibrium, the observable quantity can be expressed in terms of the populations of states as

$$<O> = P_A o_A + P_B o_B + P_C o_C$$

$$= \frac{1}{1 + K_B + K_C} o_A + \frac{K_B}{1 + K_B + K_C} o_B + \frac{K_C}{1 + K_B + K_C} o_C \qquad (6.26)$$

where the variables have the same meaning as before. For instance, K_B measures the free energy difference between the intermediate state and

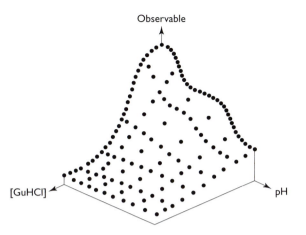

Observable

[GuHCl]

pH

Fig. 6.13 Stability surface. The figure shows how an observable quantity like fluorescence or ellipticity or chemical shift or heat capacity might depend on two independent variables, in this case pH and urea. There are at least three accessible states, as one can see from the pH axis. The set of ordered triples (pH, urea, observable) define a surface. Further analysis would take into account baseline values, as in Fig. 6.10, and enable determination of the free energy difference between states under any combination of pH and urea at the temperature of the experiments. A plot of ΔG against two independent variables is called a stability surface. (Compare Fig. 5.27.)

the folded state at the temperature of the experiment. Fitting such a model to chemical denaturant data would require a minimum four adjustable parameters (if the baselines are known): ΔG_A, ΔG_B, m_A, and m_B. Another approach to the measurement of the populations would be to plot the values of specific probes of the folded and unfolded states as a function of denaturant concentration. Circular dichroism spectroscopy, for instance, is commonly used to monitor the unfolding of proteins. As in Fig. 6.12 ellipticity in the 'far UV' (200–240 nm) can be used as a probe of secondary structure content, while ellipticity in the 'near UV' (260–310 nm) provides information on tertiary structure. When the change in ellipticity with independent variable in the far UV coincides with that in the near UV, unfolding is cooperative. If the curves are not coincident, unfolding is not all-or-one (Fig. 6.12). The difference between the probes of the folded and unfolded states measures the fraction of molecules in intermediate states. Figure 6.13 shows the value of an observable quantity for two different independent variables, pH and [urea]. Note that unfolding involves three states one axis and two along the other; the cooperativity of the unfolding transition can depend on the medium in which unfolding occurs.

Let's look at a qualitatively different example: a protein with multiple subunits. Suppose there are four identical subunits, and let each one have two accessible states, x and y. The enthalpy of any one subunit is H_x or H_y. There is only one way in which all subunits can be in enthalpy state H_x, and the enthalpy of the reference state, H_0, is $4H_x$. There are four possible ways in which just one subunit can have enthalpy state H_y ($H_1 = 3H_x + H_y$), six possible ways of having two subunits in enthalpy state H_y ($H_2 = 2H_x + 2H_y$), four possible ways of three subunits being in state H_y ($H_3 = H_x + 3H_y$), and only one way of having all four subunits in state H_y ($H_4 = 4H_y$). The total number of states is $1 + 4 + 6 + 4 + 1 = 16$, but the enthalpic properties of some cannot be distinguished from the others. In the context of thermodynamic measurements, the number of *experimentally distinguishable* enthalpy states is 5, and they are H_0, H_1, H_2, H_3, and H_4. The degeneracy of each enthalpy state is given by Pascal's triangle, from probability theory (Fig. 6.14). If there were three identical subunits, each in one of two

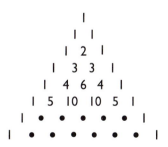

Fig. 6.14 Pascal's triangle. This triangular array of binomial coefficients (we saw the binomial theorem in Chapter 4) was taught as early as the thirteenth century by Persian philosopher Naṣir ad-Dīn al-Ṭusi. The triangle's discovery several centuries later by Pascal was apparently independent of earlier work.

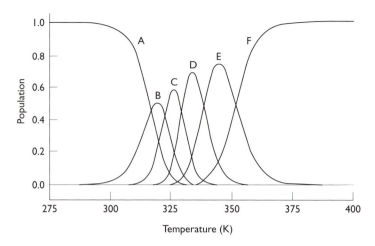

Fig. 6.15 Population of states as a function of temperature. The population of an accessible state is never zero. Depending on conditions, however, the population of some states could be practically negligible. For example, whenever states A, B and C of tRNA^{Phe} are stable (populated), the population of state F is small. Despite the jump in energy between states, the populations are smooth functions of temperature, as long as the number of molecules in the system is so large that fluctuations are extremely small.

possible states, then according to Pascal's triangle there would be a total of four enthalpy states, and the degeneracies would be 1, 3, 3 and 1.

By Eqn. 6.17 and the definition of the partition function,

$$Q = \sum_{j=1}^{16} \exp\left(\frac{-(\varepsilon_j - \varepsilon_1)}{k_B T}\right) \tag{6.27}$$

The subscript runs from 1 to 16 because there are 16 states in our multi-subunit protein, as we said above. When this equation is rewritten in terms of the five enthalpy states, we have

$$Q = \sum_{i=0}^{4} \omega_i \exp\left(\frac{-(H_i - H_0)}{RT}\right) = \sum_{i=0}^{4} \exp\left(\frac{-(G_i - G_0)}{RT}\right) = \sum_{i=0}^{4} \exp\left(\frac{-\Delta G_i}{RT}\right) \tag{6.28}$$

As always, the partition function is the sum of the Boltzmann factors. One need not be concerned that the index has changed on going from Eqn. 6.27 to Eqn. 6.28. That's because the index is just a dummy variable, a handy device that helps to distinguish one state from another.

And now for the final example of this section. The thermal unfolding of tRNA^{Phe} has been studied by scanning calorimetry. This instrument measures **the heat capacity function**, the heat absorbed as a function of temperature. Analysis of the heat capacity function (next section) has suggested that there are six states populated in the course of thermal denaturation of the macromolecule. A plot of the population of states as a function of temperature is given in Fig. 6.15. In the six-state case, Eqn. 6.26 is written as

$$<O> = \sum_{i=1}^{6} P_i o_i \tag{6.29}$$

where the sum is over all the accessible states. If our experimental observable is enthalpy, Eqn. 6.29 has the following appearance

$$<\Delta H> = \sum_{i=1}^{6} P_i \Delta H_i$$

$$= \sum_{i=1}^{6} \frac{\exp(-\Delta G_i/RT)}{Q} \Delta H_i \tag{6.30}$$

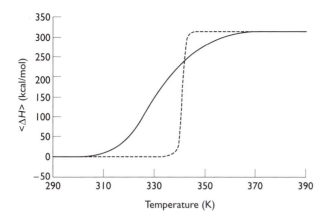

$<\Delta H>$ is called the average enthalpy. A plot of Eqn. 6.30 as a function of temperature for the populations of states given in Fig. 6.15 is depicted in Fig. 6.16. The way in which $<\Delta H>$ changes with temperature is called the heat capacity function.

G. | Protein heat capacity functions

Recall from Chapter 2 that $\Delta C_p = \Delta\Delta H/\Delta T$. When ΔH is our observable quantity – when we measure the average heat absorbed or evolved at a given temperature ($<\Delta H>$) – and if we make the measurements as a function of temperature, we measure $\Delta<\Delta H>/\Delta T = <\Delta C_p>$. This is exactly what a scanning calorimeter does. The rest of the name of instrument, 'differential,' refers to how the measurement is made (Fig. 2.10A, Appendix B). Being punctilious about it, a *differential* scanning calorimeter measures the *partial* heat capacity of a sample. The heat capacity of the solution of macromolecules is measured relative to the heat capacity of pure buffer, so only *part* of what could be measured (the *difference* between sample and reference) is actually measured. Calorimetry is the only means by which one can make a direct measurement of the heat absorbed or evolved in a process. What follows is a qualitative description of how to rationalize scanning calorimetry data.

In principle, it does not matter whether the sample is a protein, tRNA, a protein–DNA complex, a protein–lipid complex, or whatever. The approach to analysis is basically the same in every case. After accounting for buffer baseline effects, which can be assessed by measuring the heat capacity of the buffer with respect to itself, one analyzes DSC data using the temperature derivative of Eqn. 6.30:

$$<\Delta C_p> = \sum_i \frac{\Delta H_i^2 K_i}{RT^2(1 + K_i)^2} + \sum_i \Delta C_{p,i}\frac{K_i}{1 + K_i} \tag{6.31}$$

This relationship can be derived from Eqn. 6.30 using differential calculus. The sums are over all states. The left-hand side of Eqn. 6.31 is the observable in a DSC experiment. The measurement is interpreted in terms of the quantities on the right-hand side. The first of these is a 'bell-shaped' heat absorption peak (see Fig. 2.10B). The second term gives the change in 'baseline' heat capacity resulting from the order–disorder transition. It the context of proteins, the second term measures the increase in exposure to solvent of hydrophobic residues as the protein is denatured. As discussed in Chapter 2, ΔC_p can be large for large proteins, as they have a large solvent-inaccessible surface in the folded state and a small solvent-inaccessible surface in the unfolded state. For nucleic acids this terms is generally small, because the heat capacity difference between the folded and unfolded forms of DNA and RNA is small.

When only two states are present, Eqn. 6.31 reduces to

$$<\Delta C_p> = \frac{\Delta H^2 K}{RT^2(1+K)^2} + \Delta C_p \frac{K}{1+K} \tag{6.32}$$

As before, the first term represents the heat absorbed during the transition, the second one the shift in baseline heat capacity (Fig. 2.10B). Ignoring the baseline shift, the area below the curve is the 'calorimetric enthalpy.' We have supposed that each macromolecule of the sample has two accessible states. The sharp heat-absorption peak occurs when the two states are equally populated and transitions between them – i.e. fluctuations – are a maximum. There is no free energy barrier between the states at the transition temperature.

The probability that any given molecule will be in the unfolded state is just the statistical weight of that state divided by the partition function, or $K/(1+K)$. The second term on the right-hand side thus makes good intuitive sense. When the population of the unfolded state is small, the contribution of this term to the change in baseline is small, practically zero. When the population of the unfolded state is large, the contribution of the term approaches ΔC_p, the heat capacity difference between the folded and unfolded states.

Scanning calorimetry is useful for studying the thermodynamics of order-disorder transitions in macromolecules when the enthalpy change between states is sufficiently large. DSC is none the less a blunt instrument, as it provides no direct or detailed information about the conformations adopted by the protein molecule, even if statistical mechanics is used to analyze the calorimetry data. The molecular interpretation of a DSC signal often depends on other techniques, e.g. fluorescence, CD, NMR, etc. One would ideally use a combination of techniques to characterize a system.

H. | Cooperative transitions

Suppose we have a dimeric protein in which both subunits unfold simultaneously by heat denaturation. Will the van't Hoff enthalpy change for denaturation be the same as the calorimetric enthalpy change, i.e. the heat absorbed in the course of the transition? No! Why not? The van't Hoff enthalpy measures the rate of change with temperature of the population of the folded dimeric state with respect to the unfolded monomeric state; the calorimetric enthalpy measures the heat absorbed on unfolding subunits. (Note that we are neglecting interaction terms between sub-

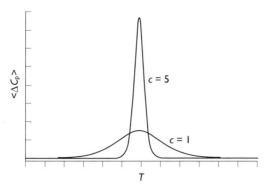

Fig. 6.17 Cooperativity and the heat capacity function. Heat capacity functions are shown for $c = 1$ and $c = 5$. The calorimetric enthalpy and melting temperature are the same in both cases.

units; these will be dealt with in the next section.) Each time a dimer unfolds, the population of folded state decreases by one while the population of unfolded state increases by two. The heat absorbed during thermal denaturation is the same whether the protein is one large monomer or a dimer of two half-sized monomers.

We can define cooperativity, c, as the ratio of the van't Hoff enthalpy change to the calorimetric enthalpy change:

$$c = \Delta H_{vH}/\Delta H_{cal} \tag{6.33}$$

In the case of a *monomeric* protein that shows cooperative unfolding, $\Delta H_{vH} = \Delta H_{cal}$ and $c = 1$; there is but one cooperative unit. A monomeric protein that exhibits multistate unfolding has $c < 1$. Systems composed of a very large number of identical subunits, for example polynucleotides and phospholipid membranes, have $10 < c < 500$.

To see this more clearly; and to give the discussion an experimental context, suppose we have an oligomeric protein that exhibits the following equilibrium:

$$c[U] \Leftrightarrow [F]_c \tag{6.34}$$

The folded state comprises c identical copies of a polypeptide. The equilibrium constant for this reaction is

$$K = \exp(-c\Delta G/RT) \tag{6.35}$$

Note the factor of c in the argument of the exponential function. ΔG is the free energy change of folding one subunit; $c\Delta G$ is the energy change of folding c subunits. We say 'folding' and not 'unfolding' here because of the direction of Eqn. 6.34. The average enthalpy is

$$<\Delta H> = \Delta H P_F = K/(1 + K) \tag{6.36}$$

where ΔH is the enthalpy difference between the folded and unfolded states of a subunit and P_F is the population of the folded state. It can be shown using Eqn. 6.36, Eqn. 6.33 and a little calculus that

$$<\Delta C_p> = \frac{K}{(1 + K)^2} \frac{c\Delta H^2}{RT^2} = \frac{K}{(1 + K)^2} \frac{\Delta H_{vH}\Delta H_{cal}}{RT^2} \tag{6.37}$$

The effect on $<\Delta C_p>$ of changing c is illustrated in Fig. 6.17. As the cooperativity increases, the curve becomes more and more sharply peaked, though the area below the curve (ΔH_{cal}) remains the same.

At the midpoint of the transition, where $\Delta G = 0$ and $K = 1$,

$$<\Delta C_p>_{T_m} = \frac{\Delta H_{vH} \Delta H_{cal}}{4RT_m^2} \qquad (6.38)$$

It can be shown that when the transition occurs far from absolute zero, the temperature at which the heat absorption peak has a maximum is practically indistinguishable from T_m. The only unknown in Eqn. 6.38 is ΔH_{vH}. The van't Hoff enthalpy can therefore be calculated from the result of a DSC experiment. The cooperativity of the transition is assessed by comparing ΔH_{vH} to ΔH_{cal}, as in Eqn. 6.33.

I. | 'Interaction' free energy

What other considerations are there if our protein has more than one domain or subunit? There are numerous examples of multidomain proteins and multisubunit proteins in nature so it will be worth our while to think about them a little while longer. For example, the extracellular matrix protein fibronectin consists of a number of repeats of a small unit called a fibronectin domain, and each domain comprises about 100 amino acids and is globular. The domains are joined together like beads on a string. In some cases, but not in all, the folded states of the individual domains are stable at room temperature in the absence of the rest of the protein. This is possible only if the stability of a domain does not depend substantially on interactions with other parts of the protein. We have already encountered the multisubunit protein hemoglobin and shall study it in greater depth in the next chapter. Whether the interactions are between domains or subunits, the interaction would not occur unless it was thermodynamically favorable. In this section we wish to study a way of modeling domain interactions.

Now, suppose we have a two-domain protein, for example yeast phosphoglycerate kinase (PGK). Our object of study need not be a protein, but it will simplify things in accounting for all the contributions to the thermodynamics to let the system be a single covalent structure. (Why?) The separate domains of phosphoglycerate kinase are not very stable; they are denatured at *room temperature*, even when combined in a 1:1 stoichiometric ratio. This means that the combined thermodynamics of

Table 6.1	Summary of the energetics of two-domain protein	
State	Free energy change	Boltzmann factor
$A_F B_F$	0, reference state	1
$A_U B_F$	$\Delta G_A + \Delta g_B$	$K_A \Phi_A$
$A_F B_U$	$\Delta G_B + \Delta g_A$	$K_B \Phi_B$
$A_U B_U$	$\Delta G_A + \Delta G_B + \Delta g_{AB}$	$K_A K_B \Phi_{AB}$

Notes:
A sum of free energy terms translates into a product of Boltzmann factors by way of a basic property of exponential functions: $e^{(x+y)} = e^x e^y$, where $x = -\Delta G/RT$

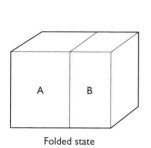

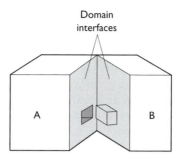

Domain
interfaces

Folded state

Fig. 6.18 Domain interface of a two-domain protein. In the folded state, interacting residues in the domain interface are inaccessible to the solvent. There is a free energy cost to exposing the domain interface to solvent, even if the domains themselves remain folded. The domain interface consists of two geometrically and electrically complementary surfaces.

the continuous polypeptide chain and the domain–domain interactions is responsible for the stability of the intact protein.

We can model this situation as follows. Let the two domains be called A and B. The free energy difference between the folded and unfolded states *of domain A alone* is ΔG_A. This energy term does *not* include contributions from *interactions* with domain B (Fig. 6.18). The free energy change of exposing the surface of domain B that is solvent inaccessible when domain A is present, is Δg_B. And so on. The states, free energy changes and Boltzmann factors are summarized in Table 6.1. Note that in general, $\Delta g_A \neq \Delta g_B$, though in practice the values might be very similar. $\Delta g_{AB} = \Delta g_A + \Delta g_B$. Table 6.1 is perfectly general, for if there is no interaction between domains, $\Delta g_A = \Delta g_B = 0$, and the corresponding Boltzmann factors are simply equal to one.

In the presence of the chemical denaturant guanidine hydrochloride, PGK unfolds reversibly on heating or cooling; it exhibits cold denaturation. The character of the transition, however, is not the same in both cases. In heat unfolding, the transition is cooperative; both domains unfold simultaneously; the individual domains are apparently too unstable to be found in appreciable abundance. In contrast, in the low temperature transition unfolding is not cooperative; the domains denature independently. This suggests that the cooperativity of the folding/unfolding transition depends on the thermodynamic properties of the domain interface. A similar phenomenon is observed for some multisubunit proteins when they dissociate into individual subunits at temperatures around $0\,°C$.

One approach to rationalizing the thermodynamic data is to examine available structural information, to see if it might provide clues to the molecular origin of the macroscopic behavior. Analysis of the crystal structure of PGK reveals that each domain has a certain amount of solvent inaccessible surface; this surface becomes solvated when the protein unfolds. On heat denaturation, both domains unfold simultaneously, and one assumes that if domain A has, say, 65% of the total buried surface, it accounts for *c.* 65% of the heat absorbed and *c.* 65% of the heat capacity change. The same sort of reasoning can be applied to estimating the enthalpy change of unfolding of domain B and both domain interfaces at the transition temperature.

Further analysis of the PGK structure shows that each domain contributes to the interface somewhat less than 500 $Å^2$ of hydrophobic surface and nine hydrogen bonds. This information can be coupled with the thermodynamics of solvation of small model compounds and used to evaluate Δg_A and Δg_B. The missing data are the reference values of Δs_A and Δs_B, and there is unfortunately no reliable way of predicting them by crystal structure analysis. To get round this obstacle, one can simply substitute in different values, simulate the heat capacity of the entire two-domain protein as a function of temperature using Eqn. 6.31, and compare the results of the simulation with the observed experimental properties. When the simulation matches the observed behavior, one can assume that the values of the thermodynamic quantities used in the simulation might be close to what one would measure if experiments to measure the thermodynamics of exposing the domain interface to solvent could actually be carried out.

J. | Helix–coil transition theory

The α-helix is a very basic structural motif in proteins; most known folded structures of proteins contain at least one α-helix (Fig. 6.19). It is generally believed that the folding pathway of a helix-containing protein will involve helix formation as an early event. On this view, the coil–helix transition plays a key role in the mechanism of folding. A qualitative description of helix–coil theory is included here for two reasons: it is of basic interest to people studying protein folding and it illustrates yet another way in which statistical mechanics can be used in biological thermodynamics research. Theoretical descriptions of the other major type of secondary structure element, the β–sheet, are not as advanced as for the α-helix and will not be discussed in detail here.

Helical peptides show thermally induced unfolding. This probably indicates that helix formation is enthalpically driven and that the sign of the heat capacity difference between the coil and helix states is positive, as in protein denaturation. ΔC_p for the helix–coil transition, though unknown, is likely to be very small. In any case, the enthalpy change on helix formation will have two major components: (1) the difference in enthalpy between (a) hydrogen bonds formed between water and polypeptide backbone donors (amide hydrogen atoms) or acceptors (carbonyl oxygen atoms) and (b) hydrogen bonds formed between backbone donors and acceptors themselves; and (2) the enthalpic effect of changes in the solvation of other parts of the peptide. It is possible but unlikely that formation of a hydrogen bond in an α-helix results in the same change in enthalpy as formation of a hydrogen bond between two water molecules. Irrespective of the precise partitioning of the enthalpy into its various components, experiments suggest that ΔH for helix formation is ~ 1 kcal mol^{-1} residue^{-1}. Hydrogen bond strength is likely to depend on the donor and acceptor involved, as well as the distance between donor and acceptor and the electronic properties of the medium in which the bond forms. But here we ignore all such details.

In the **Zimm–Bragg model** (B. H. Zimm and J. K. Bragg, 1959), the helix formation process is generally considered to involve two-steps:

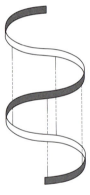

Fig. 6.19 The α-helix. Helices in proteins are right-handed. That is, the thumb of one's right hand points from N to C if the fingers of the right hand curl around the helix in the direction of N to C. This very low resolution diagram is intended to show the helical structure of this element of secondary structure and how it involves the hydrogen bonds that stabilize it. The hydrogen bonds are formed between N–H groups and C = O groups that are four residues back along the polypeptide chain. There are 3.6 peptide units per turn in the α-helix. In other words, each peptide unit corresponds to a rotation about the helix axis of 100°.

nucleation and **propagation**. Nucleation is the process of forming one hydrogen bond characteristic of an α-helix between two amino acid residues when no other such bonds are present. Propagation of the helix, or elongation, depends entirely on nucleation already having occurred. Although nucleation can occur randomly at multiple locations along the polypeptide chain, each nucleation event is relatively improbable; nucleation is energetically unfavorable. This is because it involves a substantial decrease in entropy, the cost of fixing the orientation of residues in space so that the first helix hydrogen bond can be formed. Once a helix has been nucleated, however, the additional decrease in entropy on fixing the geometry of the polypeptide chain is more than offset by the energetically favorable formation of hydrogen bonds.

The thermostability of existing helical structure is explained as follows. After initiation, a helix would extend indefinitely and encompass all amino acids in a polypeptide chain, if it were not entropically favorable for there to be several helical regions in a polypeptide instead of one. In view of this, the helical content of a homopolypeptide depends primarily on its length: nucleation is unfavorable, so more than one nucleation site in a short polypeptide is improbable. Given an ensemble of identical polypeptides, the specific location of helices and helix content will vary from molecule to molecule. This is because from an entropic point of view it is favorable for molecules with the same percentage helix to differ in one way or another. It follows that one cannot use a two-state model to describe the helix–coil transition, unless the polypeptides involved are as short as the helices found in proteins.

Measurement of helix content of an ensemble of identical polypeptides estimates not the percentage helix in each polypeptide, but the average helix content in all the polypeptides in the ensemble. In other words, there is a distribution of helix content in a collection of identical polypeptides. In addition to this, the stabilization of helical structure will depend on the character of the amino acids involved, as some amino acids have a higher **helix propensity** than others. The helix-propensity or helix-forming tendency of an amino acid type can be rationalized in terms of its structure (see below).

In the Zimm–Bragg theory of the helix–coil transition, two parameters are used to give a quantitative account for the helical content of polypeptides. These are σ, the helix nucleation parameter, and s, the helix propagation parameter. σ and s are called parameters and not variables because their values must be determined by optimizing the 'fit' of experimental data to the model. The cooperativity of the helix–coil transition depends on the value of σ. When s is large (~ 1), cooperativity is low; when s is small ($\sim 10^{-4}$), cooperativity is high. For real peptides, $\sigma \sim 10^{-3}$ and $s \sim 1$.

The first helical residue (i) in a Zimm–Bragg helix has a Boltzmann factor of $\sigma s = \exp(-\Delta G_{\text{initiation}}/RT)$. That of the second helical residue ($i + 1$) is $s = \exp(-\Delta G_{\text{propagation}}/RT)$. The Boltzmann factor is 1 for either a coil residue following a helical one and a coil residue following a coil one. Note that because the Boltzmann factors refer to α-helical hydrogen bonds formed between peptide groups, σ is a property of more than one

Table 6.2 | Helix propensity scale

Amino acid residue	Relative contribution to stability of α-helical conformation (kcal mol^{-1})
Ala	−0.77
Arg	−0.68
Lys	−0.65
Leu	−0.62
Met	−0.50
Trp	−0.45
Phe	−0.41
Ser	−0.35
Gln	−0.33
Glu	−0.27
Cys	−0.23
Ile	−0.23
Tyr	−0.17
Asp	−0.15
Val	−0.14
Thr	−0.11
Asn	−0.07
His	−0.06
Gly	0
Pro	≈3

Source: Data from O'Neil & DeGrado (1990) *Science*, **250**:646–651. Compare these data with the table in Appendix C.A.

peptide group and several residues. The equilibrium constant between the helical state (H_n) and coil state (C) for n peptide groups is $K_n = [H_n]/[C] = \sigma s^n$. In the first approximation, neither σ nor s depends on residue type, excluding proline and glycine, which have very different polypeptide backbone characteristics from the other amino acid residues. This can easily be seen by comparing the Ramachandran plots for the different amino acids.

Another approach to analysis of the helix-coil transition is that given by Lifson and Roig (1961). In the **Lifson–Roig model**, the Boltzmann factors correspond to amino acid residues, not peptide groups. This facilitates accounting for effects at the ends of helices, which do not contribute to helix stability to the same extent as residues in the middle of a helix. The ends of helices in helical peptides are 'frayed,' and the percentage of time that helical hydrogen bonds are formed at the end of a helix is lower than in the middle of a helix. In the Lifson–Roig model, the equilibrium constant of the conformation ccccchhhhhccchhcc is *uuuuuvwwwwvuuuvvuu*. The coil–helix junction (ch) is represented by a Boltzmann factor of *v*, hh by *w*, and cc by *u*.

The propensity of a given amino acid type to form helix structure has been measured by so-called host-guest experiments. A water-soluble polypeptide serves as the host, and amino acid replacement is carried out at a specific site in the polypeptide chain. One can generate a set of

20 different peptides by chemical synthesis, measure helix content, and rank the various amino acids according to their ability to promote helix formation. Helix content is usually assessed by a technique called circular dichroism spectroscopy, which measures the difference in absorption of right- and left-circularly polarized light. The technique is particularly sensitive to helical structure in the far UV.

A typical helix propensity scale is shown in Table 6.2. The experimental results have been normalized to the result for Gly. This amino acid has no chiral center, and because it has no side chain, the bonds of the polypeptide backbone on either side of the alpha carbon are able to rotate freely. There is therefore a large entropic cost to placing severe restrictions on the motion of these atoms, as would be necessary to form an α helix. In contrast, the next most complicated amino acid, Ala, stabilizes helical structure more than any other amino acid type. Why is that? Ala has a very small side chain, just a methyl group. Formation of helix from the coil state results in a relatively small decrease in the motions of the Ala side chain (rotations of the methyl group about the C_α–C_β bond and of the C_β–H bonds; Ala has no γ substituent); there is a smaller decrease in the entropy of the side chain of Ala than for other amino acid residues. This is the physical basis of the high helix propensity of Ala. Pro has such a low helix propensity because of the restrictions it places on polypeptide backbone geometry.

A question we should be asking ourselves is whether helical peptides tell us anything at all about helices in proteins or protein folding. Another approach to sorting out the role of individual amino acids in stabilizing α-helix structure is to make mutations at a solvent-exposed site in the middle of a helix in a well-studied protein. In most cases, mutations of this sort cause no significant distortion of helix geometry relative to the wild-type protein, effectively ruling out one possible origin of helix stability. Instead, the effect of a mutation on overall protein stability correlates with the change in the difference in hydrophobic surface area exposed to solvent of the mutated residue in the folded state of the protein (crystal structure) and in the unfolded state (fully extended chain). In symbols, $\Delta T_m = T_{m,\ mutant} - T_{m,\ wild\text{-}type} \propto \Delta(A_{unfolded} - A_{folded})$, where A is solvent-exposed hydrophobic surface area. In general, there is relatively good agreement between various helix propensity scales. Differences exist; however, these are likely to be attributable more to structural changes that are propagated throughout the host than to experimental error. This tells us that analysis of the experimental thermodynamic properties of model helices will probably not be able to tell us substantially more about what stabilizes proteins than what is known already.

K. | References and further reading

Atkins, P. W. (1998). *Physical Chemistry*, 6th edn, cc. 19.1–19.6. Oxford: Oxford University Press.

Atkins, P. W. (1994). *The Second Law: Energy, Chaos, and Form*. New York: Scientific American.

Ben-Naim, A. (1991). The role of hydrogen bonds in protein folding and protein association. *Journal of Physical Chemistry*, **95**, 1437–1444.

Ben-Naim, A. (1992). *Statistical Thermodynamics for Chemists and Biologists*. New York: Plenum.

Bergethon, P. R. (1998). *The Physical Basis of Biochemistry: the Foundations of Molecular Biophysics*, ch. 3 & 13. New York: Springer-Verlag.

Beveridge, D. L. & Dicapua, F. M. (1989). Free energy via molecular simulation: application to chemical and bimolecular systems, *Annual Review of Biophysics and Biophysical Chemistry*, **18**, 431–492.

Britannica CD 98, 'Boltzmann Constant,' 'Dalton's Law,' 'Diffusion,' 'Principles of Thermodynamics,' and 'Statistical Mechanics.'

Chakrabartty, A. & Baldwin, R. L. (1995). Stability of α-helices. *Advances in Protein Chemistry*, **46**, 141–176.

Chakrabartty, A., Schellman, J. A. & Baldwin, R. L. (1991). Large differences in the helix propensities of alanine and glycine. *Nature*, **351**, 586–588.

Cooper, A., Eyles, S. J., Radford, S. E. & Dobson, C. M. (1991). Thermodynamic consequences of the removal of a disulfide bridge from hen lysozyme. *Journal of Molecular Biology*, **225**, 939–943.

Creighton, T. E. (1993). *Proteins: Structures and Molecular Properties*, 2nd edn, ch. 5.3.1. New York: W. H. Freeman.

Einstein, A. (1956). *Investigations on the Theory of the Brownian Movement*. New York: Dover.

Encyclopædia Britannica CD 98, 'Binomial Theorem.'

Fersht, A. R. (1999). *Structure and Mechanism in Protein Science: a Guide to Enzyme Catalysis and Protein Folding*. New York: W. H. Freeman.

Feynman, R. P., Leighton, R. B. & Sands, M. (1963). *Lectures on Physics*, vol I, cc. 1, 6, 40-1–40-4, 43-1, 43-2 & 43-5. Reading, Massachusetts: Addison-Wesley.

Freire, E. & Biltonen, R. I. (1978). Statistical mechanical deconvolution of thermal transitions in macromolecules. I. Theory and application to homogeneous systems, *Biopolymers*, **17**, 463–479.

Freire, E., Murphy, K. P., Sanchez-Ruiz, J. M., Galisteo, M. L. & Privalov, P. L. (1992). The molecular basis of cooperativity in protein folding. Thermodynamic dissection of interdomain interactions in phosphoglycerate kinase. *Biochemistry*, **31**, 250–256.

Fruton, J. S. (1999). *Proteins, Enzymes, Genes: the Interplay of Chemistry and Biology*. New Haven: Yale University Press.

Gasser, R. P. H. & Richards, W. G. (1995). *Introduction to Statistical Thermodynamics*. Singapore: World Scientific.

Gurney, R. W. (1949). *Introduction to Statistical Mechanics*. New York: McGraw-Hill.

Hamada, D., Kidokoro, S. I., Fukada, H., Takahashi, K. & Goto, Y. (1994). Salt-induced formation of the molten globule state of cytochrome c studied by isothermal titration calorimetry. *Proceedings of the National Academy of Sciences of the United States of America*, **91**, 10325–10329.

Haynie, D. T. (1993). *The Structural Thermodynamics of Protein Folding*, cc. 4 & 6. Ph.D. thesis, The Johns Hopkins University.

Haynie, D. T. & Freire, E. (1993). Structural energetics of the molten globule state. *Proteins: Structure, Function and Genetics*, **16**, 115–140.

Haynie, D. T. & Freire, E. (1994). Thermodynamic strategies for stabilizing intermediate states of proteins. *Biopolymers*, **34**, 261–272.

Haynie, D. T. & Freire, E. (1994). Estimation of the folding/unfolding energetics of marginally stable proteins using differential scanning calorimetry. *Analytical Biochemistry*, **216**, 33–41.

Hill, C. P., Johnston, N. L. & Cohen, R. E. (1993). Crystal-structure of a ubiquitin-dependent degradation substrate – a 3-disulfide form of lysozyme. *Proceedings of the National Academy of Sciences of the United States of America*, **90**, 4136–4140.

Hill, T. L. (1986). *An Introduction to Statistical Mechanics*. New York: Dover.

Kim, P. S. & Baldwin, R. L. (1990). Intermediates in the folding reactions of small proteins. *Annual Review of Biochemistry*, **59**, 631–660.

Kittel, C. & Kroemer, H. (1980). *Thermal Physics*, 2nd edn, ch. 1. San Francisco: W. H. Freeman.

Klotz, I. M. (1986). *Introduction to Biomolecular Energetics*, ch. 8. Orlando: Academic Press.

Kuwajima, K., Nitta, K., Yoneyama, M. & Sugai, S. (1976). Three-state denaturation of α-lactalbumin by guanidine hydrochloride. *Journal of Molecular Biology*, **106**, 359–373.

Kuwajima, K. (1989). The molten globule state as a clue for understanding the folding and cooperativity of globular-protein structure. *Proteins: Structure, Function and Genetics*, **6**, 87–103.

Lazarides, T., Archontis, G. & Karplus, M. (1995) Enthalpic contribution to protein stability: insights from atom-based calculations and statistical mechanics. *Advances in Protein Chemistry*, **47**, 231–306.

Lifson, S. & Roig, A. (1961) The theory of helix-coil transitions in polypeptides. *Journal of Chemical Physics*, **34**, 1963–1974.

Lumry, R., Biltonen, R. I. & Brandts, J. (1966). Validity of the 'two-state' hypothesis for conformational transitions of proteins. *Biopolymers*, **4**, 917–944.

McKenzie, H. & White, F. (1991). Lysozyme and α-lactalbumin: structure, function and interrelationships. *Advances in Protein Chemistry*, **41**, 174–315.

Minor, D. L., Jr & Kim, P. S. (1994) Measurement of the β-sheet-forming propensities of amino acids. *Nature*, **367**, 660–663.

Murphy, K. P. & Freire, E. (1992). Thermodynamics of structural stability and co-operative folding behavior in proteins. *Advances in Protein Chemistry*, **43**, 313–361.

Microsoft Encarta 96 Encyclopedia, 'Thermodynamics.'

Nelson, P. G. (1988). Derivation of the second law of thermodynamics from Boltzmann's distribution law. *Journal of Chemistry Education*, **65**, 390–392.

Nelson, P. G. (1994). Statistical mechanical interpretation of entropy. *Journal of Chemical Education*, **71**, 103–104.

O'Neil, K. T. & DeGrado, W. F. (1990). A thermodynamic scale for the helix-forming tendencies of the commonly occurring amino acids. *Science*, **250**, 646–651.

Ohgushi, M & Wada, A. (1983). 'Molten globule:' a compact form of globular proteins with mobile side chains. *Federation of European Biochemical Societies Letters*, **164**, 21–24.

Ogasahara, K., Matsushita, E. & Yutani, K. (1993). Further examination of the intermediate state in the denaturation of the tryptophan synthase subunit. Evidence that the equilibrium denaturation intermediate is a molten globule. *Journal of Molecular Biology*, **234**, 1197–1206.

Peusner, L. (1974). *Concepts in Bioenergetics*, ch. 4. Englewood Cliffs: Prentice-Hall.

Privalov, P. L. (1979). Stability of proteins: small globular proteins. *Advances in Protein Chemistry*, **33**, 167–239.

Privalov, P. L. & Gill, S. J. (1988). Stability of protein structure and hydrophobic interaction. *Advances in Protein Chemistry*, **39**, 191–234.

Ptitsyn, O. B. (1995). Molten globule and protein folding. *Advances in Protein Chemistry*, **47**, 83–229.

Richards, F. M. (1991). The protein folding problem. *Scientific American*, **264**, no. 1, 54–63.

Scholtz, J. M. & Baldwin, R. L. (1992). The mechanism of α-helix formation by peptides. *Annual Review of Biophysics and Biomolecular Structure*, **21**, 95–118.

Shortle, D. (1993). Denatured states of proteins and their roles in folding and stability. *Current Opinion in Structural Biology*, **3**, 66–74.

Tolman, R. C. (1938). *The Principles of Statistical Mechanics*. Oxford: Oxford University Press.

van Holde, K. E. (1985). *Physical Biochemistry*, 2nd edn, cc 1.2, 1.3 & 3.4. Englewood Cliffs: Prentice-Hall, 1985.

Voet, D. & Voet, J. G. (1995). *Biochemistry*, 2nd edn, cc. 2-2, 3-2A & 3-2B. New York: John Wiley.

Williams, T. I. (ed.) (1969). *A Biographical Dictionary of Scientists*. London: Adam & Charles Black.

Wrigglesworth, J. (1997). *Energy and Life*, cc. 1.1 & 2.1. London: Taylor & Francis.

Zimm, B. H. & Bragg, J. K. (1959). Theory of the phase transition between helix and random coil in polypeptide chains, *Journal of Chemical Physics*, **31**, 526–535.

L. | Exercises

1. Explain in statistical mechanical terms why it is easier to remove a book from a specific place on a bookshelf than it is to put the book in a specific place on a bookshelf.

2. Suppose you two have glass bulbs of equal volume that are connected by a stopcock. Initially, one bulb contains N identical molecules of an inert gas, and the other bulb is evacuated. When the stopcock is open, there is an equal probability that a given molecule will occupy either bulb. How many equally probable ways are there that the N molecules can be distributed among the two bulbs? The gas molecules are, of course, indistinguishable from each other. How many different states of the system are there? Write down a formula for W_L, the number of (indistinguishable) ways of placing L of the N molecules in the left bulb. The probability of such a state occurring is its fraction of the total number of states. What is that probability? What is the value of W_L for the most probable state? Calculate $W_{N-2 \pm 1}/2^N$ for $N = 10^1, 10^3, 10^5, 10^{10}$ and 10^{23}. Explain the significance of W as a function of N.

3. Given a system with configuration $\{0, 1, 5, 0, 8, 0, 3, 2, 0, 1\}$, calculate Ω.

4. Given a system in which $N = 20$, give the configuration that maximizes Ω. Show how you arrived at your answer.

5. The 17th June 1999 issue of *Nature*, the international weekly journal of science, published a research on chimpanzee 'cultures.' There is a famous joke, attributed to renowned British physicist Sir James Hopwood Jeans (1877–1946), that goes something like this. A population of chimps typing randomly but diligently at a computer keyboard would eventually produce Shakespeare's greatest work,

'Hamlet.' Calculate how long, on the average, it would take 107 chimps to type the phrase 'to be or not to be, that is the question'? Assume that each chimp has a 45-key keyboard plus a space bar (assume no shift key). How long, on the average, would it take one chimp to type this phrase if its computer were programmed to shift to the right after each correct keyboard entry? Compare these numbers and comment on their possible significance for theories of the origin of life and chimp cultures.

6. The protein folding/unfolding example described in section D, 'Measurement,' involves changes in the urea concentration. Is the system open or closed? Why? Can the results be analyzed using equilibrium thermodynamics? Why or why not? Give a protocol for carrying out such an experiment that is independent of the instrument used and the observable monitored.

7. Polyacrylamide gel electrophoresis is used to separate proteins on the basis of size. Fig. 5.28 shows the pattern one finds when a monomeric protein that unfolds cooperatively in the presence of denaturants is studied by urea gradient polyacrylamide gel electrophoresis. The gel has been stained with a protein-sensitive dye. Rationalize the result in terms of urea concentration and migration behavior of proteins in a polyacrylamide gel. Assuming that there is no geometrical distortion in the sigmoidal curve, describe a method of determining the free energy of unfolding in the absence of denaturant and the urea concentration at the midpoint of the transition. Would you expect this method to give very accurate results in practice? Why or why not?

8. Explain how disulfide bonds stabilize the folded states of proteins.

9. Sickle-cell hemoglobin differs from normal wild-type hemoglobin by one amino acid change. This results in aggregation of hemoglobin molecules under certain conditions. Sickle-cell hemoglobin filaments formed at 37 °C can be disaggregated by cooling to 0 °C. Rationalize this behavior.

10. Write down the partition function for the three-state unfolding of a protein in a chemical denaturant.

11. Note that the phase transition shown in Fig. 2.10B is not 'infinitely' sharp. The transition occurs over a rage of temperatures, not at a single temperature, as one might expect for a pure substance. We have claimed, however, that this transition is essentially no different from what finds for the cooperative thermal denaturation of a crystal. What is the origin of the behavior shown in Fig. 2.10B?

12. Suppose you are using differential scanning calorimetry to study a two-domain protein. Suppose also that thermal denaturation shows a heat absorption peak for which $\Delta H_{vH}/\Delta H_{cal} = 1$. Does it follow necessarily that thermal denaturation is cooperative? Why or why not?

13. In Eqn. 6.6, why is the speed of the cyclist irrelevant to the number of collisions? Is the same true if the 'collisions' are with raindrops? Why?

14. At 310 K, $\Delta G°$ for ATP hydrolysis is -30.5 kJ mol^{-1}, and $\Delta H° = -20.1$ kJ mol^{-1}. Calculate $\Delta S°$ for this process and explain the result in molecular terms.

15. How many possible different tetrapeptides can be made using the 20 standard amino acids? How many possible different tetrapeptides could be made from a pool of twenty different amino acids? How many possible different tetrapeptides of the same composition could be made from a pool of twenty amino acids. How many possible different tetranucleotides are there?

16. Pascal's triangle. Consider Fig. 6.14. Give the next three rows of Pascal's triangle.

17. Given a two-domain protein with a binding site in one of the domains, enumerate and describe the various accessible states.

18. Devise a test for assessing whether or not the inclusion of additional fitting parameters actually improves the goodness-of-fit of a model to experimental data.

19. Given a three-state system, describe the population of the intermediate state relative to the other two states for an independent variable that results in a transition from the completely ordered state to the completely disordered state.

20. Consider a molecule of ethane. How many different ways are there of arranging the position of one methyl group relative to the other, accounting only for rotations about the sigma bond between the carbon atoms? What is the entropy of this bond rotation?

21. The thermodynamics of hen egg white lysozyme at pH 4.5 and 25 °C are as follows: $\Delta H = 50\,000$ cal/mol, $\Delta S = 132$ cal/mol-K, $\Delta C_p = 1500$ cal/mol-K. Calculate ΔG at 60 °C. Evaluate the partition function, P_F, and P_U at this temperature. The thermodynamics of *apo α-lactalbumin* at pH 8.0 and 25 °C are $\Delta H = 30\,000$ cal/mol, $\Delta S = 94$ cal/mol-K. Calculate ΔG at 60 °C. Assume that the protein exhibits a two-state transition under the stated conditions. Evaluate the partition function, P_F, and P_U at this temperature.

22. Helix–coil transition theory. Give the statistical weight of cchhhhh-hhhcchhchhhccc.

For solutions, see http://chem.nich.edu/homework

Chapter 7

Binding equilibria

A. Introduction

In earlier chapters we saw how the First and Second Laws are combined in a thermodynamic potential function called the Gibbs free energy. We have also seen how the Gibbs energy can be used to predict the direction of spontaneous change in a wide variety of systems under the constraints of constant temperature and pressure. One type of application of the Gibbs energy which we studied in Chapter 5 is the binding of a ligand to a macromolecule. In Chapter 6 we were introduced to statistical thermodynamics, a mathematical formalism that permits a molecular interpretation of thermodynamic quantities. The present chapter combines and extends all the main ideas we have encountered thus far. **Binding** is an extremely common and immensely important phenomenon of biochemistry. While binding can be considered just one of many different types of equilibrium process, the central role it plays in the physiology of biological macromolecules makes the present chapter one of the most important ones of this book.

Before plunging into a sea of complex mathematical equations, let's do a brief survey of areas in biology where binding plays a role. In our discussion of the First Law (Chapter 2) we encountered RNase A, a digestive enzyme that can bind tightly to a nucleotide inhibitor. Then in Chapter 5 we looked at receptor–ligand interactions and an introductory analysis of oxygen binding to hemoglobin. And an exercise in Chapter 5 involved the binding of a regulatory protein to DNA (Fig. 7.1). All these intermolecular interactions are involved in different biological processes and have quite different biological effects. From the perspective of physical chemistry, however, they all bear a distinctly similar mark. In every case, binding is made specific by steric complementarity between ligand and macromolecule (hand fits in glove), complementary charges on macromolecule and ligand (opposites attract), complementary polar and non-polar surfaces on macromolecule and ligand (likes attract), and so on. In a word, complementarity!

Binding occurs when a hormone, e.g. insulin, interacts with the extracellular portion its membrane-spanning receptor protein (Fig.

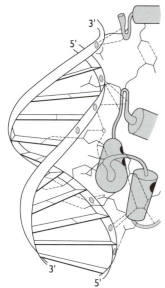

3'
5'
3'
5'

Fig. 7.1 Binding. There are many different kinds of biomolecular interaction, as discussed in the text. One class of interaction is that of protein–DNA binding. The figure shows one helix-turn-helix motif of the 434 repressor protein interacting with its DNA binding site. Binding of the repressor to the operator site in DNA inhibits transcription of the gene regulated by the operator. 434 repressor is a dimeric protein, and it associates with DNA in a twofold symmetric manner. The conformation of the protein in the protein–DNA interface closely resembles that of duplex DNA itself. The protein interacts with both paired bases and the sugar–phosphate backbone through a complex network of hydrogen bonds, salt bridges, and van der Waals contacts. The DNA bends around the protein in the complex. The energetic cost of the strain induced by bending must be offset by the other interactions made between the DNA and protein. *All binding interactions are essentially electrical in origin.* After A. K. Aggarwal *et al.* (1988).

5.18). The interaction elicits a cascade of events inside a cell, the ultimate result depending on cell type. In muscle cells, for instance, insulin binding leads to an increase in glucose metabolism. In fibroblasts, insulin acts as a growth factor. And in liver cells, insulin stimulates the activity of enzymes that synthesize glycogen, a polymeric form of the energy molecule glucose. A major aim of this chapter is to learn ways of quantifying the interaction between a ligand and a receptor.

Many enzymes require a bound ion or multi-atom cofactor in order to perform their catalytic function. For instance, the folded state of the milk protein α-lactalbumin (Chapter 6) is stabilized by calcium, and it is the *holo* (ligand-bound) form of the protein that interacts with galacto-syltransferase to form galactose synthetase, a heterodimer. The DNA polymerases from hyperthermophilic bacteria that are so important in PCR require bound divalent cations for activity (Chapter 5). And several of the enzymes involved in the citric acid cycle (Chapter 5) bind a flavin (FAD^+) for use in redox reactions (Chapter 4). These are just a few more examples of the importance of binding in the living organism.

Having seen the panoramic view, let's zoom in on the direct interaction of molecular oxygen with the blood protein hemoglobin (Hb) and the muscle protein myoglobin (Mb), two closely related molecular machines. In humans, Hb plays a vitally important role in the transfer of oxygen from the lungs to cells situated throughout the body. Hb does this by being confined to red blood cells, a.k.a. 'bags of hemoglobin.' Oxygen is loaded onto Hb in the lungs, where the partial pressure of oxygen is high, and unloaded in the extremities of the vasculature, where the partial pressure of oxygen is low. Oxygen simply moves down its concentration gradient. In striated muscle, the offloaded oxygen is picked up by the protein Mb and stored until needed for respiration. This 'macroscopic' description, while true, does not provide any sense of the marvelous complexity of the proteins involved, nor of how their complex structure enables them to do their jobs.

Though research in this area had begun decades earlier, it was not until the 1920s that Hb was found to be a tetramer with an average of about one oxygen-binding site per subunit. The structure of Hb was solved at low resolution by the British crystallographer Max Ferdinand Perutz (1914–) and co-workers at Cambridge University in 1959 (Perutz and John Cowdery Kendrew were awarded the Nobel Prize in Chemistry in 1962). This work showed that the tetramer is held together by ion pairs, hydrogen bonds, and hydrophobic interactions. Detailed analysis of binding data implies that the binding equilibrium involves not just H_2O, O_2, Hb, and $Hb(O_2)_4$, but several other molecular species as well: $Hb(O_2)$, $Hb(O_2)_2$, $Hb(O_2)_3$, $Hb(O_2)_3$. Importantly, there are multiple forms of these partially ligated species, since there are multiple permutations of the ligands bound. At the same time, each intact species of Hb is in equilibrium with subunit dimers and monomers. And as though that were not a high enough level of complexity, the binding affinity of oxygen to a subunit depends not only on whether it is in the monomeric, dimeric, or tetrameric state, but also on whether other oxygen molecules are bound! Hb is an extremely complex protein.

Determination of its structure was aided by the existing structure of Mb, which has been solved at low resolution in 1957 by John Kendrew (1917–1997) and colleagues. This was possible because there are clear structural similarities between the α-chain of Hb and Mb; a reflection of their similar functions. It is probable that hemoglobin and myoglobin genes originate come from the same proto-gene. The 'original' Mb/Hb gene is certainly very ancient because myoglobin-like heme-containing proteins are also found not only in vertebrates but also in plants, eubacteria, and archaea. And yet, as we shall see, Hb and Mb are also very different. One aim of this chapter is to provide a means of discussing such differences in terms of relatively straightforward mathematics.

As we shall see, one of the happier aspects of using mathematics to describe binding is that a relatively small number of equations can be applied very effectively to a tremendous variety of situations. Some of the questions we shall examine in this context are: How many ligand-binding sites are there per macromolecule? How strongly does the ligand bind? If there is more than one binding site per macromolecule, is the binding of a ligand to one site independent of binding to another site on the same macromolecule?

B. | Single-site model

We have already discussed the single-site binding model in moderate detail in Chapter 5, in the section on molecular pharmacology. (Before getting into the thick of things here, you might find it helpful to return to Chapter 5 for a quick review of basic concepts.) Eqn. 5.37 says that the *average* number of moles of ligand bound per mole of macromolecule (the fraction of sites occupied or **fractional saturation**), ϕ, is

$$\phi = [M \bullet L]/[M]_T = [L]/(K_d + [L]) \tag{7.1}$$

where M is the macromolecule (e.g. DNA or protein), $[M]_T$ the total concentration of macromolecule, [L] the concentration of *free* ligand, and K_d the *dissociation* constant. Note that when $[L] = K_d$, $\phi = 1/2$. That is, the dissociation constant measures the concentration of free ligand at which the binding sites are half-saturated. In terms of the *association* constant, $K_a = K_d^{-1}$, the fraction of sites bound is

$$\phi = K_a[L]/(1 + K_a[L]) \tag{7.2}$$

where $K_a = \exp(-\Delta G_b/RT)$, and ΔG_b, the free energy of binding, is the free energy difference between the bound state and the unbound state. We can see from this that if K_a is large (if binding is 'tight'), even small concentrations of free ligand will give $\phi \sim 1$. Rearrangement of Eqn. 7.2 in terms of $K_a[L]$ gives

$$K_a[L] = \phi/(1 - \phi) \tag{7.3}$$

This is the **Langmuir adsorption isotherm** (Fig. 7.2). The Langmuir isotherm gets its name from Irving Langmuir, an American physical

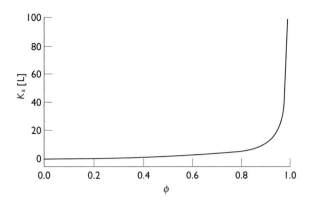

Fig. 7.2 Langmuir isotherm. $K_a[L]$ is plotted against ϕ. As K_a becomes infinitely large, $K_a[L] / (1 + K_a[L])$ approaches 1.

chemist who lived 1881–1957. He was awarded the Nobel Prize in Chemistry in 1932. 'Adsorption' because bound ligand molecules can be thought of as being adsorbed onto macromolecules, and 'isotherm' because the temperature is constant.

Equations 7.1 and 7.2, simple though they are, can nevertheless be extremely useful to the biochemist; they describe a tremendous variety of biochemical phenomena. Consider, for instance, the role of phosphorylated tyrosine residues in signal transduction: if a certain phosphotyrosine recognition module interacts with a target molecule at just one site, then these equations apply. It just so happens that this is indeed the case with a large number of known phosphotyrosine recognition motifs, including the SH2 domain and the PTB (phosphotyrosine binding) domain. PTB domains interact with phosphorylated tyrosine residues in the cytoplasmic domain of growth factor receptors with K_d ~1 μM. Some proteins, for instance the large cytoskeleton-associated signaling protein tensin, have multiple phosphotyrosine recognition domains, enabling them to interact with several proteins at once.

C. Multiple independent sites

The situation here is obviously more complicated than the one-site model! Suppose that there are n sites per macromolecule. The concentration of bound ligand is

$$[L]_b = [M \bullet L] + 2[M \bullet L_2] + 3[M \bullet L_3] + \ldots \sum_{i=1}^{n} i[M \bullet L_i] \tag{7.4}$$

The integral coefficients $\{1, 2, 3, \ldots\}$ tell us how many ligand molecules are bound to the n binding sites per macromolecule. The average fraction bound is

$$\phi = \frac{\sum_{i=1}^{n} i[M \bullet L_i]}{\sum_{i=0}^{n} [M \bullet L_i]} \tag{7.5}$$

Note the differences between numerator and denominator of Eqn. 7.5. In the denominator there is no coefficient (there is no weighting of a term's contribution to the sum), and the sum starts at zero (to include the macromolecule with no ligands bound).

Now let's think about the various ligand-bound states in terms of equilibrium constants. The relevant equilibria, which represent the time-average situation, look like this:

$$M + L \Leftrightarrow M \bullet L \qquad K_1 = [M \bullet L]/[M][L]$$

$$M + 2L \Leftrightarrow M \bullet L_2 \qquad K_2 = [M \bullet L_2]/[M][L]^2 \qquad (7.6)$$

$$M + 3L \Leftrightarrow M \bullet L_3 \qquad K_3 = [M \bullet L_3]/[M][L]^3$$

and so forth, where each K is a **macroscopic** *association* **constant**, or apparent association constant. Note that each K represents an all-or-none association of ligands and macromolecule, not the sequential addition of a single ligand; a convenient mathematical device but not an accurate model of the physical process! Because the free energy of a bound state is lower than the free energy of an unbound state, the more ligands bound the more stabilized the complex. In other words, binding stabilizes – *always*.

Substituting the relationships in the right-hand column into Eqn. 7.5, we have

$$\phi = \frac{\sum_{i=1}^{n} iK_i[M][L]^i}{\sum_{i=0}^{n} K_i[M][L]^i} = \frac{\sum_{i=1}^{n} iK_i[L]^i}{\sum_{i=0}^{n} K_i[L]^i} \qquad (7.7)$$

This is the famous **Adair equation**, first described in 1924 by the British physiologist Gilbert Smithson Adair (1896–1979). Though complex, the equation is nice because it does not depend on assumptions about the type of binding that occurs. When there is just one binding site, the Adair equation reduces to $\phi = K_1[L]/(K_0[L]^0 + K_1[L]) = K_1[L]/(1 + K_1[L])$, which is Eqn. 7.2. $K_0 = 1$ because the unbound macromolecule is the reference state, i.e. $\Delta G_0 = 0$.

Note that Eqn. 7.7 makes no distinction between binding to the folded state and binding to the unfolded state. Generally speaking, though, the geometry of the binding site will be suitable for specific binding only in the folded state of a protein. In the unfolded conformation, the constellation of chemical groups that coordinate the ligand is rarely suitable for binding, and usually specific intermolecular association does not occur. The situation is different, however, if a ligand binds non-specifically.

Consider, for example, the chemical denaturants urea and guanidine hydrochloride. The exact mode of binding of these small molecules to proteins is not entirely clear, and they probably alter the structure of solvent as much as or more than they bind to proteins at the concentrations required for denaturation. But the fact of the matter is that many proteins unfold cooperatively as the chemical denaturant concentration goes up. Regardless of whether denaturant–protein interactions are the principal cause of denaturation, the situation can be *modeled* as though binding were the cause of unfolding. In this context, the *difference* in the number of denaturant binding sites between the folded and unfolded state is a quantity of interest. This number is purely phenomenological, not only because it is a thermodynamic quantity and all thermodynamic quantities are phenomenological in the absence of a unique structural interpretation, but also because the number of binding sites determined experimentally is an effective number not necessarily the number actually believed to bind. In short, if you plug the phenomenological quantity into the right equation, your theoretical curve will closely resemble the date, but you do not necessarily believe that the phenomenological quantity gives an accurate description of the situation on the physical level. In any case, because the binding of a chemical denaturant to a protein is non-specific, there *must* be more binding sites in the unfolded state than in the folded one. As the denaturant concentration increases, the folded state becomes progressively stabilized through interacting non-specifically with ligand molecules, but the unfolded state becomes even more stabilized, because it has many more non-specific 'binding sites' than the folded state. In view of this, to a first approximation the phenomenological *m*-values discussed in the section on protein denaturation in Chapters 5 and 6 are proportional to the solvent-*inaccessible* surface area of the protein in the *folded* state. The solvent-accessible surface area in the unfolded state is the total surface area: the solvent-accessible surface in the folded state plus the solvent-inaccessible in the folded state. An *m*-value should also be roughly proportional to the difference in number of denaturant binding sites between the unfolded state and folded state of a protein.

To some readers Eqn. 7.7 will have a rather frightening appearance, possibly even sending chills up and down the spine. Take heart: the good news is that we can simplify things a bit, not merely because this is an introductory text but because that's what's usually done in practice! We'd like to be able to see, for example, how ϕ has a maximum value of n when we impose the condition that the n binding sites per macromolecule be *independent* and *identical*. If we require the binding of one ligand *not* to affect the affinity of ligands at the other binding sites, and if each site bound the same type of ligand with the same affinity, there is a simpler way of writing down Eqn. 7.7. By way of a few mathematical tricks (some hairy algebra), one can show that

$$\phi = nk[L]/(1 + k[L]) \qquad (7.8)$$

where k is the so-called **microscopic *association* constant**, or intrinsic association constant. This binding constant is defined as $k_i = [ML_i]/[ML_{i-1}][L]$, and it represents the free energy change on binding of the ith ligand to a macromolecule with $i-1$ ligands already bound. It is assumed that $k_i = k$ for all i; in other words, every ligand binds with exactly the same affinity. Note how k differs from K, the equilibrium constant for the free energy difference between the completely unbound state of the macromolecule and the state in which i ligands are bound. Note also that ϕ is now measured in units of moles. In fact, ϕ has the same units in Eqn. 7.2, but it less obvious there because $n = 1$. **The microscopic and macroscopic approaches, though they might seem quite**

different, describe binding equally well. The approach one chooses to describe a particular situation will be governed by personal preference and research objectives.

We now wish to look at the mathematics of ligand binding in the context of **titration**, the gradual filling up of sites. Such experiments are very common in biochemistry, so the odds that this knowledge will be of *practical* use to you at some point in your life in science are fairly good. You might even have call for this knowledge someday when analyzing the data of an isothermal titration calorimetry experiment!

We are interested in the general reaction

$$M + nL \Leftrightarrow M \bullet L_n \tag{7.9}$$

where the symbols have the same meaning as before. The total ligand concentration is

$$[L]_T = [L] + [L]_b \tag{7.10}$$

$$= [L] + [M]\phi \tag{7.11}$$

$$= [L] + [M]nk[L]/(1 + k[L]) \tag{7.12}$$

where we have used Eqn. 7.8 to get from Eqn. 7.11 to Eqn. 7.12. Multiplying both sides of Eqn. 7.12 by $(1 + k[L])$ gives

$$(1 + k[L])[L]_T = (1 + k[L])[L] + [M]nk[L] \tag{7.13}$$

This can be rearranged to

$$[L]_T + k[L][L]_T = [L] + k[L]^2 + nk[M][L] \tag{7.14}$$

or, in a more useful form,

$$k[L]^2 + (nk[M] - k[L]_T + 1)[L] - [L]_T = 0 \tag{7.15}$$

This equation is *quadratic* in [L] (i.e. the highest power of [L] is 2), so we can simply plug the coefficients of [L] into the solution of a standard quadratic equation in order to solve for [L]. The standard form of a quadratic equation is $ax^2 + bx + c = 0$. The two solutions to this equation: $x = (-b \pm [b^2 - 4ac]^{1/2})/2a$. Prove to youself that the result in the present case is

$$[L] = \frac{-(1 + nk[M] - k[L]_T) \pm \sqrt{(1 + nk[M] - k[L]_T)^2 + 4k[L]_T}}{2k} \tag{7.15}$$

This is the *free* ligand concentration (Fig. 7.3). Only the positive solution is physically meaningful, so '$\pm$' can be replaced with '+.'

We can connect this discussion of theory to concrete measurements made using isothermal titration calorimetry. What we would like to know is how to relate the heat released upon binding, q_b, a macroscopic quantity, to molecular interactions (Fig. 7.4). At constant pressure, this will be

$$q_b = \phi \Delta H_b \tag{7.17}$$

$$q_b = V_{cell}[L]_b \Delta H_b \tag{7.18}$$

Fig. 7.3 Variation of the free ligand concentration. The curves show the effect on the variation of [L] with [M] of changing the number of binding sites (n) or the microscopic binding constant (k). As expected, as [M] goes up, [L] falls; the macromolecule 'mops up' the ligand, reducing the free concentration for a fixed quantity of ligand. k is in units of $(\mu M)^{-1}$.

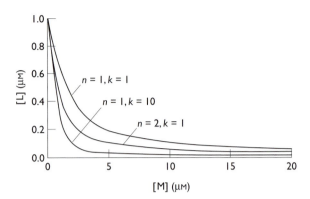

Fig. 7.4 Isothermal titration calorimetry. An experiment consists of equal-volume injections of a ligand solution into a macromolecule solution. Each peak results from a single injection and corresponds to an exothermic process. (Binding can also be endothermic.) The first few peaks are very similar in size, indicating that nearly all of the injected ligand is binding. The injection peaks become smaller as the fraction of injected ligand that becomes bound decreases. The last few peaks are about the same size: binding is effectively complete at the end of the experiment. Fig. 2.8 shows a schematic diagram of an ITC instrument.

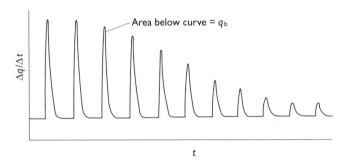

where V_{cell} is the volume of the reaction cell (Fig. 2.8). Note that $V_{cell} \times [L]_b = \phi$ has dimensions [volume][moles/volume] = [moles], as required. Substituting Eqns 7.10 and 7.16 into Eqn. 7.18 gives

$$q_b = V_{cell}\left([L]_T + \frac{(1 + nk[M] - k[L]) - \sqrt{(1 + nk[M] - k[L]_T)^2 + 4k[L]_T}}{2k}\right)\Delta H_b$$
(7.19)

When the macromolecule is saturated with ligand, the concentration of bound ligand is equal to the concentration of macromolecule times the number of binding sites per macromolecule:

$$q_{b,sat} = V_{cell}n[M]\Delta H_b$$
(7.20)

which, when solved for the binding enthalpy, is

$$\Delta H_b = \frac{q_{b,sat}}{V_{cell}n[M]}$$
(7.21)

If the number of binding sites per macromolecule and the cell volume are known, and if the macromolecule concentration and heat of binding can be measured, the enthalpy of binding can be calculated from Eqn. 7.21.

What if there is more than one class of binding site *for the same ligand*? That is, what if all the ligands are the same but the ks are different? A more general form of Eqn. 7.8 is needed, and it is:

$$\phi = \Sigma n_i k_i[L]/(1 + k_i[L])$$
(7.22)

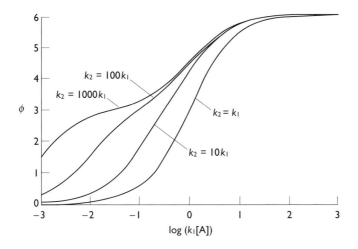

Fig. 7.5 Multiple classes of binding site. Here there are two different types of site, one with k_1 and another with k_2. $n_1 = n_2 = 3$. When $k_1 = k_2$, the curve is indistinguishable from that for six identical sites. As k_2 becomes increasingly different from k_1, however, the curve becomes less and less sigmoidal. ϕ/n, which varies between 0 and 1, could be replaced by the percentage change in any observable quantity. $(n = n_1 + n_2.)$

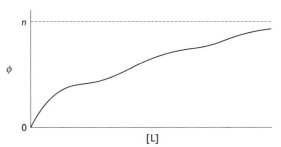

Fig. 7.6 Complex binding process. It would be very difficult to say in such a case how many different classes of binding site are present, but there is no doubt that *something* is happening. Compare Fig. 4.12.

where the sum is over the various different kinds of binding site. For example, if there are two types of binding site, Eqn. 7.22 becomes

$$\phi = \frac{n_1 k_1[L]}{1 + k_1[L]} + \frac{n_2 k_2[L]}{1 + k_2[L]} \tag{7.23}$$

There are n_1 sites of with binding constant k_1 and n_2 sites with constant k_2. Figure 7.5 shows how ϕ varies with $k_1[L]$ for different values of k_2/k_1. As we should expect, when $k_2 = k_1$ the curve is indistinguishable from Eqn. 7.8 with $n = n_1 + n_2$. Marked deviations from Eqn. 7.8 occur if k_2/k_1 is relatively large (>50) or small (<0.02). When k_2/k_1 lies between these values, unless great care is taken in collecting and handling experimental data it might be difficult to distinguish between some large number of identical sites with $k_2 < k < k_1$ and two smaller numbers of two classes of site. Such difficulties in data analysis are common in biochemistry, and a lot of clever thinking might be needed to sort out how many binding sites there are. A plot of ϕ versus [L] for a very complex binding process is shown in Fig. 7.6.

D. Oxygen transport

In Chapter 5 and in the Introduction to the present chapter we looked at how Hb plays a key role in oxygen transport. Here we wish to build on the previous discussions and make a more detailed comparison of Hb and Mb

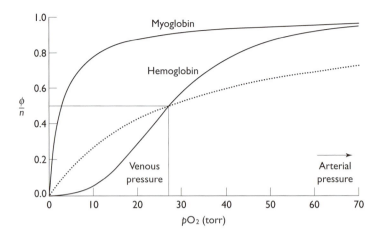

Fig. 7.7 Oxygen-binding properties of myoglobin and hemoglobin. The myoglobin oxygen-binding curve is a rectangular hyperbola. The hemoglobin curve is sigmoidal. The shape of the hemoglobin curve arises from interactions between subunits. The broken line represents the percentage saturation of hemoglobin in the absence of inter-subunit cooperativity, assuming that the partial pressure of oxygen required for 50 percent binding (p_{50}) is the same as when cooperative interactions are present. At low partial pressures of oxygen, the affinity of oxygen is much higher for myoglobin than for hemoglobin. This helps to explain why oxygen is unloaded from hemoglobin in capillaries.

in the context of binding. Our approach will be to set aside the biological *origin* of the proteins for a moment and to think about them as mere objects of physical chemistry. Then we'll see what additional insight accounting for biological function can add to the mathematical development. Our overall aim is to see how the physicochemical properties of these proteins underlie their biological functions.

As discussed in the Introduction, Hb and Mb are similar and different. They are similar in that the three-dimensional shape of native Mb, known as the myoglobin fold, closely resembles the native structure of the α-subunits of Hb. Both proteins bind **heme** (Greek: *hyma*, blood) in much the same way, and the bound iron atom (Fig. 5.1) helps to coordinate diatomic oxygen or carbon dioxide (or carbon monoxide!). Hb and Mb differ in that the former is tetrameric (two α-chains and two β-chains) while the latter is monomeric under physiological conditions. As we shall see, this structural difference has profound consequences for regulation of protein function. Indeed, the macroscopic oxygen-binding properties of these two proteins are very different (Fig. 7.7). Oxygen binding to Hb can be regulated by more than just the partial pressure of oxygen: it is also sensitive to the Cl^- concentration, pH, and a small ionic compound called bisphosphoglycerate (a.k.a. BPG. If the name of this compound sounds suspiciously like something you read about in Chapter 5 in the section on glycolysis, that's because BPG is made out of 3PG.).

Analysis of the structural and thermodynamic properties of Hb has shed much light on how protein assemblies 'self-regulate' in response to environmental 'signals.' The 'signals' are changes in the concentration of solutes and temperature, and 'self-regulation' is the effect such changes have on protein conformation and ligand binding capability. The structural properties of Hb shift subtly with changes in the chemical environment, in order to minimize free energy. The energy changes can be large, and although the associated shifts in structure are often small, they can have dramatic effects on protein biochemistry. In short, Hb is an extremely sophisticated transport system. In-depth analysis of Hb and comparison with other biological macromolecules shows that it provides a useful model for general theories of regulation of biochemical processes at the molecular level. Hb can be thought of as a type of

molecular switch. Its switch-like character 'arises' from the *interactions* between subunits.

And now for a more mathematical look at the oxygen binding capabilities of Hb and Mb. The binding of oxygen to Mb can be described as

$$Mb + O_2 \Leftrightarrow Mb \bullet O_2 \tag{7.24}$$

The dissociation constant for this reaction is

$$K_d = [Mb][O_2]/[Mb \bullet O_2] \tag{7.25}$$

and, by Eqn. 7.1, the fraction bound is

$$\phi = [Mb \bullet O_2]/[Mb]_T = [O_2]/(K_d + [O_2]) \tag{7.26}$$

When this relationship is written in terms of the partial pressure of oxygen (pO_2) and the pressure at which half of the binding sites are occupied (p_{50}), it looks like this:

$$\phi = pO_2/(K_d + pO_2) = pO_2/(p_{50} + pO_2) \tag{7.27}$$

The shape of Eqn. 7.27 is a rectangular hyperbola (Fig. 7.7). In humans, p_{50} ≈ 2.8 torr. (1 torr = 1 mm Hg at 0 °C = 0.133 kPa; 760 torr = 1 atm. The torr is named after Evangelista Torricelli (1608–1647), the Italian physicist and mathematician who invented the barometer.) Because this is well below venous pressure, it would be thermodynamically unfavorable for O_2 to be released back into the blood once it had become bound to Mb.

So much for Mb. As for Hb, experimental studies have shown that the variation of the fractional saturation with pO_2 differs considerably from that of Mb. The binding curve of Hb is *not* a rectangular hyperbola. The earliest attempt to rationalize the oxygen-binding characteristics of Hb mathematically was put forth by Archibald Vivian Hill in 1911. Hill (1886–1977), a Briton, was awarded the Nobel Prize in Physiology or Medicine in 1922 for his discovery concerning the production of heat in muscle. In general terms, for a ligand L and macromolecule M with n binding sites,

$$M + nL \Leftrightarrow M \bullet L_n \tag{7.28}$$

This is just Eqn. 7.9. If the ligand molecules are assumed to bind in an all-or-none fashion (an oversimplification that will be dealt with shortly), there is but one (macroscopic) *dissociation* constant:

$$K_d = [M][L]^n/[M \bullet L_n] \tag{7.29}$$

The degree of saturation is

$$\phi = \frac{n[M \bullet L_n]}{[M] + [M \bullet L_n]} \tag{7.30}$$

The factor n appears in the numerator on the right-hand side because there are n ligands bound to each macromolecule at saturation; ϕ is the average number of ligands bound per macromolecule. Rearrangement of Eqn. 7.30 gives

$$\frac{\phi}{n} = \frac{[M \bullet L_n]}{[M] + [M \bullet L_n]} \tag{7.31}$$

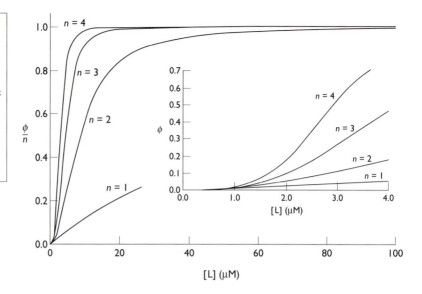

Fig. 7.8 The Hill equation. Fractional saturation is plotted as a function of ligand concentration and n. As n increases, the concentration of ligand required for saturation decreases, if the magnitude of the binding constant is maintained. The inset helps to reveal the increasingly sigmoidal shape of the curve as n increases from 1. Compare the curves in Fig. 7.7. $K_d = 1$ μM.

which could be called the fractional saturation, since it varies between 0 and 1. On substituting in Eqn. 7.29,

$$\frac{\phi}{n} = \frac{[L]^n}{K_d + [L]^n} \tag{7.32}$$

This is the famous **Hill equation**. Figure 7.8 shows plots of the Hill equation for different values of n. Note that although one would not ordinarily think of a non-integral number of ligands bound, n can in principle take on any real value. In other words, n is a *phenomenological representation* of the *average* number of ligands bound, or the *effective* number of ligands bound. Using statistical mechanics to analyze binding data does not necessarily get around the phenomenological quality of binding data, particularly when the situation is the slightest bit ambiguous, as it is with the interaction of chemical denaturants with proteins. When binding is very specific, however, the measured number of binding sites will ordinarily be close to the number one would 'see' on visualizing the structure of the macromolecule at atomic resolution. Note that n can be negative as well as positive. When n is positive, the binding cooperativity is positive; when it is negative, cooperativity is negative. 'Negative binding' (of water) occurs in the context of exclusion of the solvent from the surface of a macromolecule in the presence of 'inert' co-solvents like sucrose, which do not bind to the protein but nevertheless influence protein stability.

Now let's look at oxygen binding to hemoglobin at increased magnification. The various binding equilibria to consider are:

$$M + L \Leftrightarrow M \cdot L \qquad K_{a,1} = 4k_{a,1} \qquad k_{d,1} = 4K_{d,1}$$

$$M \cdot L + L \Leftrightarrow M \cdot L_2 \qquad K_{a,2} = 3k_{a,2}/2 \qquad k_{d,2} = 3K_{d,2}/2$$

$$M \cdot L_2 + L \Leftrightarrow M \cdot L_3 \qquad K_{a,3} = 2k_{a,3}/3 \qquad k_{d,3} = 2K_{d,3}/3$$

$$M \cdot L_3 + L \Leftrightarrow M \cdot L_4 \qquad K_{a,4} = k_{a,3}/4 \qquad k_{d,4} = K_{d,3}/4 \tag{7.33}$$

Note the relationships between the microscopic *association* constants (the $k_{a,i}$s) and the macroscopic dissociation constants (the $K_{d,i}$s). The coefficients of the microscopic binding constants, called **statistical factors** (Chapter 6), arise as follows. There is only one way in which the ligand can dissociate from a singly ligated molecule, but four possible ways in which it can bind in the first place; hence $K_{a,1} = 4k_{a,1}$. When one site is occupied, there are three possible ways in which a second ligand can bind, and two ways in which a doubly ligated molecule can lose ligands, so $K_{a,2} = 3k_{a,2}/2$. And so forth.

The concentrations of the various bound states are:

$$[M \cdot L] = [M][L]/k_{d,1}$$

$$[M \cdot L_2] = [M \cdot L][L]/k_{d,2} \quad = \quad [M][L]^2/k_{d,1}k_{d,2}$$

$$[M \cdot L_3] = [M \cdot L_2][L]/k_{d,3} \quad = \quad [M][L]^3/k_{d,1}k_{d,2}k_{d,3}$$

$$[M \cdot L_4] = [M \cdot L_3][L]/k_{d,4} \quad = \quad [M][L]^4/k_{d,1}k_{d,2}k_{d,3}k_{d,4} \qquad (7.34)$$

where dissociation constants have been used. The fractional saturation of the macromolecule is

$$\frac{\phi}{4} = \frac{[M \cdot L] + 2[M \cdot L_2] + 3[M \cdot L_3] + 4[M \cdot L_4]}{4([M] + [M \cdot L] + [M \cdot L_2] + [M \cdot L_3] + [M \cdot L_4])} \qquad (7.35)$$

The coefficients in the numerator refer to the number of ligands bound to each species. Substituting in Eqns. 7.34 gives

$$\frac{\phi}{4} = \frac{\begin{array}{l}[M][L]/k_{d,1} + 2[M][L]^2/k_{d,1}k_{d,2} \\ + 3[M][L]^3/k_{d,1}k_{d,2}k_{d,3} + 4[M][L]^4/k_{d,1}k_{d,2}k_{d,3}k_{d,4}\end{array}}{4([M] + [M][L]/k_{d,1} + [M][L]^2/k_{d,1}k_{d,2} \\ \qquad + [M][L]^3/k_{d,1}k_{d,2}k_{d,3} + [M][L]^4/k_{d,1}k_{d,2}k_{d,3}k_{d,4})} \qquad (7.36)$$

$$\frac{\phi}{4} = \frac{[L]/k_{d,1} + 2[L]^2/k_{d,1}k_{d,2} + 3[L]^3/k_{d,1}k_{d,2}k_{d,3} + 4[L]^4/k_{d,1}k_{d,2}k_{d,3}k_{d,4}}{4(1 + [L]/k_{d,1} + [L]^2/k_{d,1}k_{d,2} + [L]^3/k_{d,1}k_{d,2}k_{d,3} + [L]^4/k_{d,1}k_{d,2}k_{d,3}k_{d,4})} \qquad (7.37)$$

In terms of the macroscopic dissociation constants, Eqn. 7.37 is

$$\frac{\phi}{4} = \frac{\begin{array}{l}[L]/K_{d,1} + 3[L]^2/K_{d,1}K_{d,2} \\ + 3[L]^3/K_{d,1}K_{d,2}K_{d,3} + [L]^4/K_{d,1}K_{d,2}K_{d,3}K_{d,4}\end{array}}{1 + 4[L]/K_{d,1} + 6[L]^2/K_{d,1}K_{d,2} \\ \qquad + 4[L]^3/4K_{d,1}K_{d,2}K_{d,3} + [L]^4/K_{d,1}K_{d,2}K_{d,3}K_{d,4}} \qquad (7.38)$$

Note that the statistical factors in the denominator are the same as line five of Pascal's triangle (Fig. 6.14), reflecting their origin in the combinations of things. We shall return to this development in the second part of the next section.

E. Scatchard plots and Hill plots

Now we turn our attention to useful ways of graphing binding data. The discussion follows on from the previous discussion. At the close of this section we'll note strengths and weaknesses of the Scatchard plot and Hill plot, and comment on the more general utility and greater value of non-linear least-squares regression methods in data analysis, despite the much more recent development of the Scatchard plot and Hill plot. Least-squares methods were developed about 1794 by Carl Friedrich Gauss (1777–1855), who with Archimedes and Newton ranks as one of the greatest mathematicians of all time. Gauss, a German, also made important contributions to astronomy, geodesy, and electromagnetism, and his treatment of capillary action contributed to the development of the principle of the conservation of energy. In least squares analysis, the best-estimated value is based on the minimum sum of squared differences between the 'best-fit' curve and experimental data points.

Rearrangement of Eqn. 7.1 gives

$$[M \cdot L] = [M]_T - K_d[M \cdot L]/[L] \tag{7.39}$$

This is a variant of the **Scatchard equation** (see Eqn. 5.38 and Fig. 5.16). The bound ligand concentration ([M • L]) is linear in [M • L]/[L], with a slope of $-K_d$; a **Scatchard plot** can be used to determine K_d graphically. Note that Eqn. 7.39 assumes that there is just one binding site per macromolecule. A more useful equation would be one that could be used to analyze data directly, whether the instrument used for experiments was a fluorimeter, circular dichroism spectrometer, NMR spectrometer, calorimeter, or whatever. Assuming that the change in the observable quantity, ΔO, is directly proportional to [M • L], the bound concentration, as is often the case, then $\mathbf{c}\Delta O = [M \cdot L]$. If ΔO_{max} is the change in observable on saturation of the binding sites, then $\mathbf{c}\Delta O_{max} = [M_T \cdot L]$. $\mathbf{c}$ is a proportionality constant whose value we do not know *a priori* (i.e. before the data have been analyzed); it depends on the protein–ligand system being studied, the solution conditions, and of course the technique used for measurements. Making the appropriate substitutions into Eqn. 7.39, we have

$$\mathbf{c}\Delta O = \mathbf{c}\Delta O_{max} - K_d\mathbf{c}\Delta O/[L] \tag{7.40}$$

or

$$\Delta O = \Delta O_{max} - K_d\Delta O/[L] \tag{7.41}$$

This tells us that ΔO measures the change in signal (e.g. fluorescence intensity, ellipticity, resonant frequency, heat uptake, . . .) when [L] is added to the solution.

From Eqns. 7.1 and 7.2,

$$\phi = [M \cdot L]/[M]_T = K_a[L]/(1 + K_a[L]) \tag{7.42}$$

Multiplication of both sides by $(1 + K_a[L])$ gives

$$\phi(1 + K_a[L]) = \phi + \phi K_a[L] = K_a[L] \tag{7.43}$$

which, when solved for ϕ and divided by [L], is

$$\phi/[L] = K_a - \phi K_a \tag{7.44}$$

This is yet another form of the Scatchard equation. $\phi/[L]$ plotted against ϕ gives a line of slope $-K_a = -1/K_d$. The curve intersects the ordinate at K_a,

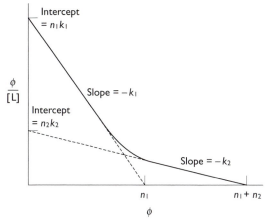

Fig. 7.9 Scatchard plot. There are two classes of ligand, $k_1 > k_2$, and $n_1 \neq n_2$. The ks are microscopic association constants. The abscissa-intercepts are particularly informative here: they provide information on the number of ligand binding sites.

the association constant, and the abscissa at 1, the number of binding sites.

The Scatchard approach can also be used in more complicated cases, for instance when there are two classes of binding site and both n_1 and n_2 are greater than 1. From Eqn. 7.23

$$\phi/[L] = n_1 k_1/(1 + k_1[L]) + n_2 k_2/(1 + k_2[L]) \tag{7.45}$$

Figure 7.9 shows the form of $\phi/[L]$ *versus* ϕ when $k_1 > k_2$. Here the ks are microscopic association constants.

A simple rearrangement of the Hill equation (Eqn. 7.32) leads to the **Hill plot**, another useful way of graphing binding data. We arrive at the Hill plot by multiplying both sides of Eqn. 7.32 by the denominator on the right-hand side ($K_d + [L]^n$), grouping terms in $[L]^n$, and solving for $[L]^n/K_d$. The result is

$$\frac{\frac{\phi}{n}}{1 - \frac{\phi}{n}} = \frac{[L]^n}{K_d} \tag{7.46}$$

This relationship, which is linear in $1/K_d$, can be cast in a somewhat more useful form by taking the logarithm of both sides. This gives

$$\log \frac{\frac{\phi}{n}}{1 - \frac{\phi}{n}} = \log[L]^n - \log K_d = n\log[L] - \log K_d \tag{7.47}$$

The left-hand side of this expression, which is admittedly complex, is nevertheless a *linear* function of $\log[L]$ with slope n and ordinate-intercept $-\log K_d$.

The usefulness of Eqn. 7.47 can be illustrated as follows. If the ligand is oxygen, as in hemoglobin, Eqn. 7.47 becomes

$$\log \frac{\frac{\phi}{n}}{1 - \frac{\phi}{n}} = n \log pO_2 - \log K_d \tag{7.48}$$

Hill plot. When the data are good enough, this type of plot can be very informative. The slopes of the upper and lower asymptotes are 1 in cooperative binding. Between the onset of binding and saturation, the slope changes dramatically. The slope of the curve at the point where it crosses the line $n = 1$ gives the *apparent* number of ligand binding sites. In this case it is about 3.0, as in hemoglobin. The actual number of binding sites differs from the apparent number because the binding process is not perfectly cooperative. The apparent number of sites is a phenomenological parameter; it does not provide an atomic resolution view of the binding process. It should go without saying that the Hill coefficient and p_{50} depend on conditions, e.g. temperature, ionic strength, etc.

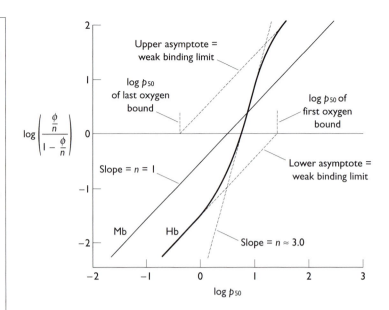

where, as before, pO_2 is the partial pressure of diatomic oxygen. What is K_d? It can be found from Eqn. 7.32. At half saturation $\phi/n = 0.5$, so $K_d = (p50)^n$. A comparison of the form of Eqn. 7.48 for $n = 1$ (myoglobin) and $n \approx 3$ (hemoglobin) is shown in Fig. 7.10.

If only things were *always* so simple! Before the advent of inexpensive desktop computers, the Scatchard plot and Hill plot were very sensible graphical approaches to the routine analysis of binding data, and they were used all the time. This is because linearized versions of the theoretical equations made determining parameters and estimating errors a cinch using linear regression, which can be carried out relatively painlessly with a no-frills hand-held calculator. Despite the apparent simplicity and neat appearance of linear equations, linearizations of complicated mathematical relationships *should not* normally be used for data analysis. Why not? Linearization (or any type of mathematical transformation for that matter) *distorts* the experimental error. Two assumptions of basic linear regression are that the errors in the experimental data points follow a *Gaussian* distribution, a 'bell-shaped' curve, (named after C. F. Gauss), that is that all errors are exclusively random, and that the variation in the error of the independent variable (pO_2 in the last example) is very small and constant throughout its range. These conditions, which can be difficult to realize in practice in the best of circumstances, are not likely to hold after data have been transformed mathematically. Moreover, mathematical transforms can change the form of the relationship between the independent variable and the dependent variable. For instance, in a Scatchard plot the dependent variable ([bound]/[free]) not only varies with the 'independent' variable ([free]), as is generally the case, but it is also multiplied by its inverse! To summarize, the algebra involved in deriving the Scatchard equation and the Hill equation is sound, but the gains in using these equations are more than offset by the losses, at least in serious data analysis. Experimental work requires a good deal of care, effort, and time, so one

Table 7.1	Outline of non-linear regression analysis

- Collect data and enter values into a spreadsheet
- Choose a model
- Guess the best-fit values of the fitting parameters
- After fitting, question whether the parameter values make good sense
- Compare the results of fitting different models to the data set
- Use various criteria to decide which model is the best one

should look for the most useful and suitable technique for analyzing data and not just one that happens to be appear in just about every biochemistry book ever published. What should be done? Non-linear regression analysis is in many cases the best way forward.

A non-linear function is just that: any function that is not a straight line. It can be a polynomial of order two or higher (e.g. $y = a + bx + cx^2 + \ldots$; the order of a polynomial is given by the highest power of the independent variable; a first-order polynomial is a straight line) or any other type of relationship that is not a straight line (e.g. $y = \sin x$, $y = e^x$, etc.). Let's take this opportunity to cover in simple terms what nonlinear regression can and cannot do. An outline of regression analysis is given in Table 7.1. One generally starts with a spreadsheet (e.g. Excel, Origin, …) into which experimental data have been entered, a mathematical model that *might* describe the data of the physical or chemical process we're interested in, and some sensible guesses as to the likely final values of the fitting parameters (adjustable quantities like **c** that are determined by the outcome of experiments under specific conditions). Making good guesses of the parameters can be very important with complicated models. This is because the computer programs used in regression analysis 'search' for the best-fit parameters by trying to minimize the deviations of the model from the experimental data, just as in linear regression. When the model is complex, many possible combinations of parameters can give a fit that appeals to the eye but does not necessarily make much sense from the physical point of view. The calculated deviations resulting from the initial guesses can sometimes be so large that the program will not be able to adjust the parameters in such a way as to home in on the 'correct' values.

That what's regression analysis does for you. What it does not do is choose your model! And for that reason regression analysis is often not a trivial task. It is not simply a matter of finding any old mathematical function that gives a good fit, which would be fine for data presentation but not necessarily of any help at all for understanding what's going on! Instead, one looks for an equation that fits the data well and that is *physically meaningful*. For when the best-fit parameters have been found, they will satisfy the mathematical criterion of minimizing the deviations of the model from the data, but it will still be necessary to question whether the values (and the model) make sense! One would like to be able to stand back and say, 'These results and analysis suggest that the underlying mechanism is x, y or z, and we can test this hypothesis by doing another experiment.' But be careful, for there is often more than one physically plausible model that will fit the data well! In such cases, more information would probably be needed to choose one model over another. Indeed, modeling can be more a matter of ruling out what seems not to work instead of choosing what is *known* to be 'right.'

A great many books have been written on data analysis, and it is hardly the main subject of the present one, so we'll end this digression presently. The important thing to remember is that knowing the ins and outs of non-linear regression is very important in modern biochemistry. Regression analysis, like all areas of scientific study beyond textbook-level knowledge, is something of an art. At the same time, though, expertise in data analysis (or information technology) does not necessarily make a person a good physical biochemist.

F. | Allosteric regulation

The metabolic and physiological processes of living cells and tissues are regulated in myriad ways. In general, regulatory biochemical mechanisms respond to change by *damping* the effect (though there are, of course, situations were *amplification* occurs). A general type of metabolic control that is common to all living things is **feedback inhibition**, where the concentration of a product of a metabolic pathway regulates an enzyme upstream by binding to it, the result being a decrease in enzyme activity. Enzymes regulated in this way are often the ones that catalyze the first step in the pathway that does not lead to other products (Fig. 7.11). This gives the cell maximum control over metabolism and conserves energy resources. Although feedback inhibition can be brought about in a variety of ways, all of the ways share the property that when the concentration of metabolite is low its production proceeds, and when the concentration is high production is inhibited. This

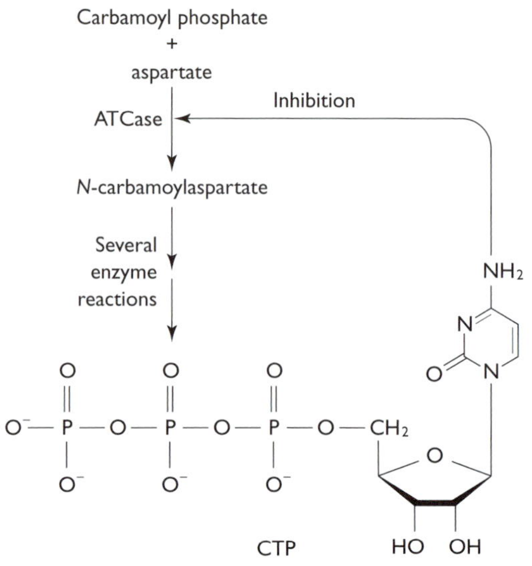

Fig. 7.11 Feedback inhibition. CTP closely resembles ATP. When the concentration of CTP is low, it does not inhibit ATCase because it does not bind. ATCase and the other enzymes of the pathway produce CTP. When [CTP] is high, however, it binds ATCase and inhibits production of CTP. Why should the energy resources of the cell be spent unnecessarily?

is yet another way in which the energy resources of the cell are spent on an as-needed basis and not willy-nilly.

The synthetic pathway of the amino acid histidine is a specific example of feedback inhibition. The pathway involves about ten different enzymes. If enough histidine is present in the cell, synthesis stops and energy resources are not depleted without reason. Enzyme inhibition comes about by the binding of histidine, the end product, to the first enzyme on the synthetic pathway. Another example is the regulation of gene expression by DNA-binding proteins. Certain protein molecules called repressors prevent the synthesis of unneeded enzymes by binding to specific locations on chromosomal DNA and preventing the transcription of the mRNA required for protein synthesis. If the substrates of these enzymes are present, however, enzyme synthesis occurs; synthesis is induced by the presence of the metabolite. For example, addition of galactose to a growth medium containing E. coli induces the synthesis of β-galactosidase. In this way synthesis of the galactose-metabolizing enzyme is regulated and the energy of the cell is not spent unnecessarily. (See Chapters 8 and 9 for additional information on 'energy conservation'.)

The regulation of ligand binding to a macromolecule is known as **allosteric regulation** (Greek, *allos*, other + *stereos*, space), and it one of the most interesting of all aspects of the function of biological macromolecules. The term *allosteric* gets its name from the influence binding at one site has on binding at a remote location in the same macromolecule, possibly the active site. Note the qualitative difference between allosteric regulation and the independent binding of different types of ligand to different sites on the same enzyme with no effect on enzyme activity. The ligand that brings about allosteric regulation of the binding of another ligand is called an **effector** or modulator. The ligands involved in allostery can be identical, as in the case of oxygen binding to hemoglobin, or different, as with aspartate transcarbamoylase (ATCase). The binding of the effector can either increase or decrease the affinity of the protein for another ligand. For instance, the binding of the first oxygen molecule to hemoglobin increases the affinity of the other sites for oxygen; this is positive allostery. And the binding of ATP to ATCase increases enzymatic activity, while binding of CTP decreases activity. CTP is a negative allosteric effector of ATCase.

Let's take a closer look at ATCase. This oligomeric enzyme of identical subunits catalyzes the formation of N-carbamoylaspartate from carbamoyl phosphate and aspartate (Fig. 7.11). Synthesis of N-carbamoylaspartate is the first step in the biosynthesis of pyrimidines, including cytosine, thymine and uracil. ATCase has at least two stable folded conformations, known as R (high substrate affinity) and T (low substrate affinity). The relative stability of these states is affected by the binding of ATP (a purine) to R and CTP (a pyrimidine) to T. Note, however, that although different nucleotides bind to different conformations, they bind to the same site on the enzyme! That is, binding is competitive. Both ATP binding and CTP binding to ATCase are examples of **homoallostery**. That is because the several binding sites in ATCase are intrinsically identical.

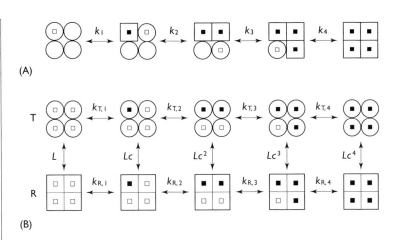

Fig. 7.12 Models of ligand binding. The ligand is not shown in the binding equilibria for the sake of clarity. (A) The KNF or sequential model. The binding affinity increases as the number of ligands bound increases. (B) The MWC or concerted model. An equilibrium constant L describes the structural change between the T state and the R state in the absence of ligand. The R conformation has a higher affinity for ligand than the T conformation by the factor c, the ratio of the microscopic binding affinities (k_{R_i}/k_{T_i}). In some cases, L and c are completely separable, but in the case of Hb they are linked, thwarting independent evaluation.

There are two basic models of allosteric regulation in proteins: the sequential model and the concerted model (Fig. 7.12). In the sequential model (**KNF model**), proposed by Daniel Edward Koshland, Jr. (1920–), G. Némethy and D. Filmer, the folded structure of a macromolecule is assumed to be sufficiently plastic for the binding of a ligand at one site to directly alter the conformation of the macromolecule at another site and thereby affect the affinity of the second site for its ligand. When cooperativity is positive, binding of the first ligand results in a conformational change that increases the affinity for a ligand at the second site, and so on. Binding affinities in allosteric systems usually vary within a range of a few kilocalories per mole (a few orders of magnitude). The binding of different ligands to the same site could have different effects on enzyme conformation at remote locations. In the concerted model (**MWC model**), proposed by Jacques Lucien Monod (1910–1976) and Jean-Pierre Changeux (1936–), both Frenchmen, and Jeffries Wyman (1901–1995), an American, each subunit of a multicomponent macromolecule has two folded conformations (T and R), the conformations are in equilibrium regardless of the presence of a ligand, the binding of a ligand to one site has no direct influence on the binding of a ligand to another site, the affinity of a ligand for a subunit depends only on the conformation of the subunit and not on the number of ligands bound, and all subunits are either in one conformation or the other. Instead of altering the conformation of the enzyme, binding shifts the equilibrium between the two conformations of the occupied subunit. The association constants include statistical factors that account for the number of ligands bound. In practice, it can difficult if not impossible to say which of these two models is a better representation of a particular physical situation.

G. Proton binding

Proton binding, seemingly insignificant, can have a dramatic impact on protein structure and function. We have seen in Chapter 2 how changes in pH affect protein thermostability and in Chapter 5 how it influences solubility. Even when it brings about no large change in

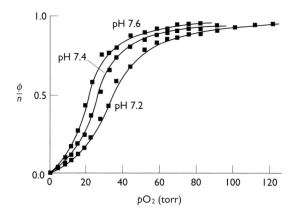

Fig. 7.13 Bohr effect. The degree of saturation of Hb depends not only on pO_2 but also on pH, as shown. The Bohr effect describes the effect of changes in pH on the number of oxygen molecules bound. The binding of oxygen is *linked* to the binding of protons. In general a thermodynamic change of the state of Hb will involve contributions from oxygen binding, proton binding, ion binding, and structural rearrangement.

structure, proton binding or release can have marked physiological consequences. Hb presents an important illustrative example. A 1904 research paper on the effect of pH on the affinity of oxygen for Hb describes what is known as the **Bohr effect** (named after the Dane Christian Bohr, father of the Nobel laureate in physics, Neils Bohr): the release of oxygen when Hb binds protons (Fig. 7.13). In the reverse process, roughly two protons are released when Hb binds thee oxygen molecules. These protons add up, and together they can result in a significant change in the pH of the local environment. We can see how the Bohr effect fits into the broader scheme of things by recalling the beginning of Chapter 5, where we discussed some aspects of respiration on a detailed level. For every mole of O_2 consumed by humans in respiration, approximately 0.8 moles of CO_2 are formed. Carbonic anhydrase, a red blood cell enzyme, catalyzes the conversion of CO_2 to carbonate and H^+, keeping the partial pressure of CO_2 low in the capillaries. As a result, the pH of the capillaries is acidic. Combination of low pH and low partial pressure of O_2 in the capillaries results in release of oxygen from Hb. The Bohr effect is particularly important in very active muscle, where the demand for O_2 is high. The role of specific inter-subunit interactions in bringing about such properties is highlighted by the finding that tetramers of the β-chain, which occur in some types of thalassemia (an inherited anemia resulting from the impaired synthesis of a hemoglobin subunit), exhibit neither cooperativity nor the Bohr effect. The unfortunate persons who suffer from some types of thalassemia require regular blood transfusions.

A deeper look at proton binding seems in order. Returning to Eqn. 7.8, and letting the ligands be protons, we have

$$\phi = nk[H^+]/(1 + k[H^+]) \tag{7.49}$$

where n is the number of dissociable protons that bind with microscopic *association* constant k. In terms of the dissociation constant, Eqn. 7.49 is

$$\phi = \frac{n[H^+]/k_d}{1 + [H^+]/k_d} \tag{7.50}$$

Defining ρ as the average number of protons dissociated from the fully protonated macromolecule,

$$\rho = n - \phi \tag{7.51}$$

because ϕ is the average number of protons bound. Substituting in Eqn. 7.50 gives

$$\rho = n - \phi = n - \frac{n[H^+]/k_d}{1 + [H^+]/k_d} = n\left(1 - \frac{[H^+]/k_d}{1 + [H^+]/k_d}\right) = \frac{nk_d/[H^+]}{1 + k_d/[H^+]} \tag{7.52}$$

Rearranging Eqn. 7.51 and substituting in Eqn. 7.52 for ρ, yields

$$n - \rho = n - \frac{nk_d/[H^+]}{1 + k_d/[H^+]} = \frac{n}{1 + k_d/[H^+]} \tag{7.53}$$

Combining Eqns. 7.52 and 7.53, we have

$$\frac{\rho}{\phi} = \frac{[A^-]}{[HA]} = \frac{\rho}{n - \rho} = \frac{\dfrac{nk_d/[H^+]}{1 + k_d/[H^+]}}{\dfrac{n}{1 + k_d/[H^+]}} = \frac{k_d}{[H^+]} \tag{7.54}$$

Dividing the numerator and denominator of the third term in Eqn. 7.54 by n, we obtain

$$\rho/(n - \rho) = \psi/(1 - \psi) = k_d/[H^+] \tag{7.55}$$

ψ is the fraction of protons *dissociated*. Taking the logarithm of both sides of Eqn. 7.55 gives

$$\log[\psi/(1 - \psi)] = \log k_d - \log[H^+] = -pk_a + pH \tag{7.56}$$

Note how closely Eqn. 7.56, a form of the Hill equation, resembles the Henderson–Hasselbalch equation (Eqn. 4.57)!

Throughout this discussion, we have assumed that there is but one type of ionizable chemical group present and that all occurrences of it have the same k_d. Real proteins are not like this, of course, not only because different acidic and basic side chains are present, but also because the specific electronic environment of a particular type of side chain can have a profound effect on its pK_a. Glutamic acid 35 of hen lysozyme, for example, has a pK_a of about 6 in the folded state, more than two pH units above the pK_a of glutamic acid side chains in other parts of the folded state of the protein! And several anomalously low pK_as of histidine residues in myoglobin play an important role in the stability of the molten globule state of apomyoglobin (myoglobin from which heme has been removed) at acidic pH. In view of this, Eqn. 7.52 is written in more general form as

$$\rho = \frac{n_1 k_{d,1}/[H^+]}{1 + k_{d,1}/[H^+]} + \frac{n_1 k_{d,2}/[H^+]}{1 + k_{d,2}/[H^+]} + \cdots \tag{7.57}$$

where there are n_1 protons with dissociation constant $k_{d,1}$, n_2 with $k_{d,2}$, and so on. In the absence of further information, each dissociation constant can be assumed to be the same as the 'intrinsic k_d,' the value for the free amino acid in the absence of 'end-effects' resulting from the partial protonation of the amino and carboxyl groups. This is the same as assuming that no titratable site is influenced by the electric field of another titratable site or by other electrons in the vicinity. This of course is an oversimplification, as the example of Glu35 proves.

As we have seen in Chapter 2 and Chapter 5, proteins can be denatured by extremes of pH. We would like to know how we can model such effects mathematically, in order to determine the electrical work done to unfold a protein. One way is to write the total free energy difference between the unbound and bound states of each proton as the sum of the intrinsic free energy (ΔG_{in}) change plus that arising from electrical effects (ΔG_e):

$$\Delta G = \Delta G_{in} + \Delta G_e \tag{7.58}$$

In terms of microscopic dissociation constants, this is

$$-RT \ln k = -RT \ln k_{in} + \Delta G_e \tag{7.59}$$

$$-RT \log k = -RT \log k_{in} + \Delta G_e / 2.303 \tag{7.60}$$

$$pk_a = pk_{a,in} + \Delta G_e / (2.303RT) \tag{7.61}$$

If $pk_{a,in}$ is known and pk_a can be determined experimentally, e.g. by NMR spectroscopy, then one can measure ΔG_e, the work done to overcome electrical effects.

H. References and further reading

Ackers, G. K. (1998) Deciphering the molecular code of hemoglobin in allostery. *Advances in Protein Chemistry*, **51**, 185–254.

Ackers, G. K., Shea, M. A. & Smith, F. R. (1983). Free energy coupling with macromolecules. The chemical work of ligand binding at the individual sites in cooperative systems. *Journal of Molecular Biology*, **170**, 223–242.

Aggarwal, A. K., Rodgers, D. W., Drottar, M., Ptashne, M. & Harrison, S. C. (1988). Recognition of a DNA operator by the repressor of phage 434: a view at high resolution. *Science*, **242**, 899–907.

Atkinson, D. E. (1965). Biological feedback control at the molecular level. *Science*, **150**, 851–857.

Attle, A. D. & Raines, R. T. (1995). Analysis of receptor-ligand interactions. *Journal of Chemical Education*, **72**, 119–124.

Baldwin, J. & Chothia, C. (1979). Hemoglobin: the structural changes related to ligand binding and its allosteric mechanism. *Journal of Molecular Biology*, **129**, 175–220.

Breslauer, K. J., Freire, E. & Straume, M. (1992) Calorimetry: a tool for DNA and ligand–DNA studies. *Methods in Enzymology*, **211**, 533–567.

Connelly, P. R., Thomson, J. A., Fitzgibbon, M. J. & Bruzzese, F. J. (1993). Probing hydration contributions to the thermodynamics of ligand binding by proteins. Enthalpy and heat capacity changes of tacrolimus and rapamycin binding to FK506 binding protein in D_2O and H_2O. *Biochemistry*, **32**, 5583–5590.

Connelly, P. R., Aldape, R. A., Bruzzese, F. J., Chambers, S. P., Fitzgibbon, M. J., Fileming, M. A., Itoh, S., Livingston, D. J., Navia, M. A. & Thomson, J. A. (1994). Enthalpy of hydrogen bond formation in a protein-ligand binding reaction. *Proceedings of the National Academy of Sciences of the United States of America*, **91**, 1964–1968.

Cooper, A. & Johnson, C. M. (1994). Introduction to microcalorimetry and bio-molecular energetics. In *Methods in Molecular Biology*, ed. C. Jones, B. Mulloy, A. H. Thomas, vol. 22, pp. 109–124. Ottowa: Humana.

Creighton, T. E. (1993). *Proteins: Structures and Molecular Properties*, 2nd edn, ch. 8.4. New York: W. H. Freeman.

Debru, C. (1990). Is symmetry conservation an unessential feature of allosteric theory? *Biophysical Chemistry*, **37**, 15–23.

Doyle, M. L., Louie, G., Dal Monte, P. R. & Sokoloski, T. D. (1995). Tight binding affinities determined from thermodynamic linkage to protons by titration calorimetry. *Methods in Enzymology*, **259**, 183–194.

Doyle, M. L. (1997). Characterization of binding interactions by isothermal titration. *Current Opinion in Biotechnology*, **8**, 31–35.

Edelstein, S. J. & Changeux, J. P. (1998) Allosteric transitions of the acetycholine receptor. *Advances in Protein Chemistry*, **51**, 121–184.

Fersht, A. R. (1999). *Structure and Mechanism in Protein Science: a Guide to Enzyme Catalysis and Protein Folding*, ch. 6.D.1. New York: W. H. Freeman.

Fruton, J. S. (1999). *Proteins, Enzymes, Genes: the Interplay of Chemistry and Biology*. New Haven: Yale University Press.

Haynie, D. T. & Ponting, C. P. (1996). The N-terminal domains of tensin and auxilin are phosphatase homologues. *Protein Science*, **5**, 2643–2646.

Honig, B. & Nicholls, A. (1995). Classical electrostatics in biology and chemistry. *Science*, **268**, 1144–1149.

Hou, S., Larsen, R. W., Boudko, D., Riley, C. W., Karatan, E., Zimmer, M., Ordal, G. W. & Alam, M. (2000). Myoglobin-like aerotaxis transducers in Archaea and Bacteria. *Nature*, **403**, 540–544.

Klotz, I. M. (1986). *Introduction to Biomolecular Energetics*, ch. 10. Orlando: Academic.

Koshland, D. E., Némethy, G. & Filmer, D. (1966). Comparison of experimental binding data and theoretical models in proteins containing subunits. *Biochemistry*, **5**, 365–385.

Kuroki, R., Kawakita, S., Nakamura, H. & Yutani, K. (1992). Entropic stabiliza-tion of a mutant human lysozyme induced by calcium binding. *Proceedings of the National Academy of Sciences of the United States of America*, **89**, 6803–6807.

Ladbury, J. E. (1995). Counting the calories to stay in the groove. *Structure*, **3**, 635–639.

Ladbury, J. E. & Chowdhry, B. Z. (1996). Sensing the heat – the application of iso-thermal titration calorimetry to thermodynamic studies of biomolecular interactions. *Chemistry and Biology*, **3**, 791–801.

Miller, K. R. & Cistola, D. P. (1993). Titration calorimetry as a binding assay for lipid-binding proteins. *Molecular and Cellular Biochemistry*, **123**, 29–37.

Monod, J., Wyman, J. & Changeux, J. P. (1965). On the nature of allosteric transi-tions: a plausible model. *Journal of Molecular Biology*, **12**, 88–118.

Morton, A., Baase, W. A. & Matthews, B. W. (1995). Energetic origins of specific-ity of ligand binding in an interior nonpolar cavity of T4 lysozyme, *American Chemical Society*, **34**, 8564–8575.

Perutz, M. F. (1978) Hemoglobin structure and respiratory transport. *Scientific American*, **239**, 92–125.

Perutz, M. F. (1989) Mechanisms of cooperativity and allosteric regulation in proteins. *Quarterly Reviews of Biophysics*, **22**, 139–236.

Schoelson, S. E. (1997). SH2 and PTB domain interactions in tyrosine kinase signal transduction. *Current Opinion in Chemical Biology*, **1**, 227–234.

Steinhardt, J. & Beychok, S. (1964). Interactions of proteins with hydrogen ions and other small ions and molecules. In *The Proteins*, 2nd edn, ed H. Neurath, vol. II, pp. 139–304. New York: Academic Press.

Timasheff, S. N. (1993). The control of protein stability with water: how do solvents affect these processes? *Annual Review of Biophysics and Biomolecular Structure*, **22**, 67–97.

van Holde, K. E. (1985). *Physical Biochemistry*, 2nd edn, cc 3.2 & 3.3. Englewood Cliffs: Prentice-Hall.

Voet, D. & Voet, J. G. (1995). *Biochemistry*, 2nd edn, cc. 9-1 & 9-4. New York: Wiley, 1995.

Weber, G. (1975). Energetics of ligand binding to proteins. *Advances in Protein Chemistry*, **29**, 1–83.

Williams, T. I. (ed.) (1969). *A Biographical Dictionary of Scientists*. London: Adam & Charles Black.

Wiseman, T., Williston, S., Brandts, J. F. & Lin, L. N. (1989). Rapid measurement of binding constants and heats of binding using a new titration calorimeter. *Analytical Biochemistry*, **179**, 131–137.

Wyman, J. (1984). Linkage graphs: a study in the thermodynamics of macromolecules. *Quarterly Reviews of Biophysics*, **17**, 453–488.

Wyman, J. & Gill, S. J. (1990). *Binding and Linkage: Functional Chemistry of Biological Macromolecules*. Mill Valley: University Science Books.

I. Exercises

1. There are four fundamental physical forces: gravity, electromagnetism, the weak nuclear force and the strong nuclear force. Electromagnetism and the weak nuclear force have recently been shown to be aspects of the same force, the electroweak force. Which of these forces mediates interactions between a ligand and a macromolecule to which it binds?

2. Find a mathematical expression for the binding free energy in terms of the dissociation constant.

3. When $[L] = K_d$, $F_b = 0.5$. Calculate the concentration of ligand required for 90% saturation and 99% saturation.

4. Describe protein folding/unfolding in terms of 'binding' of heat.

5. Once O_2 has become bound to Mb it is not released back into the blood, because p_{50} is below venous pressure and venous pressure is lower than arterial pressure. Describe the association of oxygen with Mb in terms of the chemical potential of O_2.

6. Write down a general equation relating k_i to K_i.

7. The following data were collected in a binding experiment.

[L]	ϕ	[L]	ϕ
0	0.0	55	0.85
5	0.33	60	0.86
10	0.50	65	0.87
15	0.60	70	0.875
20	0.67	75	0.88
25	0.71	80	0.89
30	0.75	85	0.89
35	0.78	90	0.90
40	0.80	95	0.90
45	0.82	100	0.91
50	0.83	—	—

Plot the data by the Scatchard method to evaluate the dissociation constant. Use a spreadsheet and the relevant equations from the text to determine the binding constant by non-linear least squares regression.

8. Show that $\phi/[L] = nK_a - \phi K_a$ for n identical binding sites.

9. Binding of a ligand stabilizes the folded conformation of macromolecule. Explain in thermodynamic terms why this must be so.

10. Given the definition of *homoallostery*, define *heterallostery*.

11. Show that

$$\frac{\phi}{n} = \frac{x(1+x)^{n-1} + Lcx(1+cx)^{n-1}}{(1+x)^n + L(1+cx)^n}$$

for homotropic allosteric interactions. Write down an expression for the fractional saturation for ligand binding. Define $x = [L]/k_R$ and $c = k_R/k_T$. Use $k_R = \{(n-i+1)/i\}[R_{i-1}][L]/[R_i]$, $i = 1, 2, 3, \ldots, n$ to show that $([R_1] + 2[R_2] + \ldots + n[R_n]) = [R_0]\{n\alpha + 2n(n-1)\alpha^2/2 + \ldots + nn!\alpha^n/n!\} = [R_0]\alpha n(1+\alpha)^{n-1}$. Show also that $([R_0] + [R_1] + \ldots + [R_n]) = [R_0]\{1 + n\alpha + \ldots + n!\alpha^n/n!\} = [R_0]\alpha n(1+\alpha)^{n-1}$, that $([T_1] + 2[T_2] + \ldots + n[T_n]) = [T_0]([L]/k_T)n(1+[L]/k_T)^{n-1} = L[R_0]c\alpha n(1+c\alpha)^{n-1}$, and that $([T_0] + [T_1] + \ldots + [T_n]) = [T_0](1+[L]/k_T)^n = L[R_0](1+cx)^n$. Note that $L \neq [L]$. (See Fig. 7.12.) Combine these terms to give the general result for homotropic allosteric interactions.

12. Which of the Scatchard plots shown below indicates that compound a binds with half the affinity as compound b but to twice the number of sites?

Figure for Exercise 12. Scatchard plots. The ratio of the bound concentration to the free concentration is plotted against the bound concentration.

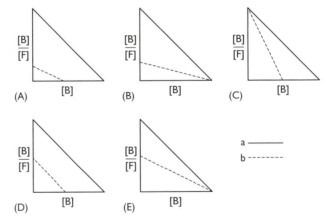

13. The Scatchard plots below compare data on the binding of a hormone (H) to the receptors (R) of five abnormal persons (solid line in plots A–E) and a normal person (broken line in each plot). [R•H] = concentration of bound hormone, [H] = concentration of free hormone. Which patient shows a decreased number of binding sites but the same receptor affinity as the normal person?

14. Consider a homotrimeric protein with three identical and independent binding sites and microscopic association constants of 10^6. Plot the fractional saturation of the protein against the free ligand

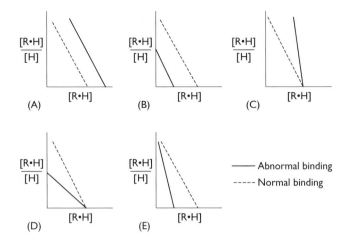

(A) (B) (C)

(D) (E)

—— Abnormal binding

- - - - Normal binding

Figure for Exercise 13. Scatchard plots. The ratio of the bound concentration of hormone to free concentration of hormone is plotted against the bound concentration of hormone.

concentration. Write down equations to describing the macroscopic binding constants (K_1, K_2, K_3) in terms of the microscopic binding constant.

15. Suppose you are studying the binding of heme to myoglobin using equilibrium dialysis (Chapter 5). The total concentration of myoglobin is 10 μM. The following data were obtained at equilibrium.

Experiment	[heme in chamber without myoglobin (μM)	[heme] in chamber with myoglobin (μM)
1	3	5.3
2	30	37.5

Calculate the concentration of bound and free myoglobin in the chamber where myoglobin is present. Use these values to calculate the fractional occupancy of myoglobin at the two ligand concentrations. Determine the affinity of heme for myoglobin using your favorite method. Can these data be used assess binding cooperativity? Explain.

16. Give three biological examples of negative feedback. Give three biological examples of positive feedback. Comment on the role of binding in each example.

For solutions, see http://chem.nich.edu/homework

Chapter 8

Reaction kinetics

A. | Introduction

The foregoing chapters have focused on practical and applied aspects of thermodynamics and statistical mechanics. These subjects provide ways of thinking about energy transformation, methods for determining the direction of spontaneous change, the magnitude of thermodynamic quantities when a system passes from one state to another, and the molecular origin of change. Useful as they are, however, thermodynamics and statistical mechanics do not tell us everything we'd like to know: they give no direct indication of the rate at which a chemical change will occur nor how the rate of change will vary with conditions.

The present chapter seeks to fill a few of the gaps remaining from a strictly thermodynamic treatment of biochemical change. It might seem counter-intuitive for this chapter to appear next to last instead of first, as one of the most basic aspects of our experience of the world is constant change. Plants grow, go to seed, and die, while animals move, eat, reproduce, and die. And the molecules of which bacteria, plants, and animals are made are *always* moving. But, the title of this book is *Biological Thermodynamics*, not *Biological Kinetics*!

As we have seen, analysis of free energy changes provides a way of answering such questions as 'Why is most of the energy of the glucose molecule obtained in the citric acid cycle and not in glycolysis?' But in order to respond intelligently to 'Does H_2CO_3 break down fast enough on its own to permit metabolic CO_2 to be excreted into the atmosphere?' we must turn to reaction kinetics. In a previous chapter we met the allosteric enzyme ATCase. This enzyme is inhibited by CTP and activated by ATP, though neither of these nucleotide triphosphates binds in the active site. ATCase provides an excellent example of how the rate of enzymatic activity can be regulated through allostery.

Back in Chapter 4 we described chemical equilibrium as the condition of equivalent forward and reverse rates of reaction. When these rates are identical, their ratio is 1: $\ln(1) = 0$, so $\Delta G = 0$, the condition for equilibrium at constant T and p. Awareness of the relative rates of the forward and reverse reactions can be useful in the biochemical laboratory in many ways. For example, the ELISA assay we discussed in Chapter 5 is based on the knowledge that the reverse ('off') rate of antibody–antigen binding is tiny in comparison with the forward ('on') rate, even though dissociation of antibody from antigen is predicted by mass action. Rates are also important to the molecular collisions that give rise to diffusion (Chapter 6). A more in-depth treatment of association and collision rates has been postponed until now, because we have wished to keep equilibrium thermodynamics as our principal focus. This chapter provides an introduction to the measurement of kinetic properties of macromolecules, include enzyme activity.

Our general approach will be to start with the phenomenological perspective. Once we've reconnoitered the lay of the land, we'll pose questions about the molecular mechanisms that underlie its topography. In other words, we're not going to start from highly polished theoretical concepts and work our way down to approximations that are actually useful. In part that's because the development of most areas of science has not followed a top-down approach. Nevertheless, most of this chapter will deal with modeling kinetic behavior, and that is necessarily a matter of mathematics. Importantly, the equations we shall derive are simple enough to be useful in practice, predictions of the models have been be tested, and in many cases the predictions correspond well enough to measured values. We shall also aim to have shown by the chapter's end how the kinetics of certain phenomena – protein folding kinetics, hydrogen exchange kinetics, muscle contraction – link to the thermodynamic quantities that have been our main concern till now.

Figure 8.1 shows an energy profile for a generic chemical reaction. As we have seen, in order for the reaction to proceed *spontaneously* at constant temperature and pressure, the Gibbs free energy of the products must be lower than the Gibbs energy of the reactants. An energy 'hill' or 'barrier' separates reactants from products, preventing immediate conversion of the one into the other, and the height of the barrier is called the **activation energy**, E_a. This is the (average) amount of energy that must be added to the reactants to convert them to products. Note that the height of the barrier depends on the direction of the reaction. On a hike in the countryside, a steep hill raises questions about whether and how to proceed: in the absence of climbing gear, a sheer cliff is practically an absolute barrier, while a gentle gradient might do little more

Fig. 8.1 Reaction profile. Energy (Gibbs free energy) is plotted against the reaction coordinate, a pseudo-coordinate that represents the progress of the reaction. The Gibbs free energy of the products is lower than the energy of the reactants, and we should therefore expect the products to form spontaneously. In order for the reactants to form products, however, a substantial energy barrier must be traversed. The height of this barrier, the activation energy (E_a), determines the kinetics of the process: the reaction is slow when the barrier is high and fast when it is low.

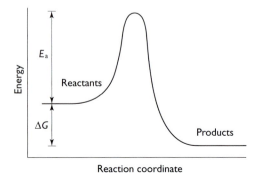

than slow things down somewhat. Similarly, if a reaction pathway is relatively slow, E_a must be large, so large in fact that its value might be very hard to determine accurately. When the rate of a reaction is relatively high, the process will make a substantial contribution to the overall reaction, and it cannot be ignored. In such case E_a can often be measured with a small error term. In the context of metabolic reactions, off-pathway reactions *always* occur, despite the exquisite specificity of enzymes, but such processes are often so slow (the energy barrier is so high) that they can be assumed not to occur. But not *always*! Indeed, some cancers are characterized by a change in the proportion of certain types off-pathway reactions, leading to abnormal cell properties. The top of the energy hill represents the free energy of the **transition state**, a transient chemical species that is unstable at equilibrium.

A couple of examples will help to illustrate how the concept of activation energy sheds light on the way the world works. Some years ago a big fire broke out in Malibu, California. To the amazement of many observers, a few houses began blazing away without having been touched by the flames. It was later surmised that the fire had generated such an intense heat that infrared radiation passed through the glass of closed windows. This heated up drapes hanging inside and they eventually ignited; the heat energy was so intense that it had exceeded the activation energy for oxidation of the curtains. In contrast, other houses nearby did not go up in flames. These dwellings were apparently protected by aluminum (in British English, aluminium) blinds hanging in the windows, which deflected enough of the infrared radiation to keep the temperature inside below the activation energy for oxidation of curtains.

Nitroglycerin is a liquid explosive. It was invented in 1846. Unlike curtains, which are usually made of cellulose or a synthetic material and take a lot of heat to combust, nitroglycerin has a low activation energy. Relatively slight agitation of the liquid is sufficient to set off very rapid formation of hot gas, despite a small free energy difference between reactants and products. Oxidation occurs rapidly because nitroglycerin is made of oxygen. Combustion releases as well as consumes oxygen, so the overall rate of reaction is not limited by access to atmospheric oxygen, as for instance in the combustion of a hydrocarbon like octane. Alfred Bernhard Nobel (1833–1896), the Swedish industrialist after whom the annual prizes in physics, chemistry, medicine or physiology, economics, literature, and peace are named, was driven to invent a safer explosive near the end of the nineteenth century, when his brother and four workers were killed in the family nitroglycerin plant. Nobel made what he called dynamite by mixing nitroglycerin, which is oily, with powdery silica. The mixture is a dry solid that is much more stable (has a higher activation energy) than nitroglycerin alone and can also be molded into shapes that stack compactly and can be carried easily – cylindrical 'sticks.'

We might suppose from these examples that chemical change in the universe as we know it is 'controlled' not so much by the laws of thermodynamics as by the laws of chemical change. The First and Second Laws are none the less 'boundary conditions' on the physically possible: all chemical changes must be consistent with these laws, but the actual rate of change will depend on both the chemistry involved and the physical conditions. This chapter provides insight into how reaction rates relate to the First and Second Laws.

Finally, we wish to say something about the breadth of the range of rates of biochemical processes. Some processes are exceedingly fast, others slow. For instance, the lifetime of the excited state of chlorophyll in the capture of electromagnetic energy is about 10^{-10} s. This is so fast that chlorophyll would be practically useless in a typical biochemical reaction. In contrast, the reduced organic molecules into which the Sun's energy is transformed have a lifetime on the order of months or years. The fastest enzymes work at a rate that is diffusion controlled; they are limited only by the amount of time it takes for a substrate molecule to come into the active site. Some polypeptides are stable in pure water for tens of thousands of years; protein turnover in cells is obviously much faster.

B. | Rate of reaction

A rate is of course a measure of how quickly something happens. For example, the rate of food intake in adult humans is about three meals per day; it's higher for babies! The rate of inflation is a measure of the change in value with time of a monetary unit, for example the yen or the peso. In chemistry, the **rate of reaction**, J, is simply a measure of how rapidly the concentration of a product or reactant changes in time. One should in general expect J itself to vary with time, as it clearly does when the amount of a reactant is limited and consumed by a process.

Recall ATP hydrolysis, which we discussed at some length in Chapter 5. The nucleotide hydrolyzes spontaneously because there is a Gibbs free energy decrease on going to ADP and P_i; hydrolysis is thermodynamically favorable. This reaction is so favorable that it is essentially irreversible at room temperature; hydrolysis results in the increasing loss of ATP. The smaller the number of ATP molecules per unit volume, the smaller the number of reactions per unit volume and the lower the rate of reaction. We can express this in symbols as follows:

$$\text{ATP} \rightarrow \text{ADP} + P_i \tag{8.1}$$

In terms of the depletion of ATP, the rate of reaction, J, is

$$J = -\Delta[\text{ATP}]/\Delta t \tag{8.2}$$

where t is time. We can see that J has dimensions [concentration][time]$^{-1}$. The rate of reaction can also be expressed in terms of the products, as

$$J = +\Delta[\text{ADP}]/\Delta t = +[P_i]/\Delta t \tag{8.3}$$

based on the stoichiometry of the reaction. Figure 8.2 shows the concentration versus time for ATP and ADP.

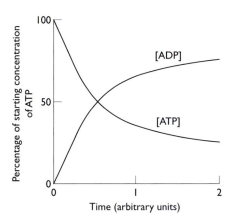

Fig. 8.2 Kinetics of nucleotide hydrolysis *in vitro*. ATP hydrolyzes spontaneously at room temperature, and the reaction is essentially irreversible. Because of the stoichiometry of the reaction the rate of ATP depletion is equal to the rate of ADP accumulation. At some stage during the reaction, the concentration of ATP will be half the starting concentration. At this time, [ATP] = [ADP].

For a more general reaction, say,

$$aA + bB \rightarrow cC + dD \tag{8.4}$$

the rate is given by

$$J = -\frac{1}{a}\frac{\Delta[A]}{\Delta t} = -\frac{1}{b}\frac{\Delta[B]}{\Delta t} = +\frac{1}{c}\frac{\Delta[C]}{\Delta t} = +\frac{1}{d}\frac{\Delta[D]}{\Delta t} \tag{8.5}$$

The reason why the rate appears as $-(\Delta[B]/\Delta t)/b$ and not $-\Delta[B]/\Delta t$ will be clear from another example. Suppose

$$A + 2B \rightarrow C \tag{8.6}$$

For every mole of A converted into C, two moles of B are consumed. So the rate of consumption of B is twice as great as that of A. Taking this into account, $-\Delta[B]/\Delta t$ must be divided by two in order to equal $-\Delta[C]/\Delta t$ or $-\Delta[A]/\Delta t$. It is important to bear in mind that J measures the rate of reaction as a whole, not the rate of consumption or production of one component of the reaction, which is given by $-\Delta[Z]/\Delta t$ for chemical species Z.

To summarize, we know from experience that change is a basic aspect of the world. We know that a chemical reaction will occur at a certain rate. And we have seen a way of describing the rate of reaction in simple mathematical terms. Time to move on to deeper things!

C. Rate constant and order of reaction

We also know from experience that the shape of a kinetic trace for a given chemical is not the same for every reaction. In other words, in one reaction, when we monitor the time rate of change of one chemical species, we get a curve of a certain shape; and in different reaction, when we monitor the rate of change of the same chemical species, we get a curve that is clearly qualitatively different from the first curve. Why? One possibility might be that the stoichiometry is different in the two situations. Another might be that the mechanism is different. To describe the various possible types of chemical behavior, we need to make our mathematical description of things a bit more sophisticated.

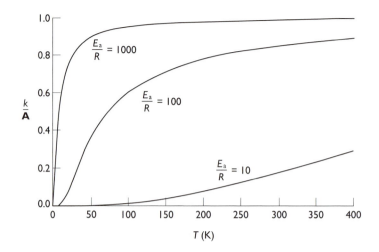

It has been found *experimentally* that, in general, the rate of a reaction is related to the concentration of a reactant as

$$J \propto [A]^n \tag{8.7}$$

n is called the **order of reaction**. According to empirical evidence, **the order of reaction for a component is *often but not always* identical to its stoichiometry; the order of a reaction order must be determined *empirically*.** We can upgrade Eqn. 8.7 to the status of an equality by inserting a constant of proportionality. This gives

$$J = k[A]^n \tag{8.8}$$

The **rate constant**, k, is a phenomenological parameter; it is the inverse of the time constant, τ (compare the average time between collisions in Chapter 6). (Note that the 'k' used here is not the microscopic binding constant of the previous chapter! In principle there is no reason why the same symbol can't be used for different things, as long as it is clear enough what the symbol means!) The value of k depends entirely on the reaction of interest and it can only be determined experimentally; the theory of reaction rates is at too primitive a stage to calculate k in the general situation. The rate constant is a measure of the activation energy, and it is expressed in terms of the activation energy as

$$k = A\exp(-E_a/RT) \tag{8.9}$$

where $\mathbf{A}$ is an empirical quantity known as the **frequency factor** (Fig. 8.3). Note how similar this equation is to $K = \exp(-\Delta G/RT)$. Eqn. 8.9 tells us that if k can be measured and T is known, then E_a can be determined. For instance, if $k = 10^{-3}\,s^{-1}$ at 300 K, and $\mathbf{A} = 1$, then $E_a = -RT\ln k = -8.314\,J\,mol^{-1}\,K^{-1} \times 300\,K \times \ln 10^{-3} = 17\,kJ\,mol^{-1}$. The exponential function is always positive.

In the general situation, where one has

$$A + B + C + \ldots \rightarrow products \tag{8.10}$$

the overall rate of reaction can be written as

$$J = k[A]^{n_A}[B]^{n_B}[C]^{n_C}\ldots \tag{8.11}$$

where n_A is the order of reaction with respect to component A, *etc.* The overall order of reaction is $n_A + n_B + n_C + \ldots$. Often, the exponents in Eqn. 8.11 will reflect the stoichiometry of the reaction, but that is certainly not always the case; the magnitudes of the orders of reaction as well as k must be determined experimentally. Because the form of the mathematical expression for J depends on the reaction, so do the units of a rate constant. For instance, in what is called a **first-order reaction**, the unit of concentration is raised to the first power and k has units of inverse time. In a **second-order reaction**, the dimensions of k are [concentration]$^{-1}$[time]$^{-1}$. If a reaction rate is independent of the concentration of a reactant, the rate is zeroth-order with respect to that reactant.

D. | First-order and second-order reactions

We want to focus on first- and second-order reactions for a couple of reasons. One is that it is often difficult to obtain data of sufficient quality that would support a more complex model (a higher-order reaction). Another is that a very large number of reactions of interest to the biochemist can be described well enough as first-order or second-order, making it hard to justify a more complex one.

Suppose we have the following first-order reaction: A→P. Combining Eqn. 8.2 and Eqn. 8.8, we obtain

$$-\Delta[A]/\Delta t = k[A] \tag{8.12}$$

This can be rearranged in terms of [A], as

$$\Delta[A]/[A] = -k\Delta t \tag{8.13}$$

Using a bit of calculus and algebra, one can show that

$$\ln[A] = \ln[A]_o - kt \tag{8.14}$$

where $[A]_o$ is the concentration of A when $t = 0$ (the beginning of the experiment). An important feature of Eqn. 8.14 is that it is *linear* in t with slope $-k$ (units of [time]$^{-1}$) and ordinate-intercept $\ln[A]_o$ (Fig. 8.4). This

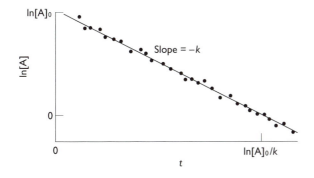

Fig. 8.4 First-order process. A plot of ln[A] against time is linear. The slope is negative, and its magnitude is the rate constant for the reaction. The curve intercepts the axes at two points: at $t = 0$, $\ln[A] = \ln[A]_o$, and $\ln[A] = 0$ when $t = \ln[A]_o/k$.

suggests that if we collected experimental rate data, plotted the logarithm of the concentration of a reactant, and found that the curve was a straight line, we could conclude that the reaction was (approximately) first-order in that reactant. One would have to do this very carefully, though, for in practice it can be difficult to distinguish between a reaction of one order and another order. We'll say more about this below.

Exponentiation of both sides of Eqn. 8.14 gives

$$[A] = \exp(\ln[A]_0 - kt) = e^{\ln[A]_0} e^{-kt} = [A]_0 e^{-kt} \tag{8.15}$$

which on rearrangement is

$$[A]/[A]_0 = e^{-kt} \tag{8.16}$$

This tells us that in a first-order reaction, the ratio of the concentration of the reactant to the starting concentration is a negative exponential function of time. When $t = 0$, the right-hand side of Eqn. 8.16 is one, and $[A] = [A]_0$, just as we should expect. As $t \to \infty$, the right-hand side becomes very small, and $[A]/[A]_0 \to 0$. The rate at which the left-hand side goes to zero depends on the size of the rate constant. When k is very large, $[A]/[A]_0 \to 0$ rapidly!

How long does it take for [A] to decrease to half its starting value? The so-called **half-time** or **half-life** ($t_{1/2}$) of a first-order reaction can be found by setting $[A]/[A]_0 = 0.5$ in Eqn. 8.16. Solving for t gives

$$t_{1/2} = -\ln(1/2)/k = \ln 2/k \tag{8.17}$$

One can describe a reaction in terms of its half-life whether the process of interest is a biochemical reaction or nuclear decay. Mathematical descriptions of things can be very general! Eqn. 8.17 shows clearly that if k is large, as in the case of the fast photosynthetic reaction mentioned in the first section of this chapter, $t_{1/2}$ is small; the reaction could go to completion in a fraction of a second. In contrast, when $t_{1/2}$ is large, as for instance with a relatively slowly decaying radioactive isotope like ^{14}C, a useful one in biochemical research, only a small fraction of a given amount of isotope would decompose in the course of a human lifetime. It is therefore advisable not to eat ^{14}C!

The simplest type of second-order reaction is $2A \to P$, where two molecules of A react with each other to form P. The analog of Eqn. 8.12 for this case is

$$-\Delta[A]/\Delta t = k[A]^2 \tag{8.18}$$

Rearrangement gives

$$\Delta[A]/[A]^2 = -k\Delta t \tag{8.19}$$

which, when transformed by a bit of jiggery-pokery (elementary calculus), yields

$$\frac{1}{[A]} = \frac{1}{[A]_0} + kt \tag{8.20}$$

As in the first-order case, we have a function of the concentration of reactant that is linear in time. Now, though, instead of the natural logarithm of the concentration of A, we have the inverse of the concen-

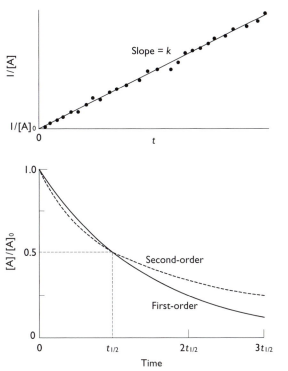

Fig. 8.6 Comparison of a first-order process and a second-order process. The fraction of reactant remaining is plotted against time. The second-order rate constant has been chosen so that $[A] = [A]_o/2$ after one time constant in the first-order reaction. Note how similar the two curves look. 'Clean' data would be needed to be able to judge whether the process was first-order or second-order on the basis of appearance alone.

tration of A, the slope of the line is $+k$ (with units of [concentration]$^{-1}$[time]$^{-1}$), and the intercept of the ordinate is $1/[A]_o$ (Fig. 8.5). Multiplying both sides of Eqn. 8.20 by $[A]_o$ and taking the inverse gives

$$\frac{[A]}{[A]_o} = \frac{1}{1 + kt} \tag{8.21}$$

Above we said that it can be difficult in practice to distinguish between a reaction of one order and another. Figure 8.6 gives some idea of this. A plot of $[A]/[A]_o$ versus time is shown for a first-order model and a second-order model when $[A]/[A]_o = 0.5$ at the half-time of the first-order model. Notice how similar the curves appear to the eye. If the data one has collected are somewhat noisy, as all real data are, it might be very difficult indeed to form a clear sense of the order of the reaction without further analysis. What one often does in such cases is *assume* that the reaction is first-order and then *calculate how well* the model fits the data, taking into account any uncertainty in the measured values. The procedure is repeated on the assumption that the reaction is second-order. A comparison is then made of quantitative measures of goodness-of-fit, and a decision is taken about which model provides a better description of the chemical process.

E. Temperature effects

Reaction rates depend on temperature! For example, because curtains don't ordinarily combust spontaneously at room temperature, k for this reaction must be fairly large when T is about 300 K. So the rate of reaction for the combustion of curtains must increase with temperature. The growth rate of E. coli depends significantly on temperature. At $-80\,°C$, water is frozen, and the growth rate of bacteria is practically nil. This is one reason why stocks are maintained at such a low temperature. At $37\,°C$, however, where the doubling time is maximal, bacterial cells divide every 30 minutes or so on the average. The human gut is a bacterium's paradise: plenty of food (in most places in the world, at least), protection from predators, and no need to pay for the heat needed for metabolic reactions to go! The rapid growth rate of bacteria at room temperature is just one of the reasons why it's a good idea to refrigerate certain foods; something cooks had discovered a very long time before anyone knew bugs existed.

The rate of reaction and therefore the rate constant increase with temperature. We can see how the rate of reaction depends on temperature by manipulating Eqn. 8.9:

$$\ln k(T) = \ln A - E_a/(RT) \tag{8.22}$$

$\ln k$ is a linear function of the inverse of T. Experiments have shown that **in many cases the rate of reaction doubles or triples for a 10 K increase in temperature**. Eqn. 8.22 is known as the Arrhenius equation, and it gets its name from the Swedish Nobel laureate. Arrhenius is also famous for his view that life on Earth arose from 'panspermia,' according to which micro-organisms or spores drifted through space by radiation pressure until finally landing on Earth. We'll come back to this extremely interesting subject in the next chapter.

Now, suppose we know by experiment that the rate of the reaction A + B→P doubles on raising the temperature from $25\,°C$ to $35\,°C$. What is E_a? Assuming that E_a is approximately independent of temperature in this range, we have

$$\ln k(T_1) = \ln A - E_a/(RT_1) \tag{8.23}$$

$$\ln k(T_2) = \ln A - E_a/(RT_2) \tag{8.24}$$

Subtraction of Eqn. 8.24 from Eqn. 8.23 gives

$$\ln k(T_1) - \ln k(T_2) = -E_a/(RT_1) + E_a/(RT_2) = E_a\{-1/RT_1 + 1/RT_2\} \tag{8.25}$$

which, when solved for E_a, is

$$E_a = \frac{\ln\{k(T_1)/k(T_2)\}}{(T_1 - T_2)/RT_1T_2} \tag{8.26}$$

Plugging in the given values and turning the crank on the calculator gives

$$E_a = 12.6 \text{ kcal mol}^{-1} \tag{8.27}$$

This is a handy rule of thumb: **a doubling of the rate of reaction near room temperature corresponds to change in activation energy of a shade under 13 kcal mol^{-1}.**

F. | Collision theory

This section and the next discuss two theories of reaction rates, collision theory and transition state theory. These theories are attempts to rationalize in molecular terms what is known from experiments. We have already encountered two basic aspects of collision theory earlier in this book. The first was in Chapter 2, during the discussion of air molecules, pressure and a bicycle tire. We saw that the pressure of a system is related to the number of collisions that the particles within make with the system boundary, and that the pressure can be increased either by raising the temperature or by pumping more particles into the system. That discussion was linked to the kinetic theory of gases, which is based on the assumption that gas particles are constantly banging into the walls of their container. Then, in Chapter 6, when looking at the molecular interpretation of thermodynamic quantities, we saw how the Brownian motion and diffusion can be explained in terms of particle collisions.

Bearing these *physical* changes in mind, we now turn our attention to chemical changes that might depend on particles colliding. Of course, this *must* have something to do with temperature, because the speed at which particles move around depends very much on temperature. Now, if a certain chemical reaction requires two particles to collide, as it does in the case of ADP and P_i coming together to form ATP, the rate of reaction must depend on concentration. This is just as we saw above in Eqn. 8.11. In other words, [ADP] and $[P_i]$ measure the likelihood that an encounter will occur between them. As we have seen, if the concentration of insects is high, a biker will collide with them relatively frequently. There must be more to collisions than this, however, because experimental studies show that in a typical reaction only about one in 10^{14} collisions leads to the formation of products! Thinking about the exquisite stereospecific nature of enzymes and enzyme-catalyzed reactions helps to throw some light on this. It's not just a collision between an enzyme and substrate that brings about an enzymatic reaction, it's a collision that brings the substrate into the active site with the correct orientation. For instance, a hand can 'collide' with a glove in a huge variety of ways, but only a relatively small number of collisions will result in a glove going on the correct hand. Things are usually less specific with non-biological molecules, but the principles involved are the same.

Another requirement for a reaction in collision theory is that the total relative kinetic energy of the colliding reactants *must* be greater than a certain amount. At a given temperature, some molecules are moving relatively rapidly, others relatively slowly, according to the kinetic theory of gases. So, in the reaction $A + B \rightarrow P$, not only do and A and B have to collide to form P, their relative orientation must be right and they must collide with sufficient energy. This energy is none other than E_a.

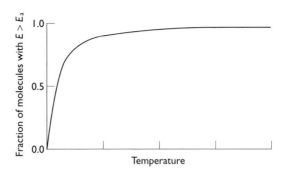

The Maxwell law predicts that the distribution of the relative kinetic energy of particles is

$$\frac{\Delta n}{n} = \frac{\Delta E}{RT} \exp(-E/RT) \qquad (8.28)$$

where n is the number of molecules of a particular type and E is the energy. It can be shown (by way of some calculus) that $n(E_a)$, the fraction of molecules with $E > E_a$, is

$$n(E_a) = n\exp(-E_a/RT) \qquad (8.29)$$

an expression that should remind us of the Boltzmann law (Chapter 6). In other words, the fraction of molecules having $E > E_a$ is $n(E_a)/n$, and these are the molecules that have enough energy to react (Fig. 8.7). We can now try to tie things together by writing

$$J \propto \text{collision rate} \times \exp(-E_a/RT) \qquad (8.30)$$

Notice how similar Eqn. 8.30 is to the rate of a first-order reaction, $J = k[A] = [A]A\exp(-E_a/RT)$. The comparison tells us that the collision rate, the number of collisions per unit volume per unit time, is $[A] \times A$ times a scaling factor. If E_a is large, $n(E_a)/n$ and $\exp(-E_a/RT)$ are small, and the rate of reaction will be low.

G. Transition state theory

As intuitive and straightforward as collision theory is, it suffers from a number of defects. The most important of these is that its predictions do not always match up with the results of experiments! This must be considered a serious flaw of the theory, for **the ultimate test of any scientific theory is how well it predicts the outcome of experiments that can actually be done.** Transition state theory was invented in the 1930s by Henry Eyring and others in order to get around some of the inadequacies of collision theory. The main device employed in this theory is the **transition state**. The rate of formation or breakdown of the transition state determines the rate of reaction.

Suppose we have the bimolecular reaction

$$A-B + C \rightarrow A + B-C \qquad (8.31)$$

At some point in the reaction, a high-energy 'intermediate' complex must be formed. This complex, which is only transiently formed, is very unstable. Moreover, its lifetime is so short that no one can observe it. We can guess, though, that it must look something like A··B··C. Whatever it looks like, we shall call this chemical species the transition state or **activated complex**, X.

Now consider the reaction

$$A \Leftrightarrow X \rightarrow P \tag{8.32}$$

Key assumptions here are that X is in *rapid equilibrium* with A and B, and that the rate of formation of X from A and B–C is so small as to be negligible. We also assume that the reaction takes place at constant temperature and pressure, for reasons that will become clear momentarily. We know from preceding sections of this chapter that

$$\Delta[P]/\Delta t = k'[X] \tag{8.33}$$

We can also write down an equilibrium constant, K, for the formation of the transition state:

$$K = [X]/[A] \tag{8.34}$$

From Chapter 4, K can be written as

$$\Delta G_a = -RT \ln K \tag{8.35}$$

where the subscript 'a' refers to activation, just as it does in E_a. Combining this equation with Eqn. 8.33 gives

$$\Delta[P]/\Delta t = k'[X] = k'K[A] = k'\exp(-\Delta G_a/RT)[A] \tag{8.36}$$

This equation tells something that makes good intuitive sense: the more energy required to form the transition state from the reactants, the smaller the exponential term on the right hand and the smaller the overall rate of reaction. This energy, ΔG_a, corresponds to the activation energy, E_a, and the chemical species it represents, X_a, is formed at the crest of the **activation barrier** or **kinetic barrier** in the energy profile in Fig. 8.1.

What about k', the rate of formation of P from X_a? Is there some way of finding a mechanistic interpretation of this rate constant? Suppose that k' is proportional to a vibrational frequency, ν, and the probability that X_a will decompose to form P. Then

$$k' = \nu\kappa \tag{8.37}$$

The probability κ is known as a **transmission coefficient**, and it is something like the odds that a salmon will be sufficiently fit to jump several feet in the air on its journey upstream to the spawning area. It often takes a salmon several tries to clear a given waterfall, and it may never make it at all. The vibrational frequency, ν, on the other hand, takes us very far back upstream in the course of this book to Planck's law, Eqn. 1.1. From statistical mechanics, the energy of an oscillator is

$$E = hc/\lambda = h\nu = k_B T \tag{8.38}$$

Combining Eqns. 8.37 and 8.38 gives

$$k' = \kappa E/h = \kappa k_B T/h \tag{8.39}$$

and inserting this into Eqn. 8.36 gives

$$J = \Delta[P]/\Delta t = \kappa k_B T \exp(-\Delta G_a/RT)[A]/h \tag{8.40}$$

We can see from comparing Eqn. 8.40 to Eqn. 8.10 that the rate constant of the forward reaction is

$$k = \kappa k_B T \exp(-\Delta G_a/RT)/h \tag{8.41}$$

When $\kappa = 1$, as is it does for many reactions, the rate is $k_B T \exp(-\Delta G_a/RT)/h$. Now we have a direct connection between reaction rate, something that can often be measured relatively easily, and the free energy of the transition state. If addition of a catalyst brings about a ten-fold *increase* in the rate of reaction, the catalyst must *reduce* the free energy barrier by $RT \ln 10 = 8.31\,\text{J mol}^{-1}\,\text{K}^{-1} \times 298\,\text{K} \times \ln 10 = 5.70\,\text{kJ mol}^{-1}$ at 25 °C. To put things in perspective, the magnitude of this energy change is roughly the same as the *enthalpy* change on forming one hydrogen bond.

We can take our analysis one step further by returning to Eqn. 8.35 and writing the activation free energy in terms of enthalpy and entropy:

$$\Delta G_a = \Delta H_a - T\Delta S_a \tag{8.42}$$

Inserting this relationship into Eqn. 8.41 and setting $\kappa = 1$ gives

$$k = \frac{k_B T}{h} \exp(-\Delta H_a/RT)\exp(\Delta S_a/RT) \tag{8.43}$$

ΔH_a and ΔS_a are known, respectively, as the enthalpy of activation and entropy of activation. Eqn. 8.43 shows very clearly that for a given activation enthalpy, the higher the activation entropy, the faster the reaction. The entropy of the activated complex, however, will generally be substantially lower than the entropy of the reactants, making $\Delta S_a < 0$. But if the formation of the activated complex can be coupled to the release of water molecules, as in the case of hexokinase (Chapter 4), the total entropy change might make the activation energy sufficiently low for the reaction to occur. By means of a little calculus, it can be shown that

$$\frac{\Delta \ln k}{\Delta T} = \frac{\Delta H_a}{RT^2} + \frac{1}{T} \tag{8.44}$$

It can be shown from this that for a *unimolecular* reaction in the *gas phase*, $E_a = \Delta H_a + RT$. Substituting this into Eqn. 8.43 gives

$$k = \frac{kT}{h} \exp(-E_a/RT)\exp(\Delta S_a/RT)\exp(1) \tag{8.45}$$

Comparison of Eqn. 8.44 with Eqn. 8.9 indicates that

$$A = \frac{kT}{h} \exp(\Delta S_a/RT)\exp(1) \tag{8.46}$$

In other words, the frequency factor measures the activation entropy.

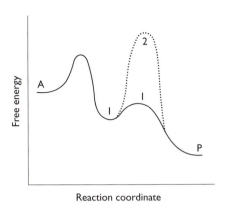

Fig. 8.8 Rate limiting step of a reaction. In reaction 1, the first step in the formation of P from A is rate-limiting; the energy barrier of this step is higher than for the second step. Consequently, the first step determines the overall rate of this reaction. In reaction 2, the second step is the rate-determining one. The rate of reaction is independent of the Gibbs free energy difference between P and A.

The preceding discussion pertained to the relatively simple case of A →P. If a stable intermediate, I, is involved, the reaction scheme looks like this:

$$A \rightarrow I \rightarrow P \tag{8.47}$$

There is no longer just one activated complex but two of them, one for each step of the reaction. The rate constants for the first step, k_1, can be either greater than or less than the rate constant for the second step, k_2, depending on the reaction. The smaller of these corresponds to the higher activation energy. The step of the overall reaction with the highest free energy barrier is called the **rate-determining step** (Fig. 8.8); something like a narrow passageway through the Alps on an automobile trip from Florence to Frankfurt. As we shall see below in the context of enzyme kinetics, catalysts speed up a reaction by reducing the energy barrier of the rate-determining step.

H. | Electron transfer kinetics

In Chapter 4 we studied electron transfer reactions in the context of redox couples. Then, in Chapter 5, we touched briefly on the role of redox reactions in the all-important processes of photosynthesis, glycolysis, and the citric acid cycle. Now we wish to take a slightly deeper look at electron transfer and consider the kinetics of the process.

In 1992 the Nobel Prize in Chemistry was awarded to Rudolph A. Marcus (1923–), an American, for his work on electron transfer. The **Marcus theory** relates the rate of electron transfer, k_{et}, to properties of the redox system involved. Specifically,

$$k_{et} = k_0 \exp(-\beta d) \tag{8.48}$$

where d is the distance between electron donor and acceptor and the coefficient β depends on the medium between them; $0.7 \text{ Å}^{-1} \geq \beta \geq 4 \text{ Å}^{-1}$, the lower value corresponding to van der Waals contact and the upper to vacuum. The less stuff between donor and acceptor, the higher β. The maximum rate of electron transfer is about 10^{13} s^{-1}. That's very fast; dinosaurs roamed Earth 10^{13} minutes ago. As it turns out, the intervening medium between donor and acceptor is not the only determinant of the transfer rate. It is also affected by ΔG, the driving force of the reaction. Notice how this situation is qualitatively different from what we saw above, where the free energy difference between products and reactants had little or no effect on the rate of reaction; what mattered most was the relative free energy of the transition state.

The dependence of rate of electron transfer on ΔG is complex. How could it not be when it must account for the size of the ions involved, the spatial orientation of neighboring dipole moments, and the number and orientation of solvent molecules? One particular arrangement of all these contributors will give the minimum free energy for transfer; all other arrangements will have a higher energy. Figure 8.9 illustrates how the Gibbs free energy varies with the reaction coordinate. In each of the three cases, two overlapping energy wells are shown. These represent the electron donor and electron acceptor. The electron is in a bound state. Electrons move on much faster timescale than do nuclei, so in the course of electron transfer there is effectively no nuclear motion. In order for transfer to occur, the energy of the acceptor must be the same as that of the donor – the point where the energy wells overlap.

According to the Marcus theory, k_{et} varies with the free energy difference between donor and acceptor ($\Delta G°$) and the energy required to reorient the nuclei so that the energy state of the electron will be the same in both donor and acceptor (λ). (Note that the meaning of 'λ' here is not the same as in Eqn. 1.1.) In polar solvents like water, the major contribution to λ is reorientation of solvent molecules resulting from the change in the charge distribution of the reactant. The magnitude of λ is reduced if electron transfer occurs in a medium of low dielectric, e.g. a lipid bilayer or the interior of a protein. λ also depends on changes in the shape and charge distribution of the electron donor as the reaction proceeds. A prediction of the theory is that the rate of electron transfer

Fig. 8.9 Electron-transfer kinetics. In contrast to rates of chemical processes, the kinetics of electron transfer do vary with free energy. Electronic motion is very fast, much faster than nuclear motion. This allows one to make a relatively clean separation of the energy into its electronic and nuclear components. The parabolas represent the energy of the nuclear component. When the distance between donor and acceptor (d_1) is constant, as in plots (A) and (B), varying the free energy difference between electron donor and acceptor affects the height of the energy barrier (the energy of the transition state). In panel (B), where $|\Delta G_2| = \lambda$, the reorganization energy, the rate of transfer is larger than when $|\Delta G| \neq \lambda$. Panel (C) shows that, although the free energy difference between donor and acceptor is the same as in panel (A), the rate is different, because a change in distance separating the donor–acceptor pair has resulted in a large increase in the energy of the transition state.

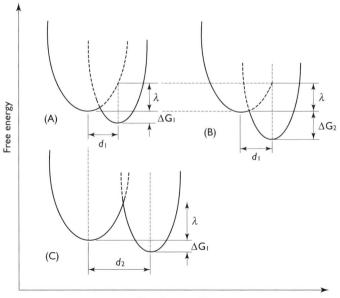

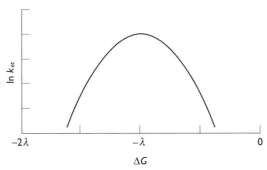

Fig. 8.10 Electron transfer rate as a function of λ. The rate is maximal when $|\Delta G| = \lambda$.

is a maximum when $\Delta G° = -\lambda$. This is shown in plot (B) of Fig. 8.9. As $\Delta G°$ becomes more negative but the distance between nuclei remains the same, as in panel (A), the rate of electron transfer decreases, because $\Delta G° \neq -\lambda$. Note that changes in the distance between nuclei (d) result in changes in the height of the energy barrier, the energy at which the curves cross. Figure 8.10 shows how k_{et} varies with $\Delta G°$.

I. Enzyme kinetics

Technical usage of the word *enzyme* is at least as ancient as the Middle Ages, when it played a part in theological discussions about whether the Eucharist should be celebrated with leavened bread (enzyme) or unleavened bread (azyme). (See Fruton (1999), p. 148.) Unlike electron transfer kinetics, the rate of enzyme (Greek, *en*, in + *zyme*, leaven) catalysis is assumed to be independent of the free energy difference between reactants and products. Instead, the rate depends on free energy of the transition state in the way outlined in the section on transition state theory. An entire section of the book is devoted to this subject because the chemical reactions of life are mediated and catalyzed by enzymes! Enzymes not only promote biochemical reactions: they effectively ensure that the reactions will proceed rapidly enough for the cell metabolism to be maintained. Though *all* catalysts speed up a reaction by reducing the free energy of the transition state (Fig. 8.11), enzymes are no ordinary catalysts. A great deal could be said about how extraordinary enzymes are, but let it suffice to mention in brief their defining characteristics. Then we'll devote the rest of the section to the mathematical relationships used to describe their behavior.

Enzymes are of course no less subject to the laws of nature than anything else made of matter. But they do nevertheless have a number of properties that set them well apart from ordinary catalysts. These are: a high rate of reaction, effectiveness under mild reaction conditions, astonishing specificity, and the capacity to be regulated. The rates of enzymatically catalyzed reactions are typically several orders of magnitude greater than those catalyzed by non-biological catalysts, and 10^6-10^{12} times greater than the uncatalyzed reaction. In other words, if a biochemical reaction occurs at a rate of about 1 day^{-1} in the absence of enzyme, when the enzyme is present it is likely to occur at a rate of 1 s^{-1} to 1000000 s^{-1}. That's amazing! Such astonishing specificity

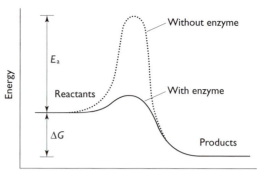

Fig. 8.11 Reaction profile for an enzyme-catalyzed reaction. The precise geometrical arrangement of atoms in the active site stabilizes the transition state of a biochemical reaction, lowering the free energy barrier between reactants and products. Because the rate of a reaction scales with E_a, a catalyzed reaction is faster than an uncatalyzed one.

probably depends in general on the induced fit of the enzyme to the substrate, as we saw with hexokinase (Chapter 4). Enzyme-catalyzed reactions work best under remarkably mild conditions: temperatures at which water is in the liquid state, near-neutral pH values, near-atmospheric pressure. Most industrial catalytic process requires extremes of these variables for reasonable efficiency. Enzymes, by virtue of being made out of chiral subunits, are themselves chiral. The sites where enzymatic reactions occur are chiral. This means that enzymes discriminate carefully between 'sheep' and 'goats,' between what can and cannot enter into the binding site, though some enzymes are clearly more specific than others. For instance hexokinase phosphorylates not just glucose but other hexoses as well, while phosphoryl group transfer by phosphoglucomutase is 10^{10} times more probable to glucose-6-phosphate than to water. Side reactions are the bane of anybody who has tried to synthesize a long polypeptide by chemical methods, and yet this is done by cellular enzymes with nearly total fidelity. Finally, enzymes can be regulated by a tremendous variety of means, depending on the enzymes. The activity of ATCase, for instance, depends on the concentration of ATP and CTP, even though neither of these is involved directly in the reaction catalyzed by the enzymes. Berzelius formulated the first general theory of chemical catalysis in 1835; Hermann Emil Fischer (1852–1919), a German who was awarded the Nobel Prize in Chemistry in 1902, discovered that glycolytic enzymes can distinguish between stereoisometric sugars in 1894. On the other side of the Channel, Adrian John Brown (1852–1919) reported on the rate of hydrolysis of sucrose by β-fructofuranosidase in 1902. About three decades later, the American John Howard Northrop (1891–1987), who was awarded the Nobel Prize in Chemistry in 1946, crystallized pepsin and demonstrated conclusively that its catalytic activity is a property of the protein. And the first enzyme structure (hen lysozyme, see below) was visualized at atomic resolution in Oxford about another three decades later, in 1965. These landmark accomplishments give a very rough idea of the pace of change in during the first half of the twentieth century.

An overall reaction scheme in which the substrate (S) forms a complex with the enzyme (E) that then decomposes into product (P) and enzyme is:

$$E + S \Leftrightarrow E \bullet S \rightarrow P + E \tag{8.49}$$

Note that the enzyme–substrate complex is *assumed* to be *in equilibrium* with free enzyme and substrate, an important simplification discussed below. Note also that when the substrate concentration is high and all of the enzyme present is in the $E \bullet S$ form, the overall rate of reaction will be independent of [S]. This is akin to the saturation of binding sites that we studied in Chapter 7. The rate or '**velocity**' of this reaction is

$$J_P = \Delta[P]/\Delta t = k_2[E \bullet S] \tag{8.50}$$

where k_2 is the rate of formation of P from $E \bullet S$. This rate constant is also known as k_{cat} and the **turnover number**, the number of substrate molecules converted to product per unit time (when all enzyme active sites are filled with substrate; see below). The rate of change in $[E \bullet S]$ is

$$J_{[E \bullet S]} = k_1[E][S] - k_{-1}[E \bullet S] - k_2[E \bullet S] \tag{8.51}$$

where k_1 is the rate of formation of $E \bullet S$ from E and S and k_{-1} is the rate of decomposition of the complex back into E and S. At steady-state, the rate of formation of the complex is equal to the rate of its decomposition, $[E \bullet S]$ is constant, and $J_{[E \bullet S]} = 0$. This is known as the **steady-state assumption**, and it was first put forth by the Englishmen John Burdon Sanderson Haldane (1892–1964) and George Edward Briggs (1893–1985) in 1925.

To make it seem more like this exercise is actually getting us somewhere, we wish to reformulate things in terms of things we can *measure*. Those things are the *total* enzyme concentration, $[E]_T$, and the concentration of substrate at the beginning of the reaction. We have assumed that the enzyme binding site can either be occupied or free, so

$$[E]_T = [E] + [E \bullet S] \tag{8.52}$$

Substitution of this equation into Eqn. 8.51 under steady-state conditions gives

$$k_1([E]_T - [E \bullet S])[S] - (k_{-1} + k_2)[E \bullet S] = 0 \tag{8.53}$$

which, when solved, for $[E \bullet S]$ is

$$[E \bullet S] = \frac{[E]_T[S]}{K_M + S} \tag{8.54}$$

where $K_M = (k_{-1} + k_2)/k_1$. The quantity K_M is called the Michaelis constant, after the German enzymologist Leonor Michaelis (1875–1949). Note that K_M is the ratio of the sum of the rates of depletion of the enzyme–substrate complex to the rate of formation of the complex. When $k_{-1} \gg k_2$, $K_M \approx k_{-1}/k_1$, and K_M is *like* a dissociation constant. Values of K_M range from about 10^{-8} and 10^{-2} M, and the value for a given enzyme depends on the conditions under which measurements are made. When K_M is small, k_1 is relatively large, the free energy barrier to complex formation is relatively small, binding of substrate to enzyme is tight, and the enzyme will catalyze reactions at very low concentrations. K_M thus reflects an enzyme's ability to bind a substrate and begin catalysis, but it is important to bear in mind that K_M is not a true binding constant.

Eqn. 8.54 can be substituted into Eqn. 8.50, giving

$$J_P = \Delta[P]/\Delta t = k_2[E \bullet S] = k_2[E]_T[S]/(K_M + [S]) \qquad (8.55)$$

This the rate of formation of the product at any time during the experiment. The equation can be used to determine K_M if $[E]_T$ and $[S]$ are known, as at $t = 0$, and J_P can be measured. This $J_P \approx J_P(0)$. Later in the experiment, although $[E]_T$ will be known, because it will not have change, $[S]$ will be unknown, because some of the substrate will have been consume, complicating determination of K_M.

The velocity of the reaction we are studying cannot be any faster than when every enzyme molecule is in a complex with a substrate molecule, i.e. when $[E \bullet S] = [E]_T$. So,

$$J_{P,max} = k_2[E]_T \qquad (8.56)$$

This equation shows how the turnover rate, $J_{P,max}$, is the ratio of maximum rate of reaction to enzyme concentration, and it usually varies from 10^5 to 10^9 molecules of product formed per molecule of enzyme per second. Substituting this relationship into Eqn. 8.55 gives

$$J_P = J_{P,max}[S]/(K_M + [S]) \qquad (8.57)$$

This is the famous **Michaelis–Menten equation**, the basic equation of enzyme kinetics (Fig. 8.12). It was developed in 1913 by Michaelis and Maud Leonora Menten (1879–1960). The Michaelis–Menten equation tells us that when $[S] = K_M$, the velocity of the reaction is half-maximal. When K_M is relatively small, the rate of enzyme catalysis is maximal at a relatively low substrate concentration; when K_M is relatively large, the concentration of substrate must be very large for the rate of enzyme catalysis to be a maximum. The fraction of active sites occupied can be found from the Michaelis–Menten equation by dividing both sides by $J_{P,max}$:

$$J_P/J_{P,max} = [S]/(K_M + [S]) \qquad (8.58)$$

When $[S]$ is relatively small, $J_P/J_{P,max}$ is small, and as $[S] \to \infty$, $J_P/J_{P,max} \to 1$, as required. Taking the reciprocal of Eqn. 8.57 leads to a popular way of plotting kinetic data:

Fig. 8.12 The relative rate of reaction as a function of substrate concentration. Note that the rate of reaction here is the *initial* rate, i.e. the rate before the concentration of substrate has change substantially, or the rate one would measure if the substrate concentration were held constant throughout the measurement. Data are shown for substrates with different values of K_M. The larger K_M, the greater the substrate required to give a half-maximal rate.

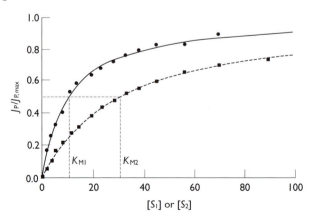

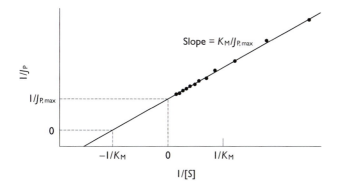

$$1/J_P = (K_M + [S])/(J_{P,max}[S]) = (K_M/J_{P,max})(1/[S]) + 1/J_{P,max} \quad (8.59)$$

This equation is linear in $1/[S]$ and has slope $K_M/J_{P,max}$, ordinate-intercept $1/J_{P,max}$, and abscissa-intercept $-1/K_M$ (Fig. 8.13). Useful as it is, this linear form of the Michaelis–Menten relationship is beset by the same difficulties discussed in Section E of the last chapter. Moreover, because most concentrations of substrate will be relatively high, most the experimental data will be clustered together in a fairly narrow range of $1/[S]$. Small errors in $[S]$ can lead to large errors in K_M and $J_{P,max}$.

Equation 8.55 can be used to determine the efficiency of an enzyme-catalyzed reaction. Suppose that $J_P = J_P(0)$. When $[S]$ is very small in comparison with K_M, $[E]_T \approx [E]$, because very little E • S can form, and Eqn. 8.55 reduces to

$$J_P(0) \approx k_2[E]_T[S]/K_M \approx k_{cat}[E][S]/K_M \quad (8.60)$$

a second-order rate equation with rate constant k_{cat}/K_M, where k_{cat} is defined as $J_{P,max}/[E]_T$. That is k_{cat} is identical to k_2 in the model discussed above; k_{cat}/K_M measures the catalytic efficiency of an enzyme. When k_{cat}/K_M is large, on the order of 10^8–10^9 M^{-1}s^{-1}, enzyme activity is effectively limited solely by diffusion of the substrate into the binding site. For example catalase, an enzyme that catalyzes the degradation of hydrogen peroxide to water and oxygen, has a k_{cat}/K_M of 10^7. This enzyme operates very close to the diffusion-controlled limit. In other words, the catalytic efficiency of this enzyme is limited by physics and by not by chemistry; no further change in the enzyme could increase its catalytic power. It is k_{cat}/K_M, not K_M, that is the generally accepted parameter for characterizing an enzyme (under given conditions). See Table 8.1.

J. Inhibition

Enzyme activity can be inhibited in various ways. In **competitive inhibition**, molecules similar to the substrate bind to the active site and prevent binding of the usual substrate. The antibiotic penicillin, for example, serves as a competitive inhibitor by blocking the active site of an enzyme that many bacteria use to construct their cell walls. In contrast, **non-competitive inhibition** results from the binding of an inhibitor to an enzyme at a location other than the active site. This can come

Fig. 8.13 A linearized version of the rate of reaction versus substrate concentration. Note how the data are clustered together, an effect of the mathematical transformation. Two points of interest are: $1/J_P = 1/J_{P,max}$, which occurs where $[S] \to \infty$, and $1/J_P = 0$, which will be the case when $[S] = -K_M$. Of course, a negative concentration is a mathematical fiction, but that does not make it any less useful for evaluating K_M. Unfortunately, such linearizations of kinetic data are often less helpful than they might seem, for reasons outlined in the text. The type of plot shown here is a Lineweaver–Burk plot.

Table 8.1 The kinetic properties of some enzymes and substrates

Enzyme	Substrate	$K_M(M)$	$k_{cat}(s^{-1})$	$k_{cat}/K_M(M^{-1}s^{-1})$
Acetylcholinesterase	Acetylcholine	9.5×10^{-5}	1.4×10^4	1.5×10^8
Carbonic anhydrase	CO_2	1.2×10^{-2}	1.0×10^6	8.3×10^7
	HCO_3^-	2.6×10^{-2}	4.0×10^5	1.5×10^7
Catalase	H_2O_2	2.5×10^{-2}	1.0×10^7	4.0×10^8
Chymotrypsin	N-Acetylglycine ethyl ester	4.4×10^{-1}	5.1×10^{-2}	1.2×10^{-1}
	N-Acetylvaline ethyl ester	8.8×10^{-2}	1.7×10^{-1}	1.9
	N-Acetyltyrosine ethyl ester	6.6×10^{-4}	1.9×10^2	2.9×10^5
Fumarase	Fumarate	5.0×10^{-6}	8.0×10^2	1.6×10^8
	Malate	2.5×10^{-5}	9.0×10^2	3.6×10^7
Urease	Urea	2.5×10^{-2}	1.0×10^4	4.0×10^5

Source: data from Table 13-1 of Voet & Voet (1995). Note k_{cat} is very high for catalase. Chymotripin is much more active against tyrosine residues than glycine or valine. Although acetylcholinesterase and fumarase have lower k_{cat} values than does catalase, all three of these enzymes have k_{cat}/K_M ratios of the same order of magnitude.

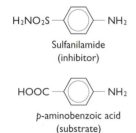

Sulfanilamide
(inhibitor)

p-aminobenzoic acid
(substrate)

Fig. 8.14 A substrate and an inhibitor. The figure shows how very similar the structure of an inhibitor and the substrate can be. In the pharmaceutical industry, medicinal and synthetic chemists look for ways of synthesizing compounds similar in structure to known substrates. The potential inhibitors are then tested in different ways. If a chemical shows the ability to inhibit an enzyme, and if it is not too toxic to cells, the chemical could become a marketable drug.

about in several ways, including deformation of the specific geometry of the active site (**allosteric inhibition**, see Chapter 7).

An example of competitive inhibition is the action of sulfanilamide (Fig. 8.14) on an enzyme involved in the metabolism of folic acid, a vitamin that is a coenzyme precursor. Sulfanilamide is sufficiently similar to *p*-aminobenzoic acid, the substrate, that it binds to the enzyme and prevents interaction with *p*-aminobenzoic acid. The enzyme in question is essential in certain disease-causing bacteria but not in humans, and this allows the chemotherapeutic use of sulfanilamide as a type of antibiotic called an anti-metabolite.

Note that both competitive inhibition and non-competitive inhibition are usually thought of as involving non-covalent interactions between inhibitor and enzyme. Other types of inhibitor, however, form covalent bonds with enzymes. For instance, the nerve gas diisopropyl fluorophosphate forms a covalent bond with an amino acid residue side chain in the active site of acetylcholinesterase and thereby prevents binding of the neurotransmitter acetylcholine and blocks nerve action. Various types of protease inhibitor are added to protein preparations to prevent digestion of the sample. Some protease inhibitors bind irreversibly to proteases by forming a covalent bond with amino acid side chains in the active site.

Having looked at qualitative aspects of inhibition, we now turn our attention to modeling it mathematically. In competitive inhibition the enzyme interacts either with the substrate, S, or with the inhibitor, I, so the total concentration of enzyme is:

$$[E]_T = [E] + [E \bullet I] + [E \bullet S] \tag{8.61}$$

As before,

$$K_M \approx [E][S]/[ES] \tag{8.62}$$

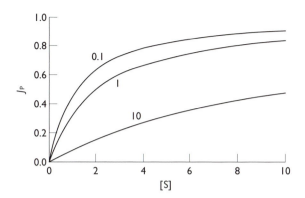

Fig. 8.15 The effect of an inhibitor on the rate of enzyme activity. The three curves represent different values of the ratio $[I]/K_I$. For a given K_I, the plots show the effect of increasing inhibitor concentration. As expected, increases in $[I]$ result in decreases in J_p for a given substrate concentration. The middle curve could also represent $[I]/K_I = 10$ and a ten-fold decrease in K_M relative to that in the lowest curve. In other words, the smaller K_M, the larger the rate of catalysis for a given concentration of substrate.

The inhibitor is *assumed* to be *in equilibrium* with the enzyme and E • I, so

$$[E][I]/[E • I] = K_I \tag{8.63}$$

Solving this equation for [E • I], and substituting Eqn. (8.62) for [E], we obtain

$$[E • I] = [E][I]/K_I = (K_M[E • S]/[S])[I]/K_I \tag{8.64}$$

Substituting this equation into Eqn. (8.61) and solving for [E • S] gives

$$[E • S] = \frac{[E]_T[S]}{K_M\left(1 + \dfrac{[I]}{K_I}\right) + [S]} \tag{8.65}$$

As before, $J_p = k_2[E • S]$, so

$$J_p = \frac{k_2[E]_T[S]}{K_M\left(1 + \dfrac{[I]}{K_I}\right) + [S]} \tag{8.66}$$

Comparison of the relationship with Eqn. 8.55 shows that the effect of increasing the concentration of competitive inhibitor is to increase the apparent magnitude of K_M. This is why the quasi-equilibrium character of the Michaelis constant must be treated with caution. Plots of Eqn. 8.66 for different values of $[I]/K_I$ are given in Fig. 8.15. $[I]/K_I = [E • I]/[E]$, approximately the proportion of the total amount of enzyme that has formed a complex with the inhibitor.

The derivation for the case of non-competitive inhibition is very similar. It is left as an exercise to show that

$$J_p = \frac{k_2[E]_T[S]}{K_M\left(1 + \dfrac{[I]}{K_I}\right) + [S]\left(1 + \dfrac{[I]}{K_I'}\right)} \tag{8.67}$$

where K_I is defined as before and $K_I' = [E • S][I]/[E • S • I]$.

K. | Reaction mechanism of lysozyme

As we have said elsewhere (Chapter 5), lysozymes hydrolyze the $\beta(1 \rightarrow 4)$ glycosidic linkages of oligosaccharides in the cell walls of bacteria. This

Table 8.2 Binding free energies of polysaccharide subunits in the active site of hen lysozyme

Site	Bound saccharide	Binding free energy (kJ mol^{-1})
1	NAG[a]	−7.5
2	NAM[b]	−12.3
3	NAG	−23.8
4	NAM	+12.1
5	NAG	−7.1
6	NAM	−7.1

Note:

[a] N-acetylglucosamine.

[b] N-acetylmuramic acid.

Lysozyme hydrolyzes the β(1→4) glycosidic linkages from NAM and NAG in the alternating NAM-NAG polysaccharide component of bacterial cell wall peptidoglycans. The data are from Chipman & Sharon (1969).

weakens the cell wall and leads to osmotic lysis. Lysozyme also digests chitin, a strong polysaccharide in the cell walls of most fungi. What we are interested in here is the mechanism of lysozyme activity.

Much of what is known about the mechanism of lysozyme is based at least in part on the structure of the enzyme at atomic resolution, which was first obtained by David Chilton Phillips (1924–) at Oxford in 1965. One of the most prominent structural features of the folded enzyme is its large active site cleft, which is capable of binding up to six residues of a polysaccharide. The rate of enzyme activity depends on the length of the sugar. Enzyme-catalyzed hydrolysis is up to 100 million times faster than the uncatalyzed reaction. When six residues are bound in the active site, hydrolysis occurs between residues four and five. This is somehow related to the experimental finding that the free energy of binding of the fourth residue to the enzyme is the weakest of the six (Table 8.2). This unfavorability is thought to reflect the distortion of the sugar at residue four that is required for tight binding of the polysaccharide to occur. Binding to the third residue is the most favorable of all.

There are two acidic groups in the active site cleft, Glu 35 and Asp 52, and they are close to the bond of the polysaccharide that is cleaved during catalysis. Replacement of Asp 52 by Ser leads to virtually complete loss of enzyme activity. There is, however, nothing particularly unusual about the pK_a of this side chain (it is *c.* 3.5), as one might expect for a side chain involved in catalysis. The carboxyl group of Asp 52 is in a polar environment. In contrast, as mentioned in Chapter 7, Glu 35 has an anomalous pK_a in the folded state of the enzyme (it is *c.* 6.3!), and this must result from the specific electronic environment of the side chain. Indeed, the carboxyl group of Glu 35 is situated in the relatively low dielectric environment of a mostly hydrophobic pocket. This has important consequences for the mechanism of enzyme action, as experimental studies have shown that activity is *maximal* at pH 5, well *below* the pK_a of Glu 35. In other words, enzyme activity is less than it could be if the Glu 35 side chain were ionized. Therefore, a plausible mechanism of lysozyme activity is that the carboxyl group Glu 35 transfers its proton to the bond between polysaccharide units four and five. The negative charge on Asp 52 must play a role in this, but it is still not entirely clear how.

L. | Hydrogen exchange

Biological macromolecules are full of covalently bound but labile hydrogen atoms, many of which can exchange readily with the hydrogen atoms in the solvent. What we have in mind here are not the ionizable protons of Glu and Asp but the amide protons of the polypeptide backbone. Because these protons undergo exchange with the solvent on a convenient time scale, hydrogen isotopes can be used to label solvent-exposed parts of a macromolecule (Fig. 8.16). Use of such labeling techniques in biochemistry, which was pioneered by Linderstrøm-Lang in the 1950s, has been extremely useful in studies of the structure, stability, dynamics, and folding properties of biological macromolecules. For instance, the rate of hydrogen exchange is an essential consideration is NMR structure determination, because if the side chain protons of aromatic residues in the core of the protein exchange too rapidly, it will not be possible to get solid information on how they interact with other protons. Such information is essential for determining the 3-D structure of a protein by NMR. In this section we study hydrogen exchange kinetics in two contexts: protein stability at equilibrium and acquisition of native-like structure on the protein folding pathway. The approaches described can also be used to study protein–ligand interactions, whether the ligand is an ion, a small organic molecule, a peptide, or a nucleic acid.

Not all hydrogen atoms exchange rapidly enough to be useful for physical biochemistry experiments. For example, aliphatic hydrogen atoms exchange extremely slowly, so these hydrogen atoms are essentially fixed. Polypeptide backbone amide hydrogen atoms (protons), however, exchange readily, making them particularly valuable for exchange experiments. Other exchangeable protons are indicated in Fig. 8.17, which shows how the rate of exchange depends on pH. This

Fig. 8.16 Exchange labeling. Proteins and other biological macromolecules comprise hydrogen atoms that exchange readily with protons in the solvent. In proteins, these protons are found in the amide groups of the polypeptide backbone, and in certain amino acid side chains (e.g. the indole NH of tryptophan); aliphatic protons do not exchange with solvent on a time scale that is suitable for protein structure studies. In general, the more exposed to solvent a labile proton, the more rapidly it exchanges. One can determine the rate of exchange at different sites in a protein by exchange labeling, in which deuterons are exchange for protons, or *vice versa*. This works because on the chemical level deuterium is identical to hydrogen: there is just a single electron and a single proton. The difference lies in the number of nucleons: hydrogen has no neutrons and deuterium has one. This changes not only the mass but also the magnetic properties of the deuterium nucleus relative to that of hydrogen, rendering deuterium 'invisible' to an NMR spectrometer set up to acquire data on proton resonances. The effect of exchange on the NMR spectrum is that the proton peaks diminish in size as deuterium gets exchanged for hydrogen, since only the remaining hydrogen at a particular site will contribute to the proton spectrum.

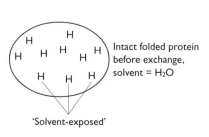

'Solvent-exposed'

Intact folded protein before exchange, solvent = H₂O

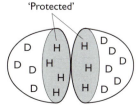

'Protected'

'Dissected' folded protein after exchange, solvent = D₂O

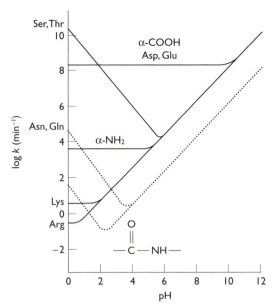

Fig. 8.17 The pH-dependence of the hydrogen exchange rate for different types of chemical group. There is a ten-fold change in reaction rate for a change in pH of one unit; exchange is both acid- and base-catalyzed, giving v-shaped curves. Peptide bond amino protons show a minimum rate of exchange at acidic pH. The rate of exchange of these protons depends not only on whether the amide group is buried or solvent exposed and on the type of secondary and tertiary structure in which it is involved, but also to some extent on the flanking residue (see Appendix D).

dependence arises from the hydrogen exchange reaction's being both acid-catalyzed and base-catalyzed (though by different mechanisms). The rate of exchange also depends on temperature, tripling every $10\,^{\circ}C$ or so. The pH and temperature determine the **intrinsic rate of exchange**, the rate of exchange of labile protons in an unstructured polypeptide (see Appendix D). Besides pH and temperature, the exchange rate varies with the electronic properties of the local environment and the exchange medium (whether the solvent water contains hydrogen or deuterium, the ionic strength, etc.). Important for studies of protein folding, the rate also varies substantially with how often the exchangeable hydrogen comes into contact with the solvent. Backbone amide protons deep in the hydrophobic core are not likely to exchange rapidly, because they rarely come into contact with the solvent when the protein is in the folded state. But interior labile protons *can* exchange! This phenomenon is usually described in terms of **local unfolding** of the protein, or 'breathing,' an interpretative image provided by Linderstrøm-Lang. There are two 'classes' of exchange, and both come from thinking of exchange as a race between the protein and the exchange mechanism. If the rate of exchange of a solvent-exposed labile hydrogen is higher than the rate at which this hydrogen ceases to be solvent-exposed (usually because of reorganization of protein structure), exchange is said to follow the **EX1 mechanism**. If, on the other hand, the rate of exchange of a solvent-exposed labile hydrogen is lower than the rate at which this hydrogen becomes 'protected,' exchange follows the **EX2 mechanism**. The second of these, in which the rate is determined by the kinetics of the exchange reaction because exchange is slower than refolding, is the type displayed by proteins under most conditions.

There are many ways in which one can make good use of protein hydrogen exchange data. Here we consider just a couple of them: NMR spectroscopy and mass spectrometry. With NMR, though there are many variations on the theme, there are two basic ways of doing hydrogen exchange (HDX) experiments. These are **equilibrium exchange** of protons for deuterons and **quenched-flow pulse labeling** of deuterated protein with protons. In either case, one uses a non-exchangeable proton as an intrinsic probe, typically a tryptophan side-chain proton: this allows normalization of the proton resonance magnitude from spectrum to spectrum. In equilibrium HDX, the protein sample is dissolved in a buffer made with D_2O instead of H_2O. As discussed in Chapter 5, even at equilibrium the protein structure fluctuates, and at any given time some fraction of the molecules will be unfolded. Fluctuations and folding/unfolding are sufficient to allow the solvent to penetrate the core of the protein, though the time-average duration of contact of protein surface with the solvent is much lower for core residues than surface ones. Exchange is generally very slow under conditions favoring the folded protein, so NMR protein peak heights, which are proportional to the number of hydrogen atoms at a given location in the protein, are measured over a course of up to several months. Between measurements, the protein sample is kept at constant temperature (and pressure).

After hydrogen peak heights have been measured and normalized using the height of a non-exchangeable hydrogen, the data for each exchangeable proton can be plotted as a function of time. Fitting procedures can then be used to determine the rate of exchange. In the usual case, exchange at a particular site is dominated by a single rate, though this is not always the case. One can then compare the rates of exchange for different protons in the same protein under the same conditions, to see how similar or different they might be. When exchange of core protons occurs via global unfolding of the protein, all such protons exchange with about the same rate. A more complicated process may be involved, however, for example if one part of the protein is particularly flexible and another is especially rigid, and when this is the case the rate data for individual protons might cluster into two or more groups. In either case, such data can be used to build models of the mechanism of protein folding/unfolding.

Equilibrium HDX data can be used in other ways. For example, the measured rate of exchange can be compared with the calculated intrinsic rate of exchange under the same conditions. The ratio of the rates is a measure of how well 'protected' against exchange a particular proton is under the conditions of the experiment, and in view of this a ratio of rates is called a **protection factor**. This quantity is similar to an equilibrium constant (see Chapter 4), and it can therefore be used to calculate the free energy difference between the folded state (for which the rate of exchange is measured) and the unfolded state (the intrinsic rate of exchange), using Eqn. 4.33. The free energy calculated in this was should compare well with that determined by calorimetry, unless the folding/unfolding mechanism deviates substantially from a two-state process.

Most single-domain proteins follow a two-state folding/unfolding mechanism, in which only the folded and unfolded states are populated to a significant extent. As we have seen in Chapters 5 and 6, some of the techniques used to monitor structural transitions in proteins are differential scanning calorimetry and spectroscopic methods like fluorescence. These techniques measure bulk properties of a sample and cannot distinguish explicitly between different states. In contrast, mass spectrometry, by which the mass of a particle can be measured to a resolution of 1 Da (dalton, named after John Dalton), allows not only the identification of different co-existing conformations but also determination of their exchange behavior. In the case of two-state behavior, only two peaks will be seen in the mass spectrum at any point in the folding/unfolding reaction: one corresponding to the folded state and the other to the unfolded state. Depending on how the data are processed, the mass spec. peak heights will reflect the relative abundance of the two states. When the folding/unfolding mechanism is more complicated, for instance when three states are present, in principle it will be possible to identify all three states as separate peaks in the mass spectrum.

A quenched-flow pulse labeling instrument can be used to label proteins during the folding process. In the typical experimental set-up, the protein is deuterated and dissolved in deuterated chemical denaturant. The experiment is begun by rapid mixing of the protein sample with a non-deuterated buffer that favors protein refolding. This comes about mainly by diluting the concentration of denaturant. Some time later, usually between just a few milliseconds and one second, a high-pH pulse is added to the refolding protein. The driving force for exchange is very high under these conditions, and deuterons that are exposed to the solvent are replaced by protons. The number of deuterons that can exchange will clearly decrease as the length of time between initiation of folding and the high-pH pulse increases. The duration of the high-pH pulse is constant. To inhibit further exchange, the high-pH pulse is followed immediately by a change back to the refolding buffer. Data collection at different time points during folding give something of a snapshot of which parts of the protein are solvent-exposed as a function of time. Using this approach, one might be able to show, for example, that one domain of a protein folds more rapidly than another, as has been done for a number of proteins.

M. | Protein folding and pathological misfolding

Protein folding is a large topic, one that has continued to develop rapidly since the early 1980s, although groundbreaking studies were done long before then (see Chapter 5). Justice could not possibly be done to this subject in the span of a few pages, so our aim is not to attempt an exhaustive analysis but just to give some sense of the place of folding within the more general subject of reaction kinetics. This section will conclude with some comments on protein misfolding and disease.

One of the first aims of a protein folding study is to measure the overall rate of attainment of native structure from an adequately defined denatured state. We say 'adequately' and not 'completely' because it can be very difficult to say just how unfolded a denatured state is. Suffice it to say here that the heat-denatured state will not always be very close in structure to the pH-denatured state or some other type of denatured form, because the way amino acids interact with each other depends on the solution conditions. In general, the form of the denatured state matters more in protein folding studies for big proteins (greater than about 130 residues) than small ones. A popular was of studying folding is to denature the protein in 6 M GuHCl and to initiate refolding by diluting one volume of denatured protein with 10 volumes of refolding buffer.

Figure 8.18 shows what one might find using optical spectroscopy to analyze the refolding of a small globular protein from a denatured state. The data were obtained by circular dichroism spectroscopy (panels (A) and (B)) and fluorescence emission spectroscopy (panel (C)). The near-UV CD signal monitors organization of specific interdigitation of side chains in the native state. We see from panel (A) that this occurs with relatively slow kinetics, and is not complete until about 1 s after initiation of folding. According to panel (B), however, in which the far-UV CD signal is shown, the secondary structure of the protein forms relatively quickly: by 200 ms into the folding reaction apparently all secondary

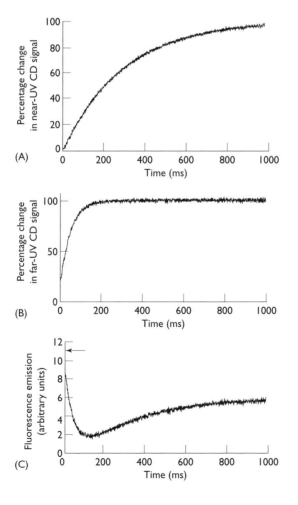

Fig. 8.18 Kinetic protein folding data. Panel (A): far-UV CD. The signal reaches the equilibrium value relatively rapidly. Panel (B): near-UV CD. Compared with the far-UV data, the signal reaches its maximum value relatively slowly. Panel (C): intrinsic fluorescence emission. This kinetic trace is more complex than either of the CD curves. The process is biphasic, the first phase being relatively fast and the second one slow. The data were acquired with instruments outfitted with stopped-flow modules, which enable rapid mixing of solutions.

structure is present. Comparison of panels (A) and (B) would suggest that there are at least two 'phases' to the folding reactions. Moreover, about one-fifth of the native far-UV signal is present by the time the first reliable signal can be acquired (about 3 ms into the experiment). This is the so-called burst phase of protein folding.

Fluorescence monitors changes in the solvent accessibility of tryptophan side chains. When exposed to the highly polar solvent, emission intensity is low; when buried in the hydrophobic core of the folded protein, emission intensity is high. The curve in panel (C) is complex. Although the curve is biphasic, distinguishing it from panels (A) and (B), the overall rate of refolding is the same as detected by near-UV CD (panel (A)). Moreover, the first phase has kinetics that closely resemble those of the far-UV CD signal (panel (B)). Thus, in a somewhat non-obvious way the fluorescence data corroborate the results shown in panels (A) and (B).

Denatured polypeptides do not always fold into native proteins under conditions favoring the folded state. In some cases, misfolding *in vivo* can be 'corrected' by other proteins, the so-called chaperones, whose general function appears to be to assist folding. This is particularly important to the life of the cell, tissue, and organism, because under some conditions misfolding can result in pathological protein aggregation.

Generally speaking, misfolding yields a partly folded structure, one that is more compact than a fully extended polypeptide chain and may contain native-like elements of secondary structure, but is less thermostable and usually more hydrophobic than the native state. The extra solvent exposure of apolar side chains in misfolded proteins makes them 'sticky,' and they tend to aggregate with kinetics that are governed by protein concentration, temperature, and solution conditions.

Why does aggregation occur when the folded state is likely to have a lower free energy than any partly folded state? Just as a large number of relatively weak individual hydrogen bonds can stabilize a folded protein, a large number of weak interactions between misfolded proteins can stabilize aggregates. Experimental work has shown that protein aggregation is often mediated by intermolecular β-strands, with the two strands of a β-sheet being contributed by two different protein molecules. Individual partly folded states are in a shallower energy well than the native state, but aggregates probably have a lower free energy than the same number of fully folded proteins, making them particularly difficult for the body to clear. If aggregates are not lower in free energy than folded states, the aggregated monomers might be 'kinetically trapped' and unable to fold because the activation energy of dissociating from the aggregate is very high.

Protein misfolding appears to be the common cause of the various amyloid diseases, which are characterized by abnormal extracellular accumulations of protein called amyloid plaques. The word *amyloid* (Greek, *amylon*, starch + *eidos*, form) was coined by Rudolf Carl Virchow (1821–1902), a German pathologist. Most of the mass of an amyloid plaque, however, is protein, not starch. Clumps of degenerating neurons surrounding deposits of protein, for example, are called neuritic plaques; twisted proteins fibers in nerve cells are known as neurofibrillary tangles.

If these descriptions conjure up images of senility, then you are right on target: the occurrence of these structures in the brain correlates with symptoms of Alzheimer's disease (named after the German psychiatrist and neuropathologist Alois Alzheimer (1864–1915)). Plaques and tangles may actually cause the disease. Neuritic plaques form around aggregates of amyloid β-protein, a proteolytic fragment of a larger molecule called amyloid precursor protein, a normal component of nerve cells. When separated from the rest of the precursor protein, amyloid β sticks to itself like glue, forming large deposits that probably interfere somehow with normal cellular activity and lead to impaired brain function. The fibers of neurofibrillary tangles consist of a different protein, called tau, which normally occurs in neurons. The tau molecules clump together and form tangles when protein processing goes awry.

Something similar happens with amyloidogenic lysozyme. In this case, however, the aggregate-forming molecules are intact; they are not the result of proteolysis or incorrect processing. The tendency of human lysozyme to aggregate in some people (an extremely small percentage of the population in Britain and probably elsewhere) comes not from misfolding or from incorrect disulfide bond connections but from native state instability brought about by a point mutation. Several amyloidogenic variants are known, and in the most pathological of these, the relative instability of the native state leads to non-cooperative protein denaturation at room temperature.

N. Polymerization

Polymerization has to do with the formation of linear polymers from subunits. In this section, our concern is polymers in which the subunit interactions are non-covalent ones, not the covalent bonds of amino acid polymers (polypeptides) or nucleic acid polymers (polynucleotides). The most famous non-covalent biological polymer is perhaps the actin filament, but in fact many biomacromolecules can polymerize. Indeed, lysozyme amyloid fibril formation stems from mutations that make the enzyme polymerize – even in the absence of additional free energy. Formation of polymers is thought to be an excluded-volume effect, one that favors compact conformations. In this sense, polymerization is similar to the folding of a single small globular protein. Another feature of polymerization is that highly elongated protein polymers (e.g. actin microfilaments) tend to form higher-ordered phases in which the elongated protein polymers associate with each other to form even more compact structures. The actin cytoskeleton consists of strong bundles of actin filaments that are highly 'cross-linked' by actin-binding proteins, not weak single filaments. Our main interest here is in actin, but before focusing on aspects of actin microfilament assembly, let's look at another example of polymerization in the living organism and the energy required to make it happen.

Microtubules are 24 nm cylindrical tubes (Fig. 8.19), and they fill the cytosol of a cell from the nucleus to the plasma membrane. These

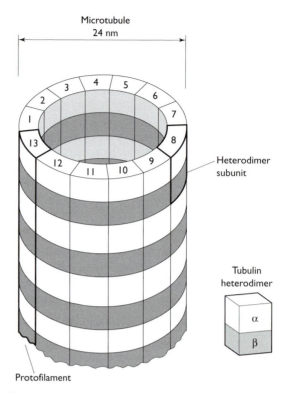

Fig. 8.19 Microtubule structure. Microtubules are composed of tubulin, of which there are two types of subunit, designated α and β. A single tubule consists of 13 protofilaments made of alternating α- and β-subunits. Note that an intact tubule, which is 24 nm wide, is helical and hollow. Microtubules have received a fair amount of popular press lately, as the noted British mathematical physicist and science popularizer Sir Roger Penrose (1931–) has suggested that they and a reworking of the foundations of physics might be keys to understanding consciousness. Is it not amazing that a human brain, which on one level can be thought of as just a highly organized collection of atoms, can investigate how those atoms are organized and indeed the structure of the atoms themselves? But why microtubules and not some other type of filamentous structure, say actin? After all, actin has more than one biochemically relevant conformational state, uses ATP for polymerization, plays an indispensable role in contractility, and serves as a conveyor for biomolecules in the cytosol. There is a great deal of disagreement about where consciousness comes from!

polymeric filaments underlie cell movement, from the beating of cilia to the transport of vesicles from the cell membrane to places within the cell, to the separation of chromosomes during cell division, to the extension of the neuronal growth cone. Microtubules also play a structural role, helping to give a cell its shape. Microtubles are composed of globular α- and β-tubulin subunits. The inherent asymmetry of the subunit gives rise to an asymmetric structure. At low temperatures (and also in the presence of calcium), microtubules dissociate into tubulin protomers (heterodimers). **Both types of tubulin subunit interact with GTP.** The α-subunit binds GTP irreversibly, the β-subunit is a GTPase. **The hydrolysis of GTP is used to add tubulin subunits at the end of a growing microtubule.** A single tube comprises 13 protofilaments. Tubes can join together to form larger diameter structures.

Actin is ubiquitous and usually the most abundant cytoplasmic protein in eukaryotic cells. A major component of the cytoskeleton, actin forms microfilaments *in vivo* and *in vitro*. The monomeric form predominates at low temperature, low salt concentration, and alkaline pH. Monomers can association and dissociate from both ends of the filament (Fig. 8.20). The kinetics of association and dissociation, however, differ at the two ends. The '**plus end,**' or '**barbed end,**' is where **ATP-bound actin monomers associate with the filament**; the '**minus end,**' or '**pointed end,**' is where **ADP-bound actin monomers dissociate from the filament**. The nicknames of the filament ends come from electron microscopic studies of actin filaments 'decorated' with the actin-binding portion of myosin (see next section). The complex has a chevron-like appearance, reflecting the underlying asymmetry of the

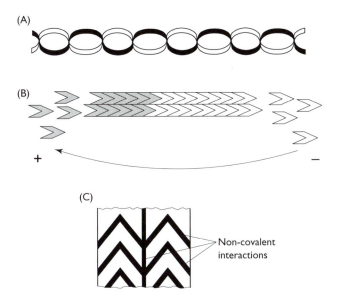

(A)

(B)

+   −

(C)

Non-covalent
interactions

Fig. 8.20 Actin filament structure. Panel (A): a single filament, also known as a microfilament, is a double helix, each strand of which is a string of actin monomers held together by *non-covalent* interactions. Panel (B): actin filaments are polar: the ends differ because subunits are asymmetrical. Preferential ATP-bound monomer association occurs at the barbed end of the filament, preferential ADP-bound monomer dissociation occurs at the point end. ATP is clearly important for actin polymerization, but actin itself is not known to be an ATPase. Panel (C): the situation is more complicated than panels (A) and (B) might suggest. For not only is there physical contact between the heads and tails of monomers, subunits interact non-covalently with each other across the helix axis.

monomers of which the actin filament is made. At some point between association of an actin monomer at the barbed end and dissociation from the pointed end, ATP hyrdolysis occurs. The precise role of ATP hydrolysis in actin polymerization is unknown, but ADP–actin polymerizes much less well than ATP–actin.

Both monomer addition at the pointed end and monomer dissociation from the barbed end are relatively improbable. Under suitable solution conditions, actin filaments *in vitro* exhibit a phenomenon known as '**treadmilling**;' the rate of monomer addition at the barbed end is about the same as the rate of monomer dissociation from the pointed end, and the average length of the filaments is approximately constant. The ability of actin to polymerize and depolymerize readily is probably very important to cell function, for when filaments are stabilized by the binding of phalloidin, a toxic component of certain poisonous mushrooms, cells look highly abnormal. Before elongation of a filament can occur, nuclei must form. Nucleation involves the thermodynamically stable association of at least three actin monomers and is therefore improbable (compare nucleation of α-helix formation in Chapter 6). In the presence of well-formed nuclei, however, once the concentration of actin monomers has exceeded the '**critical concentration**' for polymerization, elongation occurs rapidly.

The kinetics of polymerization can be determined by fluorescence spectroscopy, if the actin monomers have been labeled with a fluorescent dye. For example, it is possible to conjugate the fluorescent probe pyrene to actin. There is an increase in fluorescence intensity when polymerization occurs, because the polarity (dielectric constant) of the environment of the dye molecules is lower in filamentous actin than in monomeric actin, and the intensity of fluorescence emission is inversely proportional to the polarity of the environment surrounding the fluorophore. Experiments involving the use of fluorescence to monitor the kinetics of actin polymerization are called **polymerization assays**.

Polymerization assays are a useful tool for studying the association of actin-binding proteins with actin. For instance, in the presence of a protein that associates with the barbed end of actin, the rate of actin polymerization will be slower than in the absence of the actin-binding protein, all other conditions being the same. Moreover, one can use the results of such experiments to determine an association constant for binding of the actin-binding protein to filamentous actin. Binding of a barbed end-binding protein to actin can also result in net depolymerization of existing actin filaments. This arises from the dissociation of monomers from the pointed end and the inhibition of monomer addition at the barbed end.

There are other types of actin-binding protein. Profilin, a well-studied one, forms a 1:1 complex with actin monomers called profilactin and thereby prevents actin polymerization. The importance of this interaction can be seen exceptionally clearly in the fertilization of a sea urchin egg. Each sperm head is loaded with a bag full of profilactin called the acrosome. Contact between sperm head and the jelly surrounding the egg sets off a reaction that increases the pH of the acrosome. This change in the net charge of actin and profilin results in dissociation of profilactin, and actin nuclei begin to form. About 2 s later, a thin bundle of actin filaments called the acrosomal process begins to protrude from the sperm head. The process penetrates the jelly, joining sperm and ovum. Once activation of the formation of the acrosomal process has occurred, elongation is rapid and long: the rate is greater than 10 μM s^{-1} for c. 6 s.

O. | Muscle contraction and molecular motors

About 2 of every 5 g of the weight of a healthy adult human is muscle. The mechanical and contractile properties of non-muscle cells, erectile tissue, and other types of muscle depend in part on the actin filaments described above. In skeletal muscle, for example, which allows the skeleton to operate as a system of levers, actin filaments form regular arrays with filaments of **myosin**. This gives muscle its shape as well as its ability to contract. The action of muscles enables animals to move wings, legs, or fins, digest food, focus eyes, circulate blood, maintain body warmth, and perform a variety of other physiological functions. Our concerns in this section are kinetic and mechanistic aspects of muscle contraction.

The interaction of actin and myosin provides a basis for molecular models of 'vectorial' force generation and contraction in living muscle. On the mechanistic perspective, the contraction of muscle results from the relative motion of actin ('thin') filaments and myosin ('thick') filaments (collectively, **actomyosin**), which are oriented parallel to each other and to the long axis of the muscle. This enables muscle to contract and stretch while maintaining structural integrity. The **sliding-filament model** was proposed in 1954 by the British molecular biologists Hugh Esmor Huxley (1924–) and Emmeline Jean Hanson (1919–1973). Later, analysis of the work of Albert Szent-Györgi

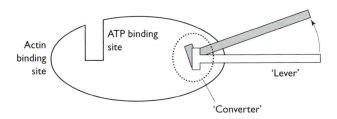

Fig. 8.21 Structure of the head group of myosin. A catalytic 'domain' binds actin and ATP, and an elongated, α-helical carboxyl-terminal 'domain' contains a variable number of proteins called calmodulin-like light chains. A plausible model of myosin function is that ATP hydrolysis leads to small intramolecular movement in the catalytic domain that is converted into a large rotation of the light-chain domain. This then acts like a lever arm in the motor mechanism. The pivot, called the 'converter,' fixes the point on which the lever rotates. One result of this process is that the end of the lever moves by several nanometers.

(1893–1986), a Hungarian biochemist, and the results of kinetics studies of the biochemical properties of myosin and actin, led others to propose models of actomyosin-mediated ATP hydrolysis.

The mode of action of myosin is very complicated. In resting muscle, the head group of a myosin subunit (Fig. 8.21) is bound to an actin filament. Release occurs upon a conformational change in the head group, which is itself induced by the binding of ATP (Fig. 8.22). Hydrolysis of bound nucleotide by the ATPase activity of myosin results in a repositioning of the head group about 6 nm down the fibril. The head group associates with the filament again, dislodging P_i. Upon release of ADP, the head group pulls the actin filament about 5–6 nm, ending the cycle about 200 ms after it began. In other words, the chemical energy of ATP hydrolysis is converted into the mechanical work of muscle contraction. The ~ 0.05 s^{-1} rate of ATP hydrolysis in isolated myosin is far lower than in contracting muscle, where the rate is $\sim 10\,s^{-1}$. The rate of hydrolysis is actually greater in the presence of actin than in its absence, because interaction between myosin and actin stimulates the release of P_i (and ADP), enabling ATP to bind.

A motor is a thing that is designed to perform a given function in a periodic fashion. For example, a motorized lawnmower depends on the combustion of gasoline (petrol) to turn both a blade for cutting grass and wheels for propelling the vehicle. An example of a biological motor is the mammalian heart. Comprising several chambers and valves, the heart is a motorized muscle that pumps blood throughout the body, delivering oxygen and food to all cells are returning carbon dioxide to the lungs. From what we've seen in this section, myosin is a molecular biological machine; it is, moreover, an example of a class of molecular machines known as protein motors. Cardiac tissue is a motor made of lots of little motors. As discussed above, the head group is the location myosin's motor function.

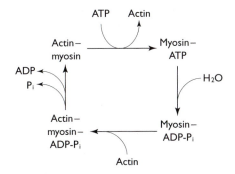

Fig. 8.22 The actomyosin ATPase reaction cycle. Compare Fig. 2.3.

Like the engines that propel cars and planes, protein motors convert chemical energy into mechanical energy. Protein motors also control the visco-elasticity of the cell cortex and undergird the transport of cell components along polymer tracks such as actin filaments, microtubules, or DNA. Some protein motors operate as individual molecules; others cooperate in large ensembles. Common to all motors, however, is energy consumption in the form of ATP hydrolysis and an associated small conformational change. 'Linear motor' molecules, for example, move along a track in nanometer-sized steps, each step corresponding to the hydrolysis of single ATP molecules. Myosin is one of the best-studied of the known linear protein motors. Analysis of the physico-chemical properties of different motor proteins is expected to shed light on general mechanisms of energy transduction in motor proteins. All myosins move along an actin filament track by hydrolysis of ATP, but at least one type of myosin can move in the opposite direction from skeletal muscle myosin. Amazing! We must stop for now, but we'll return to molecular motors in Chapter 9.

P. | References and further reading

Bai, Y., Milne, J. S., Mayne, L. & Englander, S. W. (1993). Primary structure effects on peptide group hydrogen exchange. *Proteins: Structure, Function, and Genetics*, **17**, 75–86.

Bai, Y., Milne, J. S., Mayne, L. & Englander, S. W. (1994). Protein stability parameters measured by hydrogen exchange. *Proteins: Structure, Function, and Genetics*, **20**, 4–14.

Baldwin, R. L. (1993). Pulsed H/D-exchange studies of folding intermediates. *Current Opinion in Structural Biology*, **3**, 84–91.

Benson, S. W. (1976). *Thermochemical Kinetics*. New York: John Wiley.

Bergethon, P. R. (1998). *The Physical Basis of Biochemistry: the Foundations of Molecular Biophysics*, cc. 1, 31 & 32. New York: Springer-Verlag.

Bluestone, S. & Yan, K. Y. (1995). A method to find the rate constants in chemical kinetics of a complex reaction. *Journal of Chemical Education*, **72**, 884–886.

Booth, D. R., Sunde, M., Bellotti, V., Robinson, C. V., Hutchinson, W. L., Fraser, P. E., Hawkins, P. N., Dobson, C. M., Radford, S. E., Blake, C. C. F., Pepys, M. B. (1997). Instability, unfolding and aggregation of human lysozyme variants underlying amyloid fibrillogenesis. *Nature*, **385**, 787–793.

Carr, R. W. (1992). Chemical kinetics. In *Encyclopedia of Applied Physics*, ed. G. L. Trigg, vol. 3, pp. 345–375. New York: VCH.

Chipman, D. M. & Sharon, N. (1969) Mechanism of lysozyme action. *Science*, **165**, 454–465.

Cornwell, J. (ed.) (1998). *Consciousness and Human Identity*. Oxford: Oxford University Press.

Creighton, T. E. (1992). *Proteins: Structures and Molecular Properties*, 2nd edn, ch. 9.3.3. New York: W. H. Freeman.

Delepierre, M., Dobson, C. M., Karplus, M., Poulsen, F. M., States, D. J. & Wedin, R. E. (1987). Electrostatic effects and hydrogen exchange behaviour in proteins. The pH dependence of exchange rates in lysozyme. *Journal of Molecular Biology*, **197**, 111–130.

Encyclopædia Britannica CD98, 'Collision theory,' 'Transition-state theory.'

Englander, S. W. & Mayne, L. (1992). Protein folding studied using hydrogen-exchange labeling and two-dimensional NMR. *Annual Review of Biophysics and Biomolecular Structure*, **21**, 243–265.

Fersht, A. R. (1985). *Enzyme Structure and Mechanism*, 2nd edn. New York: W. H. Freeman.

Fersht, A. R. (1999). *Structure and Mechanism in Protein Science: a Guide to Enzyme Catalysis and Protein Folding*. New York: W. H. Freeman.

Fruton, J. S. (1999). *Proteins, Enzymes, Genes: the Interplay of Chemistry and Biology*. New Haven: Yale University Press.

Gerlt, J. A. (1987). Relationships between enzymatic catalysis and active site structure revealed by applications of site-directed mutagenesis. *Chemical Reviews*, **87**, 1079–1105.

Gillespie, R. J., Spencer, J. N. & Moog, R. S. (1996). An approach to reaction thermodynamics through enthalpies, entropies and free energies of atomization. *Journal of Chemical Education*, **73**, 631–637.

Glasstone, S., Laidler, K. J. & Eyring, H. (1941). *The Theory of Rate Processes*. New York: McGraw-Hill.

Harris, D. A. (1995). *Bioenergetics at a Glance*, cc. 1 and 37. Oxford: Blackwell Science.

Hecht, C. E. (1990). *Statistical Mechanics and Kinetic Theory*. New York: W. H. Freeman.

Klots, C. E. (1988). The reaction coordinate and its limitations: an experimental perspective. *Accounts of Chemical Research*, **21**, 16–21.

Klotz, I. M. (1986). *Introduction to Biomolecular Energetics*, ch. 9. Orlando: Academic Press.

Kondepudi, D. & Prigogine, I. (1998). *Modern Thermodynamics: from Heat Engines to Dissipative Structures*, ch. 9. Chichester: John Wiley.

Laidler, K. J. (1988). Just what is a transition state? *Journal of Chemical Education*, **65**, 540–542.

Laidler, K. J. (1988). Rate-controlling step: a necessary or useful concept? *Journal of Chemical Education*, **65**, 250–254.

Lipscomb, W. N. (1983). Structures and catalysis of enzymes. *Annual Review of Biochemistry*, **52**, 17–34.

Logan, S. R. (1986). The meaning and significance of 'the activation energy' of a chemical reaction. *Education in Chemistry*, **23**, 148–150.

Maier, C. S., Schimerlik, M. I. & Deinzer, M. L. (1997) Thermal denaturation of *Escherichia coli* thioredoxin studied by hydrogen/deuterium exchange and electrospray ionization mass spectrometry. *Biochemistry*, **38**, 1136–1143.

Maskill, H. (1984). The extent of reaction and chemical kinetics. *Education in Chemistry*, **21**, 122–123.

Maskill, H. (1990). The Arrhenius equation. *Education in Chemistry*, **27**, 111–114.

Mata-Perez, F. & Perez-Benito, J. F. (1987). The kinetic rate law for autocatalytic reactions. *Journal of Chemical Education*, **64**, 925–927.

Millar, D., Millar, I., Millar, J. & Millar, M. (1989). *Chambers Concise Dictionary of Scientists*. Cambridge: Chambers.

Morozova-Roche, L., Arico-Muendel, C., Haynie, D., Emelyanenko, V., van Dael, H. & Dobson, C. (1997). Structural characterisation and comparison of the native and A-states of equine lysozyme. *Journal of Molecular Biology*, **268**, 903–921.

Oyama, S. T. & Samorjai, G. A. (1988). Homogeneous, heterogeneous, and enzymatic catalysis. *Journal of Chemical Education*, **65**, 765–769.

Penrose, R. (1994). *Shadows of the Mind*. Oxford: Oxford University Press.

Pepys, M. B., Hawkins, P. N., Booth, D. R., Vigushin, D. M., Tennent,, G. A., Soutar, A. K., Totty, N., Nguyent, O., Blake, C. C. F., Terry, C. J., Feest, T. G., Zalin, A. M. & Hsuan, J. J. (1993). Human lysozyme gene mutations cause hereditary systemic amyloidosis. *Nature*, **362**, 553–557.

Pilling, M. J. & Seakins, P. W. (1995). *Reaction Kinetics*. Oxford: Oxford University Press.

Raines, R. T. & Hansen, D. E. (1988). An intuitive approach to steady-state kinetics. *Journal of Chemical Education*, **65**, 757–759.

Rashin, A. A. (1987). Correlation between calculated local stability and hydrogen exchange rates in proteins. *Journal of Molecular Biology*, **198**, 339–349.

Rayment, I. & Holden, H. (1993). Title. *Current Opinion in Structural Biology*, **3**, 949.

Roberts, T. J., Marsh, R. L., Weyland, P. G., & Taylor, C. R. (1997). Muscular force in running turkeys: the economy of minimizing work, *Science*, **275**, 1113–1115.

Shaw, H. E. & Avery, D. J. (1989). *Physical Chemistry*, ch. 4. Houndmills: Macmillan.

Siegel, I. H. (1993). *Enzyme Kinetics*. New York: Wiley-Interscience.

Stossel, T. P. (1994). The machinery of cell crawling, *Scientific American*, **271**, no. 3, 54–63.

Taubes, G. (1996). Misfolding the way to disease. *Science*, **271**, 1493–1495.

Voet, D. & Voet, J. G. (1995). *Biochemistry*, 2nd edn, cc. 12 & 13. New York: John Wiley.

Wagner, C. R. & Benkovic, S. J. (1990). Site directed mutagenesis: a tool for enzyme mechanism dissection. *Trends in Biotechnology*, **8**, 263–270.

Wells, A. L., Lin, A. W., Chen, L. Q., Safer, D., Cain, S. M., Hasson, T., Carragher, B. O., Milligan, R. A. & Sweeney, H. L. (1999). Myosin VI is an actin-based motor that moves backwards. *Nature*, **401**, 505–508.

Woodward, C. K. (1994). Hydrogen exchange rates and protein folding. *Current Opinion in Structural Biology*, **4**, 112–116.

Wrigglesworth, J. (1997). *Energy and Life*, ch. 4.2.1. London: Taylor & Francis.

Q. | Exercises

1. How is E_a overcome in the oxidation of materials in a bomb calorimeter in Chapter 1? (See Fig. 1.11.)

2. The following statements pertain to energy transfer within a cell and between a cell and its surroundings.
   (a) A cell can convert energy into a useful form by allowing carbon and hydrogen to combine with oxygen.
   (b) Chemical energy is converted by a cell to heat, where the energy is transferred into a more ordered form.
   (c) A cell obeys the Second Law of Thermodynamics by acting like a closed system.
   (d) Enzymes are important for cell catabolism because they lower the change in free energy of the reaction.
   Which of these statements are true? Which are false? Explain.

3. Which of the following are true?
   (a) Resting cells do not produce any heat.
   (b) Growing cells release less heat to the environment than do resting cells because they use more energy.
   (c) Life is a thermodynamically spontaneous process.

(d) Enzymes that couple unfavorable reactions to favorable reactions cause a decrease in total entropy.
Explain.

4. What are the units of k in a fourth-order reaction?

5. Does the Arrhenius equation hold for enzymes? If yes, under what conditions? If no, why not?

6. Prove that if k_2/k_1 is small in comparison with $[E][S]/[E \cdot S]$, K_M is a measure of the affinity of an enzyme for a substrate.

7. Referring to panel (B) of Fig. 8.9, describe the effect on the rate of electron transfer of increasing $\Delta G°$.

8. The rate of ATP hydrolysis to ADP and P_i is influenced by the muscle protein myosin. The following data are tabulated at 25 °C and pH 7.0.

Velocity of reaction in μmoles inorganic phosphate produced $l^{-1} s^{-1}$	[ATP] in μM
0.067	7.5
0.095	12.5
0.119	20.0
0.149	32.5
0.185	62.5
0.191	155.0
0.195	320.0

Find the Michaelis constant of myosin.

9. Show that

$$J = \frac{\dfrac{J^f_{max}[S]}{K^S_M} - \dfrac{J^r_{max}[P]}{K^P_M}}{1 + \dfrac{[S]}{K^S_M} + \dfrac{[P]}{K^P_M}} \qquad (8.68)$$

for a reversible enzymatic reaction. The reaction scheme might look like this

$$E + S \Leftrightarrow ES \Leftrightarrow P + E$$

and

$$J^f_{max} = k_2[E]_T \qquad J^r_{max} = k_{-1}[E]_T$$

$$k^S_M = \frac{k_{-1} + k_2}{k_1} \qquad k^S_M = \frac{k_{-1} + k_2}{k_{-2}}$$

10. The rate of hydrogen exchange is a function of temperature. Assuming that the rate increases three-fold for every increase in temperature of 10 °C, calculate the activation energy for exchange.

11. Suppose you have a bimolecular reaction in which $2A \rightarrow P$. Using standard methods of calculus, it can be shown that $[P(t)] = [A]_o kt/(1 + 2[A]_o kt)$, where $[A]_o = [A(t = 0)]$ and k is the rate constant. Compare $[P(t)]$ for a unimolecular reaction and a bimolecular reaction in the form of a graph.

12. Skeletal muscle is involved in maintaining body warmth. Explain how this might occur.

13. At low temperatures, addition of heat increases enzyme activity. The trend usually begins to reverse at about 55–60 °C. Why? (Hint: see Chapter 5.)

14. Would life be possible if the rates of biochemical reactions were not determined by activation energies? Why or why not?

15. Why does plant life tend to be more robust in tropical climates than closer to the poles?

16. Cellular respiration involves the oxidation of glucose to gluconic acid. The reaction is catalyzed by glucose oxidase. Suggest a means of measuring the rate of reaction.

17. List four variables that can affect the rate of a reaction.

18. Urea is converted into ammonia and carbon dioxide by the enzyme urease. An increase in the concentration of urea increases the rate of reaction. Explain. (Hint: see Chapter 6B.)

19. Suggest several ways in which an enzyme inhibitor might be used therapeutically.

20. Outline an experimental program by which site-directed mutagenesis could be used to study properties of the transition state of a protein folding reaction.

21. Equine lysozyme is an unusual lysozyme in that it has a calcium-binding site. The location of the ion-binding site is identical to that in α-lactalbumin (see Chapter 6). Unlike the α-lactalbumins, equine lysozyme has Glu35 and Asp53 (Asp52 in hen lysozyme), important for lysozyme activity. Morozova-Roche *et al.* (1997) have measured rates of exchange of polypeptide backbone amide protons in equine lysozyme under different conditions. Determine the protection factors of the residues shown in the table below. Helix B encompasses residues 24–36, and residues 40–60 form an anti-parallel β-sheet in the native protein. Comment on the calculated protection factors in the light of this structural information.

Residue	pH 4.5, 25 °C k_{ex}	pH 4.5, 25 °C k_{in}	pH 2.0, 25 °C k_{ex}	pH 2.0, 25 °C k_{in}
Asn27	1.7×10^{-7}	9.5×10^{-7}	3.6×10^{-4}	1.5×10^{-3}
Trp28	2.0×10^{-7}	1.0×10^{-6}	8.4×10^{-4}	8.3×10^{-3}
Val29	5.6×10^{-8}	2.7×10^{-7}	5.3×10^{-5}	3.0×10^{-4}
Met31	1.4×10^{-7}	8.0×10^{-7}	2.2×10^{-4}	6.4×10^{-4}
Ala32	1.3×10^{-8}	8.2×10^{-8}	2.1×10^{-4}	2.3×10^{-3}
Glu33	6.1×10^{-8}	2.5×10^{-7}	—	—
Tyr34	1.5×10^{-7}	7.2×10^{-7}	2.7×10^{-4}	1.0×10^{-3}
Glu35	6.1×10^{-7}	2.5×10^{-6}	9.9×10^{-5}	7.1×10^{-4}
Ser36	7.5×10^{-7}	3.5×10^{-6}	1.1×10^{-3}	1.5×10^{-3}
Thr40	2.6×10^{-7}	1.1×10^{-6}	—	—
Ala42	1.6×10^{-6}	7.0×10^{-6}	3.7×10^{-3}	2.2×10^{-3}
Lys46	4.7×10^{-6}	1.8×10^{-5}	1.2×10^{-3}	2.2×10^{-3}
Ser52	5.2×10^{-5}	1.7×10^{-4}	—	—

Residue	pH 4.5, 25 °C	pH 4.5, 25 °C	pH 2.0, 25 °C	pH 2.0, 25 °C
	k_{ex}	k_{in}	k_{ex}	k_{in}
Asp53	1.2×10^{-5}	4.7×10^{-5}	—	—
Tyr54	3.5×10^{-7}	1.5×10^{-6}	1.3×10^{-3}	1.6×10^{-3}
Phe57	4.3×10^{-7}	1.8×10^{-6}	1.4×10^{-3}	1.1×10^{-3}
Gln58	3.7×10^{-6}	1.5×10^{-5}	3.3×10^{-3}	6.6×10^{-4}
Leu59	6.3×10^{-7}	2.6×10^{-6}	1.2×10^{-3}	7.2×10^{-4}

22. Derive Eqn. 8.67.

For solutions, see http://chem.nich.edu/homework

Chapter 9

The frontier of biological thermodynamics

A. | Introduction

Thus far this book has principally been concerned with fairly well established aspects of energy transformation in living organisms, the macromolecules they're made of, and the environments in which living things flourish. Much of the discussion has had a decidedly practical slant to it, in order to show how thermodynamic and kinetic concepts can be useful in today's biochemistry laboratory. In the present chapter, we change our tack and set sail for waters less well charted. The exploration will aim to locate the material covered thus far in the broader scheme of things, and also to see how the development of topics of considerable current interest must conform to the strictures of the laws of thermodynamics. Our course might prove to be somewhat off the mark, as the questions addressed here are speculative; often no right answer is known. But the journey will nevertheless prove worth the effort, as it will show what a lively subject biological thermodynamics is and draw attention to a few of the areas where there is still much work to be done. Throughout the chapter, we shall bear in mind a program proposed over a century ago by the great British physicist Lord Kelvin (eponym of the absolute temperature scale; he lived 1824–1907): to explain *all* phenomena of the world, both natural and manmade, in terms of energy transformations.

B. | What *is* energy?

Many students find the concept of energy difficult to grasp. Definitions tend to be very abstract and are often framed in mathematical terms that may seem far-removed from our everyday experience of the world. Physics textbooks don't always help much here, as *energy* is usually defined as 'the capacity to do work.' Even when interpreted strictly mechanically, such 'operational' definitions are more open-ended than one might expect of the most basic concept in science. After all, *science* comes to us from the Latin verb *scire*, meaning 'to know.' Digging deeper doesn't always help either, because in some cases knowing more can make the basic outline of the object of one's study seem all the more enigmatic or obscure. When, for instance, we turn for guidance to that august and scholarly compendium of knowledge, the *Encyclopædia Britannica*, we find that '**the term** *energy* **is difficult to define precisely, but one possible definition might be the capacity to produce an effect**.' How's that for scientific clarity, precision, and certainty!

Nevertheless, headway can be made, and one should avoid making the genuinely mysterious seem mundane or the plainly obvious appear privileged knowledge. From the exercises of Chapter 1, Einstein's famous formula, $E = mc^2$, tells us that the energy of a thing is equal to the amount of matter it comprises times the speed of light in vacuum squared. In other words, **energy is a property of matter**. Moreover, energy is a *universal* property of matter, because **all material things possess energy**. Although this does not provide a very specific idea of what energy is, it does at least give a sense of its nature. We can also see that the Einstein energy relation says less than it might. That's because, as written, it does not apply to photons, the massless particle-like entities that are the main source of free energy in photosynthesis. An important conclusion to be drawn from this is that **matter considered generally is not the same as mass**. To describe the energy of a photon we need Eqn. 1.1. This relationship describes photon energies very well over an extremely wide range of wavelengths. The form of the equation, however, suggests that there is no upper bound on the energy of a photon. What might that mean? Does Eqn. 1.1 make sense if the universe is finite, as suggested by Big Bang cosmology? For if the universe is finite, then surely the energy of a photon cannot be arbitrarily large. Or is Eqn. 1.1 just a convenient model for photons of energies that are neither smaller nor larger than we have encountered thus far in our investigation of the universe? Are we able to know the answers to these questions?

As we have seen throughout this book, energy can be described qualitatively and mathematically in seemingly very different ways. This can make the abstract concept that much more difficult to pin down. For example, we know from an earlier chapter that heat energy has liquid-like properties, and advanced analysis shows that this form of energy can be modelled using mathematical equations developed to describe fluid flow. And yet it was demonstrated experimentally hundreds of years ago that heat energy is not a fluid in the sense that liquid water is a fluid. We have also mentioned that energy conservation, a basic principle with a simple mathematical expression, is related to the time-symmetry of physical law, a very basic principle. It is not at all clear, however, how the concepts of fluid-likeness and time-symmetry link up.

In the first few chapters of this book our discussion of energy concentrated mainly on such questions as '*How* can a biological process be described in energetic terms?' and '*What* can be done with energy?' The harder question, '*What is* energy?' was raised but not answered explicitly. This approach was adopted to take account of both the puzzling and difficult nature of energy and **the marked tendency of *modern* science to be more concerned with mechanism and utility than with *being as such*.** In a fast-pace world where what counts is 'results,' it can be difficult to see the importance of pondering the very foundations of a science, of going beyond what can obviously be applied in a concrete situation. But it is important to question and probe and reflect; indeed, it is *necessary* for the development of science. Recalling the first few chapters, we say – with confidence! – that energy is conserved, citing the First Law of Thermodynamics. We say – with even greater confidence! – that the Second Law tells us that the extent to which the energy of the universe is organized can only decrease. We can *define* energy as the capacity to do work, and write down precise mathematical equations for energy transformation. We can solve mathematical equations that give accurate predictions of what can be measured. We begin to think that we know something. And indeed, *something* is known. But one should not lose sight of the fact that such knowledge does not tell us what energy *is*.

We should not feel too bad about this, though, because **no one can say what energy is**. 'Ah,' you say, 'but don't we *know* that the kinetic energy of a body like a comet is its mass times its velocity squared divided by two, and hasn't this relationship been tested many, many times with objects under our direct control?' (Kinetic energy $= \frac{1}{2}mv^2$.) 'Indeed,' I reply, 'but do we *know* what *mass* is? Einstein tells us that mass and energy are, in some sense, equivalent, and we quickly find our reasoning becoming circular. We know that energy exists, and we know that the total amount of it doesn't change, but we cannot say what energy is.' This suggests that **although we certainly can know something about the world and describe its properties both qualitatively and quantitatively, it is by no means certain that we can know or describe its most basic aspects**. This is something like the situation with numbers and the extremely important concept of infinity. We accept that if we add or subtract a finite quantity from infinity we get infinity back again. Infinity must of course be a number, but we do not know whether that number is even or odd. No one can say what infinity *is*. In other words, we *know* that the infinite exists even though we cannot comprehend its nature. The extremely important concept of energy is something like that.

Energy sure does seem to have a mysterious quality to it. *Why* is that? Is it a pesky relic of a pre-modern age in dire want of further scientific research and clarification? Or is it a deep insight into the basic nature of the universe? Whatever one thinks about this or however one responds, it is at least clear that **it is not *necessary* to be able to say what energy *is* for the concept to be *useful*** in science, technology, or everyday life. This should not come as a great surprise. After all, for most purposes it is hardly necessary to *know* technical details of how a computer works in order to use one, or what paper is made of in order to press it into service in the form of a book, cereal box, or money. Neither should the difficulty we have in 'understanding' energy necessarily be seen as a failure of scientific research: for we should at least admit the possibility that what is known now might actually be a glimpse of the basic character of reality and that it might not be possible to go deeper in the sense of being able to say with greater specificity and clarity what energy is. In any case, the lack of certainty we might have about such things should not lead us into cozy contentment but spur us on to test anew whether ideas inherited from the previous generation are incomplete or possibly wrong (including basic thermodynamic relationships), to seek a more definite or complete awareness of the *nature* of the world, and to consider what it all means.

C. | The laws of thermodynamics and our universe

In the Preface to this book, Einstein is cited as saying that of all the laws of physics the First and Second Laws of Thermodynamics are the ones least likely to be overturned or superseded. Why are these laws so special? We have mentioned that they are *extremely* general, and anything purporting to be a law should be general. The First Law meets this criterion, as there are good reasons to believe that it applies just as well to the entire universe as a purified solution of DNA in a test tube. Similarly, the Second Law pertains not only of the entropy of a system of interest but indeed the entire universe.

Perhaps just as important as the *universality* of these Laws is their *simplicity*. From a mathematical point of view, the First Law is of the form $a + b = c$, and the Second Law $x / y \leq z$. Summing two numbers and comparing the magnitudes of two numbers are among the most basic mathematical operations conceivable. The First Law says that for any chemical reaction the sum of the energy changes must not result in a change in the total energy, while the Second Law says that for there to be a net change in a system the ratio of the entropy to the temperature of the system must increase. That's it. **The principles of thermodynamics are profound not merely because they work and are universal, but because they are so simple**.

The simplicity of thermodynamic relationships helps to make them of great practical importance in science and engineering. Their utility also stems from the severe restrictions they place on what is possible in principle. The limitations imply that a large class of conceivable machines, e.g. machines that produce more energy than they consume, cannot exist (in our universe). Now, let's suppose that the First and Second Laws provide a fundamentally correct description of our universe, as Einstein, Eddington, and many others have believed. We can conceive of the set of all possible universes in which energy can be transformed from one type to another (P), and represent this in the form of a Venn diagram (named after the British logician John Venn, who lived 1834–1923) (Fig. 9.1). Subsets of P are the set of all universes in which the First Law holds (A) and the set of all universes in which the Second Law holds (B). The intersection of A and B, which includes all universes in which both the First Law and the Second Law apply, is where our universe is found. Twenty billion years old and vast in extent, our universe occupies but a single dimensionless point in the diagram. This helps to

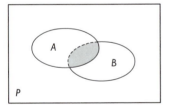

Fig. 9.1 Possible universes and the laws of thermodynamics. The Venn diagram shows all possible universes in which energy transformation can occur (P), those in which the First Law is required to hold (A), those in which the Second Law is required to hold (B), and the intersection of A and B, where our universe is a dimensionless point. Even given the constraints of the First and Second Laws, there is no limit on the number of different universes that are possible; not different successive states of the same universe, but different universes. Ours is but one of the infinite number of possible universes.

put things in perspective, including our very lives! And any theory claiming to describe how living organisms originated and continue to exist by natural causes must be compatible with the First and Second Laws of Thermodynamics.

The Second Law tells us that as time goes on the entropy of the universe will increase indefinitely. A rather curious aspect of this is that although the law demands irreversible change, the law itself does not change. That is, the Law seems not to apply to itself. However, all the ways in which we have thought about the Second Law thus far have concerned the behavior of *particles and energy*, and while the Law describes what particles must do, at least in general terms, the Law itself is not made of matter; it is a *relationship*.

Some people say that the universality of thermodynamic relationships eliminates an 'artificial' distinction between the animate and the inanimate, and thereby enables one to see the world as a single entity. A piece of granite, however, is no more a willow tree than a plant is a kangaroo. **That the world is one does not necessarily imply that qualitative differences of universal proportion will not be found in it.** For this reason (and others, see below) a cell is considered the smallest structure that can be considered alive.

D. | Thermodynamics of small systems (e.g. molecular motors)

The relationships of classical thermodynamics describe the properties of *macroscopic* systems – systems of large numbers of particles. In a typical differential scanning calorimetry experiment, for example, the protein concentration might be 1 mg ml^{-1}. Given a cell volume on the order of 1 ml and a protein molecular mass of 20 000 Da, the number of molecules in an experiment is of the order 10^{16}. This number is big, approximately 1% of the age of the universe in seconds! In contrast, a bacterial cell might contain only a small number of copies of a given macromolecule, rarely more than 1000 (excluding strains engineered for expression of recombinant protein).

So what? After all, average values of thermodynamic quantities of small systems must be the same as for large systems. But, just as the uncertainty of a measured value is inversely related to the number of measurements made, deviations from the average value are large when the concentration is low, because each molecule plays a relatively important role in determining the properties of the system. The Boltzmann distribution says that the population of an energy level scales as the negative exponential of the energy (Chapter 6). **The Boltzmann distribution is the most probable distribution in the limit that the system consists of a large number of particles, but it is not necessarily the most probable distribution in a small system.** In other words, the lower the concentration of molecules, the less well the Boltzmann distribution will describe how the particle energies are distributed.

A single molecule can play no greater role in determining the thermodynamics of a system than when it is the *only* molecule of that type in the system. This is pertinent here because much remains to be known about biological processes at the mechanistic level, and there is increasing interest in techniques that can be used to interrogate single complex biomolecular machines and the roles they play. Consider, for example, DNA replication in the cell. This occurs when regions of single-stranded DNA become available for binding by DNA polymerase, a type of molecular motor that replicates DNA for transmission to daughter cells. Polymerase catalyzes DNA synthesis. The enzyme is believed to move itself along the template DNA by making use of the favorable free energy change of incorporating nucleotides into the growing nucleotide strand, which involves hydrolysis of phosphodiester bonds. It has been found by single-molecule experiments that the catalytic activity of polymerase depends on the tension in DNA. Activity is highest when the tension is about 6 pN;[1] below or above this value, activity is less than maximal. This suggests that entropic properties of single-stranded DNA play a role in determining how quickly polymerase does its job. If the chain is not flexible enough, polymerase doesn't work too well; if the chain is too flexible, polymerase is also less effective than it could be.

Single-molecule systems are interesting and important in other ways as well. There are certain biological machines known as molecular switches, whose movement is driven by chemical, electrochemical, or photochemical forces. As we have seen, an example of a molecular switch is hemoglobin. Molecular switches are of great interest now because of hopes that their properties can be exploited in molecular-scale information processing. One can envision, for example, protein- or nucleic acid-based computers, and interesting work in this area is now underway.

Experiments involving single molecules raise questions of fairly basic importance. How does one cope with fluctuations on the level of a single molecule? How meaningful are the results of single-molecule experiments for describing the properties of molecules generally? A basic condition of credibility in experimental science is repeatability. This is obviously more easily achieved when one is investigating properties of a large collection of molecules than a single molecule. Will analysis of the fluctuations exhibited by single molecules provide insight into how macroscopic properties develop out of them? How do thermodynamic relationships apply to macromolecules when quantities are relatively limited? Can the behavior of single molecules tell us anything about how life got going?

E. | Formation of the first biological macromolecules

Microfossils in ancient rocks in Australia and South Africa strongly suggest that the first organisms flourished on Earth as early as 3.5 billion

---

[1] $1 N = 1 \text{ kg m s}^{-2}$. $1 N = 1$ newton. This unit of force is named after Isaac Newton.

years ago. Even older rocks in Greenland contain features that seem to have resulted from living organisms. So, about 100 million years after the earliest time when Earth could safely have supported life, living organisms had probably already established a toehold. (To put this in perspective, the dinosaurs met their demise a mere 65 million years ago.) The timing of life's earliest appearance on Earth suggests that there may have been 'help' from space; for it is hardly overstating the case to say that going from a bacterium to a baboon is easier than going from a mixture of amino acids to a bacterium. Ignoring the odd meteorite, Earth is a closed system (Chapter 1), and it is by no means certain that the mixture of chemicals present early on in the history of our planet was sufficiently complex to permit the spontaneous organization of the first cell. Although such reasoning might 'explain' the origin of life on Earth, it obviously would not explain how life began in the first place. In this section we shall consider several aspects of the origin of life as we know it, all of which may well pertain regardless of where life began.

Abiotic synthesis of biopolymers

Could bacteria have formed from non-living chemicals in a single step? Most scientists think not. It is hard to imagine how there could not have been 'intermediate forms,' perhaps aggregates of organic polymers. Candidates are the four major classes of biological macromolecule: proteins, nucleic acids, polysaccharides, and lipids. Such molecules constitute the bulk of tissues, cellular components, and combinations thereof. Knowing something about the biosynthetic pathways and physical properties of these macromolecules might provide clues as to which ones were important in 'pre-biotic life,' so let's take a look at some important features.

Cell component biosynthesis occurs in two main stages. In the first, intermediate chemical compounds of the main thoroughfares of metabolism are shunted to pathways that lead to the formation of the building blocks, or **precursors**, of macromolecules. These reactions are wholly dependent on the extraordinary functional specificity and speed of a broad range of enzymes and other proteins. In the second stage of biosynthesis, precursors are joined to form a protein, nucleic acid, polysaccharide, or lipid, as the case may be. Through biosynthesis of certain macromolecules, principally chromosomal DNA and DNA-binding proteins, the **biological information** specifying the identity of the cell, tissue, and entire organism is both expressed and maintained.

Though the cell makes biological polymers with relative ease – consider how readily bacteria are engineered nowadays to produce huge quantities of recombinant polypeptide – the **abiotic synthesis** of biological polymers from precursors is rather difficult. From the physical point of view, such synthesis is hard because two monomers must be in the right spatial orientation in the same place at the same time, and there is no enzyme binding pocket present to facilitate this. Synthesis is difficult from the chemical point of view as well, because the collision between precursor and growing polymer must be sufficiently energetic to overcome the activation energy barrier, and there are no enzymatic chemical groups present to reduce the height of the barrier. Moreover, the underlying mechanisms of synthesis, known as dehydrations (a molecule of water is lost in joining two subunits), require the presence of dehydrating agents or condensing agents (e.g. cyanamide). In other words, there are rather severe restrictions on the conditions that could have supported abiotic synthesis of the polymers of life. This is particularly true of enzymes, which have distinct catalytic properties; the polyamino acids of living things simply cannot be random polymers. In view of this, it is most unclear how the first biopolymers came into being billions of years ago.

A British X-ray crystallographer named John Desmond Bernal (1901–1971) has suggested that prebiotic aqueous solutions of molecular intermediates of biopolymers were perhaps concentrated and protected from degradation by adsorption onto clay or some mineral. One possibility is that **phosphates** were involved, as this would help explain the apparently preferential incorporation of phosphorous into organic molecules (nucleotides) at a time when the extremely complex biological concentration mechanisms we know about today – e.g. membranes and ion pumps – did not exist. The primitive oceans in which life is commonly thought to have begun may have contained high concentrations of organic molecules, and evaporation or freezing of pools may also have helped to concentrate precursors of biopolymers. In any case, **the means by which the first biological polymers were formed is still very much an open question**.

Proteins

One proposal regarding Earth's first biopolymers is that they were not synthesized here but delivered – by **meteorites**. These bits of debris from space are made mostly of metal and rock, but some of them contain complex organic compounds like nucleobases, ketones, quinones, carboxylic acids, amines, and amides. Of all these molecules, the amino acids command the most attention. Why? There are several reasons. One is that all proteins are made of them. The genetic code, which is nearly identical in all forms of life, specifies the 20 'standard' amino acids employed by cells to build proteins. Eight of these amino acids have been found in meteorites. And although nucleic acids are the molecules of genetic inheritance, proteins are necessary for (almost) all of the specific chemical reactions in cells. Another reason is that amino acids (and other biological molecules, e.g. sugars) exist in mirror-image pairs, a molecular quality called **chirality**, or handedness, and ribosomes in all known organisms manufacture proteins with just one of them. More specifically, individual amino acids are either left-handed (L-form) or right-handed (D-form), depending on the placement of atoms that are singly-bonded to a centrally located α-carbon. The energies of formation of the two forms are identical, so there is no apparent thermodynamic advantage in making and using one type instead of the other. **Despite the absence of a thermodynamic criterion, all the amino acids in proteins are left-handed**. (More accurately, D-amino acid residues are found in some short bacterial polypeptides that are synthesized enzymatically instead of on ribosomes. Such peptides are found in the bacterial cell wall, and this may help to protect the bugs against proteolytic attack. D-amino acid peptides are also found in bacterially produced antibiotics. It is currently believed that D-amino acids are synthesized enzymatically from L-amino acid precursors.)

Some people believe that the peculiar handedness of proteins is a matter of chance, the result of an unpredictable combination of 'blind' pre-biotic processes that may have involved polarized light. Others think that primitive life forms may have incorporated both L- and D-amino acids in polypeptides, but that the mechanisms for D-amino acid synthesis were lost long ago, presumably in order to conserve the energy resources of the cell. If extraterrestrial starting ingredients were involved in the origin of life on Earth, they could be responsible for the chiral bias, but convincing evidence of this based on the composition of comets or meteorites has not been found. And, as stated previously, **although sufficient proof of an extraterrestrial origin would answer an important question about life on Earth, it would not tell us how the biased distribution of amino acids types arose in the first place.** Related questions on this topic are: What would be the effect of changing the handedness of a protein but keeping its chemical make-up the same? Would the information content of the folded state of the protein change? Is it important that water is a non-chiral solvent? Does the universe as a whole have a hand? If so, what is its origin? Is there a thermodynamic reason for this?

Although the energies of formation of D- and L-amino acids are the same, could thermodynamics nevertheless have played a role in their asymmetric importance in life on Earth? Research has shown that far-from-equilibrium chemical systems can both generate and maintain chiral asymmetry *spontaneously*. The dominance of L-amino acids in living organisms might be caused by a small but significant chiral asymmetry that has its origin in the so-called **electroweak interactions** of electromagnetism and the weak nuclear force. Such interactions give rise to effects at the atomic and molecular levels, spin-polarized electrons in radioactive decay, and polarized radiation emitted by certain stars. It has been estimated that the chiral asymmetry of the electroweak interaction could result in a difference in the concentration of chemical mirror pairs (enantiomers) on the order of one part in 10^{-17}, a very small difference! Calculations suggest, however, that if the production of chiral molecules were fast enough and maintained long enough, the enantiomer favored by the electroweak force could dominate. **As of now, though, convincing experimental support for the hypothesis that electroweak interactions in far-from-equilibrium systems underpin the origin of biomolecular chiral asymmetry is lacking.**

Nucleic acids

Proteins were not necessarily required to catalyze all the biochemical reactions that have been important to life since its advent. Indeed, nucleic acids and not protein may have been the first biopolymers. In support of this, some RNAs go beyond their protein-encoding function and exhibit enzymatic properties. Moreover, RNA can serve as a template for DNA synthesis, the reverse of the normal transcription process, as it does in retroviruses like HIV. Taken together, these facts suggest that RNA could have been the type of macromolecule that got things going.

On this view, polymeric RNA molecules came into being spontaneously in a 'nucleotide soup,' assembling themselves into something 'proto-biological.' Importantly, RNA can self-replicate, and just as in DNA incorrect base incorporation leads to new sequences and possibly new enzymatic properties. tRNA-like adaptor molecules must eventually have appeared, and these would have associated directly with amino acids, which must have been available by this time, and ribosomal RNA would have self-assembled with proteins into ribosomes, on which others proteins were synthesized. The resulting polypeptides would eventually 'acquire' the enzymatic properties required for synthesizing the necessary components of a membrane, metabolizing foodstuffs, and sustaining growth. And DNA, which is more stable chemically than RNA and thus better for storing genetic information, would eventually replace RNA as the storage molecule of genetic information.

In support of this 'RNA world,' (a term apparently coined in 1986 by Walter Gilbert (1932–), an American molecular biologist) it has been found that highly active RNA ligases – enzymes that link RNA monomers together – can be derived from *random* RNA sequences. This suggests that biomolecular functionality *can* arise out of randomness. And recently a polymeric RNA lacking cytidine was found to exhibit enzymatic activity, implying that no more than three subunit types were necessary for catalytic activity in 'prebiotic life.' Though these findings are perfectly credible, it must be appreciated that they depend on the design of the experiments and the functional selection processes involved. A *person* designed an assay to *select* molecules exhibiting a certain property. Moreover, it is hardly irrelevant that outside the controlled environment of the laboratory one would be extremely hard-pressed to find biologically meaningful RNA strands *of any size*. This is because in addition to being difficult to synthesize abiotically, RNA is chemically very unstable. Just as important, the known range of catalytic activities exhibited by RNA is rather narrow. In other words, **we are far from having proof that the spontaneous appearance of RNA catalysts was the means by which life originated on Earth**. This view is corroborated by that of British Nobel Laureate Sir Francis Crick, who recently said that 'the gap from the primal "soup" to the first RNA system capable of natural selection looks forbiddingly wide.' Ilya Prigogine (1917–), a Russian-Belgian chemist, who won the Nobel Prize in Chemistry in 1977 for his contributions to nonequilibrium thermodynamics (see Chapter 5), has expressed a similar view, though one not necessarily intended as a comment on the RNA world: 'The probability that at ordinary temperatures a macroscopic number of molecules is assembled to give rise to the highly ordered structures and to the co-ordinated functions characterizing living organisms is vanishingly small. The idea of spontaneous genesis of life in its present form is therefore highly improbable, even on the scale on billions of years during which prebiotic evolution occurred.' In view of this, **the idea that life on Earth originated from an extra-planetary source looks very attractive indeed. And yet, should Arrhenius's panspermia hypothesis continue to prove the most plausible one, we still would not necessarily know when, where, or how life began.**

F. | Bacteria

Leaving aside the really hard questions, let's just take the existence of bacteria for granted and move on to other topics. We begin with bacteria because they are the simplest living things known to mankind. **All known types of bacteria exhibit all the essential feature of a living organism: the ability to capture, transform, and store energy of various kinds in accordance with information encoded in their genetic material. Viruses** and **prions** (infections proteins), on the other hand, which exist on the fringe of life, are themselves not alive. True, viruses (but not prions) contain genetic information in the form of DNA or RNA. Viruses, like cats, are open systems. Viruses, like cats, reproduce. And both viruses and cats change from generation to generation by way of alterations to their genetic material. What viruses cannot do, however, is self-subsist: they require the metabolic machinery of a host like a cat to produce the energy molecules required for their replication. Though some enzymes are encoded by some viral genomes, such genes are almost always readily identifiable as mutated forms of normal cellular enzymes produced by the host. And, after their constituent molecules have been synthesized, viruses *assemble spontaneously* into highly symmetrical structures similar to inorganic crystals; they do not *develop*.

Like cats, most known species of bacteria require oxygen to synthesize the energy molecules required for replicating DNA, making proteins, growing, and reproducing. Other types of bacteria, however, for instance sulfate-reducing ones, are strict anaerobes, and culturing them in the presence of oxygen is a sure-fire way of putting them to death. **Sulfate-reducing bacteria** use sulfate, not oxygen, as the terminal electron acceptor in respiration, and they 'generate' energy from a variety of simple organic molecules and molecular hydrogen. Sulfate is reduced to hydrogen sulfide, a substance that smells like rotten eggs. Both mesophilic and thermophilic species of sulfate-reducing eubacteria are known. At hyperthermophilic temperatures (85–110 °C), however, only certain archaebacteria are known to thrive by sulfate-reduction. These bacteria are called '**extremophiles.**'

Other extremophilic archaebacteria are the **methanogens**. Enclaves of these organisms have been detected *thousands* of meters below the surface of Earth, in crystalline rock aquifers within the Columbia River basalt group in North America. Archaebacteria are the only organisms known to live under such 'harsh' conditions. Methanogens are strict **anaerobic autotrophs**, meaning that they can synthesize *all* their cellular constituents from simple molecules such as H_2O, CO_2, NH_3, and H_2S in the absence of oxygen. Methanogens appear not to depend on photosynthesis at all; they use **chemosynthesis** to produce methane and water from carbon dioxide and hydrogen gas.

To come to the point. Though some species of archaebacteria live under extreme conditions, their basic metabolic machinery nevertheless closely resembles that of eubacteria and eukaryotes. This strongly suggests, though by no means requires, that archaebacteria, eubacteria, and eukaryotes had a single origin. Going further, **some people speculate that archaebacteria, being able to exist under conditions more closely resembling early Earth than the atmosphere of today, are the probable precursors of eubacteria and eukaryotes.** The properties of archaebacteria have also been interpreted as suggesting that life on Earth may have begun deep within instead of on the stormy, oxygen deficient surface. On this view, the heat of Earth's core, not the Sun, would have provided the free energy necessary to sustain the first-formed living organisms and bring about their origin in the first place. An argument in support of this hypothesis is the following. In general, the less energy required to carry out a process, the more probable it will be, as for instance when an enzyme lowers the activation energy of a reaction and thereby speeds it up. The chemical equations for photosynthetic fixation of carbon dioxide are:

$$6CO_2 + 12H_2S \rightarrow C_6H_{12}O_6 + 6H_2O + 12S \quad \Delta G^{\circ\prime} = +406 \text{ kJ mol}^{-1} \quad (9.1)$$

$$6CO_2 + 12H_2O \rightarrow C_6H_{12}O_6 + 6H_2O + 6O_2 \quad \Delta G^{\circ\prime} = +469 \text{ kJ mol}^{-1} \quad (9.2)$$

These equations show that anaerobic synthesis of sugar from carbon dioxide, in which hydrogen sulfide is the terminal electron acceptor, has a smaller energy demand than the corresponding aerobic process. *Ergo, ...*

At this stage, however, too little is known to say whether archaebacteria gave rise to life that depends on the Sun or branched off from earlier photosynthetic bacteria and then adapted to a variety of more extreme environments. For one can easily imagine that the widespread and frequent volcanic eruptions characteristic of earlier stages of Earth's history may have led to the isolation of a population of photosynthesizing bacteria that were able to make good use other sources of free energy. If an isolated population did not have access to sunlight, it is possible that with time the genes encoding proteins required for photosynthesis were lost, transformed by mutations, or simply expressed with increasing improbability, as they were no longer required for continued existence.

G. | Energy, information, and life

Regardless of when, where, or how life began, it is clear that good internal energy resource management is important to the life of an organism, whatever it might be. As we have seen in a previous chapter, energy management on the level of the entire organism determines whether weight is lost or gained. And on the cellular level, **perhaps the most 'successful' organisms (ants, bacteria, cockroaches, . . .) are the ones that utilize their energy resources the *most efficiently*.** Consider a metabolic pathway. In the usual case this will involve a number of different enzymes, each encoded by a separate gene. Functional genome analysis has shown, however, that the genes encoding enzymes of a particular pathway are often adjacent to each other in chromosomal DNA. Moreover, such genes are often turned on or off by the same molecular switch – not the same *type* of switch but the *same* switch. Such switches often come in the form of a repressor protein binding to a particular site on DNA, inhibiting gene transcription. Synthesis of mRNA and protein production are energy consuming, so 'successful' organisms might be ones in which these processes are as very efficient; or, perhaps, as efficient as possible. From an entropic point of view, it would seem most probable for the genes of a pathway to be distributed randomly in the genome. As this is often not the case, the organism must expend energy to prevent it from happening. Are the demands of the Second Law met by the entropy increase resulting from maintaining the structural integrity of a genome being even greater than the entropy increase that would result from locating all genes in a genome at random locations? Could this not be tested by experiments with bacteria or yeast?

Maintenance of cell structure and growth require a huge number of metabolic and synthetic reactions; a range of complex mechanisms is needed to regulate the highly heterogeneous distributions and flows of matter and energy within the cell. In general, the biological macromolecules that carry out the biochemical reactions of metabolism, synthesis, and active (i.e. energy-consuming) transport are the proteins. This class of machine-like biomolecules is tremendously diverse, not only in terms of monomer composition but also with regard to biochemical activity. It is not much of an exaggeration to say that proteins are what really matter to the physicochemical properties and existence of a cell, because proteins do all the work. In addition to catalyzing a plethora of biochemical reactions, proteins also give a cell its shape and determine its mechanical properties.

Proteins are similar to books. A simple calculation shows that there are 720 (6!) different ways of arranging six books on a shelf in the side-by-side and upright position. And yet, only a few of these will be *meaningful* to the person doing the arranging (alphabetical by author, subject, or title, size, and so on). The mind selects these possibilities intuitively or by force of habit and does not consider explicitly the remaining universe of arrangements (of which there are about 700). Small proteins, for example hen lysozyme, are about 100 amino acids long. Given 20 different amino acids, there are $20^{100} \approx 10^{130}$ different sequences of this size. Even if a protein must have a sequence composition that matches the *average* protein, there are still some 10^{113} *possible* different sequences of this size. In other words, **effectively unlimited variety is possible at the level of the primary structure of a small protein. The total number of 100 residue-long sequences that encode functional proteins, however, is extremely small in comparison**, as we shall see presently. Is the relatively small number of sequences one finds in nature the result of some sort of optimization process?

Thermodynamic optimization and biology

Another question we might ask is whether any given protein sequence one finds in nature is optimized is some way, for example for certain thermodynamic properties? Is the sequence of an enzyme as random as possible without loss of enzymatic activity? As we have seen, thermodynamic stability is a measurable property of a protein. Moreover, in some cases, notably relatively small proteins, all the information required for folding into something biologically functional is encoded in the amino acid sequence. The native states of such proteins represent either a global free energy minima (under conditions favoring the native state) or energy wells that are sufficiently deep not to allow structure attainment to proceed to beyond the 'kinetically trapped' native state.

Protein thermostability is related to structure, structure to function, and function to information and the ability process information. A particularly clear example of this is *arc* repressor of bacteriophage P22. Wild-type repressor interacts with DNA and regulates the expression of genes, and the native protein contains a stabilizing electrostatic interaction involving amino acid residues Arg31, Glu36, and Arg40. In folded repressor, Glu36 is inaccessible to the solvent. Site-directed mutagenesis has been used to make all 8000 combinations of the 20 amino acids at positions 31, 36 and 40 ($20^3 = 2^3 \times 10^3 = 8000$), and all variants have been tested for *arc* repressor activity in P22.

Remarkably, only four of the variants (0.05%) are as active as the wild-type enzyme: Met–Tyr–Leu, Ile–Tyr–Leu, Val–Tyr–Ile and Val–Tyr–Val. Another 16 (0.2%) are partially active. Six of the 20 active variants are more thermostable than the wild-type protein, by as much as 20 kJ mol^{-1}. These are Met–Tyr–Leu, Val–Tyr–Ile, Ile–Tyr–Val, Met–Trp–Leu, Leu–Met–Ile, and Gln–Tyr–Val; all decidedly hydrophobic combinations. Analysis of the crystal structure of the Met–Tyr–Leu variant shows that it is practically identical to the wild-type protein, excluding the replaced side chains, which pack against each as well as might be expected in the core of a wild-type protein. In contrast, variant Ala–Ala–Ala is about 16 kJ mol^{-1} less stable than the wild-type under usual conditions, though it folds under native conditions.

What do the data tell us? There would appear to be no loss of biological information in an amino acid replacement that does not impair biological activity. In all of the stabilizing mutants the favorable electrostatic interaction, which is fairly specific (low entropy), is absent. Hence, that interaction cannot be the main source of native stability in the wild-type protein. The mutated side chains in the stable variants have large hydrophobic surfaces in close contact, so such interactions, which are relatively non-specific (high entropy), must help to stabilize folded structure. It is likely that some of the inactive variants have folded states that are more stable than the wild-type protein, but there is no experimental proof of this as just the active variants were studied thoroughly. And, importantly, the thermostability of the biologically functional form of a wild-type protein can be increased without impairing functionality. That is, **the native states of natural proteins are not necessarily optimized for thermostability.**

The lack of a requirement for thermodynamic optimization is intriguing. For **most natural processes occur in such a way that some physical quantity is 'extremized.'** Water flows downhill, and in doing so its gravitational potential energy is a minimized. A closed system tends to equilibrium, and in doing so its free energy (entropy) is minimized (maximized). It is likely that there is a thermodynamic explanation for the 'minimum' size for a protein: if the polypeptide chain is not long enough, the stabilizing interactions between residues will not be great enough to overcome the energetic cost of restricting side chain motion on folding. Proteins are generally no smaller than about 50 amino acids long. This is not to say that smaller peptides cannot be bioactive. Indeed, some peptide hormones are but a few amino acids long. In much more general terms, the dynamical behavior of objects can be formulated in several logically equivalent ways, and one of these involves the minimization of a quantity called the *action*, the sum over time of the kinetic energy minus the potential energy. Something obviously similar seems not to be true of proteins, which as the example above proves are not necessarily extremized for thermostability. In other words, if existing biological macromolecules are extremized for anything, that something need not be free energy. Are biological macromolecules optimized instead for something *biological*? As we have seen, **the catalytic rate of some protein enzymes is as large as it possibly could be, since their rate is limited by the rate of diffusion of the substrate,** though most enzymes operate with sub-maximal efficiency.

Is there a biological advantage to a protein's not being optimized for stability? If a protein were extremely stable when it did not need to be, protein metabolism or programmed protein degradation might severely tax the energy resources of the cell (see below for a more quantitative discussion of this). An extremely thermostable viral protein might be particularly harmful to an organism! Less-than-maximal thermostability allows molecular chaperones and other proteins to assist in folding, facilitate protein translocation across membranes, and eliminate incorrectly folded proteins. Assuming that proteins must be optimized for something, maybe that something is **compatibility** with all aspects of its existence as a biological entity, from the availability of amino acids to folding on the ribosome, to transport, biological function, and degradation. This raises the following point: **if the distinction between complex, highly organized living things and less complex, less organized inanimate things is artificial, it is at least *unclear* how it is artificial.**

Information theory and biology

Having come thus far, we find ourselves not at the end of a quest to understanding the origin of cells and their ability to transform energy, but at the beginning. For in organisms as we know them, and despite all the work they do, proteins are not the most basic repository of biological information: the instructions for making a protein are stored in

DNA. At the current stage of development of the field of molecular biology it is clear enough how the structure of DNA translates into the amino acid sequence of a protein, but there is no known way in which the sequence of a protein could be used as a template for the synthesis of a corresponding DNA sequence.

The biological information stored in genes can be analyzed in a variety of ways. The one we wish to focus on now is **information theory**, a subject that developed out of the work in the communications industry by an American named Claude Elwood Shannon (1916–). Information theory has found its main application in electrical engineering, in the form of optimizing the information communicated per unit time or energy. But because it is so general, the theory has proved valuable in the analysis of phenomena in other areas of inquiry. As we shall see below, information theory is closely akin to statistical thermodynamics and therefore to physical properties of biological macromolecules, and it gives insight into information storage in DNA and the conversion of instructions embedded in a genome into functional proteins.

According to information theory, **the essential aspects of communication are a message encoded by a set of symbols, a transmitter, a medium through which the information is transmitted, a receiver, and noise.** *Information* stands for messages occurring in any of the standard communications media, such as radio or television, and the electrical signals in computers, servomechanisms, and data-processing devices. But it can also be used to describe signals in the nervous systems of animals. Indeed, a sense organ may be said to gather information from the organism and its environment: the central nervous system integrates this information and translates it into a response involving the whole organism, and the brain can store information previously received and initiate action without obvious external stimulation. Ideas about how the mammalian central nervous system might work were the basis on which early electronic computer development proceeded. In the biological expression of genetic information, the transmitter is the genome, the message is messenger RNA, and the receiver is the cell cytoplasm and its constituents (e.g. ribosomes).

In the context of information theory, the **information content** of a message has a more precise *meaning* than in 'ordinary' human communication. Information theory aims to be quantitative and unambiguous. The remarkable subtleties of human communication no longer exist, and there is no such thing as the **inherent meaning** of a message in information theory; unless *inherent meaning* means nothing more than a semi-objective quantitative measure of the degree of order, or non-randomness, of information. In so far as it can be treated mathematically, information is similar to mass, energy, and other physical quantities. Nevertheless, information is not a physical quantity in the sense that mass is a physical quantity: information concerns arrangements of **symbols**, which of themselves need not having any particular meaning (consider, for example, the individual letters by which the word *symbol* is symbolized), and any **meaning** said to be encoded by a specific combination of symbols can only be conferred by an intelligent **observer**.

Information theory resembles thermodynamics in a number of ways. As we have seen in Chapter 6, in the statistical mechanical theory of thermodynamic state variables define the *macroscopic* state of a system. The 'external,' macroscopic view places constraints on what is happening 'inside' the system, but it does not determine the state of the system. In other words, in general many different microscopic states correspond to the same macroscopic state. For example, a fixed current drawn by a house can correspond to a variety of different combinations of lights on inside, as we noted in Chapter 6. We saw something of a biological parallel of this in our discussion of the Adair equation in Chapter 7. To be sure, the average number of ligand molecules bound is a measurable quantity, but in the absence of additional information the average number bound does not tell us which sites are occupied in a given macromolecule; many possible combinations of occupied sites would give rise to the same measured value.

The connection between information theory and thermodynamics can be further elaborated as follows. When the entropy of a system is low, as for instance in the crystalline state, the information content (or determinacy content) is high, because it is *possible* to provide a very accurate description of the system on the microscopic level. One could say, for example, that all the atoms form a regular array with specific geometrical properties. In contrast, when the entropy of a system is high, as it is in a gas, it is impossible to describe the arrangement of particles on the microscopic level. We see from this that **the information content ascribed to a system is inversely related to our uncertainty of the microscopic state of the system**. There is as well a relationship between information and work. A liquid can be converted into a solid by doing work on it, often by using a motor to extract heat in order to lower the temperature. So the expenditure of work can result in an increase in information, albeit at the expense of an increase in the entropy of the universe. And as discussed in Chapter 4, a system at equilibrium undergoes no net change. In the context of this chapter, this means that an equilibrium system can neither gather information nor respond to it. We shall return to this theme below.

Taking a word in English as the system, the macroscopic state could be the number of letters of each type and the microscopic state their specific arrangement. In some cases, knowledge of the composition of a word is enough to determine arrangement, since only one of the many possible strings would represent a word in English. An illustrative example is *example*, for which there is no anagram. In other cases, multiple arrangements of letters could make sense, and the 'correct' one would have to be inferred from analysis of a 'higher' **semantic level** of the message. For instance, the two very different words *state* and *teats* have the same macroscopic state, and the one that is actually a constituent of a message could only be known by considering the broader context in which the word occurred. In the case of *state*, there would be the additional difficulty of deciding whether it was employed as a noun

(Plato's *Republic* discusses the ideal state, one for which the paramount concern is justice), verb (Aristotle, who in many respects surpassed his mentor, Plato, might say that the best state was the one in which justice benefited the man who is just), or an adjective (were he alive today, Plato's mentor Socrates might well agree that state-supported higher education and graduate study was a worthwhile investment – if it could be freed from the political agendas of government bureaucrats, university administrators, and of course department heads, many of whom have no better a grasp of management than phrases they have memorized from a 'how to' book.) This shows how the observer or experimenter plays a key role in determining the information content, or meaning, of a message; the observer *decides* how the message will be evaluated. Owing to such subjectivity, **there is no absolute value that can be placed on the information content of a message.**

There is, nevertheless a difference between **actual information** and **potential information**, depending on the macroscopic state and microscopic state of a particular message. Consider, for example, the relatively simple binary alphabet (a, b) and the set of all possible 'words' that are 12 letters long. There are $2^{12} = 4096$ different words of this size. In the simplest case, where the message is a string of identical letters, e.g. *aaaaaaaaaaaa*, the actual information is maximal, because determination of the microscopic state adds nothing to what is known at the macroscopic level. In contrast, the message *aabbaabbaabb* contains considerably more information than the macroscopic state, which of itself tells us only that the 'word' consists of an identical number of letters a and b. (The significance of this choice of message is as follows. If the message represents the order of bases in DNA, the sequence of the complementary strand is identical to the message itself (the coding strand and its complement read in opposite directions). Messages that read the same way forward and backward are called palindromes, and many of the known DNA sequences to which proteins bind to regulate gene expression are palindromic. Regulatory DNA-binding proteins are typically dimeric, for obvious reasons.) An intermediate semantic level might characterize the message as '*aab* followed by its mirror image, and the resultant followed by its inverse.' The greatest possible potential information is encoded by words in which the number of letters a equals the number of letters b, as this allows for the greatest *number of ways* of arranging the available symbols. As we have seen in Chapter 6, the number of ways of arranging the particles of a system is related to the entropy function of classical thermodynamics, and in the case of *aabbaabbaabb* that number is $\omega = 12!/(6!6!) = 924$ (Fig. 9.2). In view of this it is said that **entropy and information are 'isomorphous'** (Greek, *iso*, same + *morphe*, form).

We can become more biological about information theory. The complete DNA sequence is known for several genomes, including that of *Haemophilus influenzae*. In the genome of this protist, which has 1727 coding regions, only 15% of the genomic DNA does not encode a protein, and over half of the genes have known counterparts in other organisms. Assuming that each nucleotide can occur at any position with the same

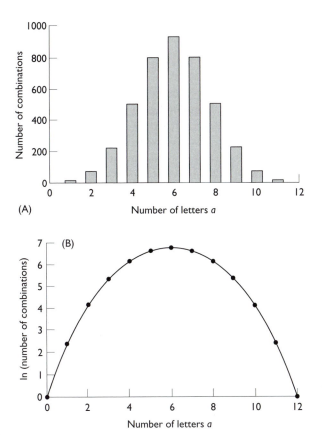

(A) Number of letters *a*

(B) Number of letters *a*

Fig. 9.2 Genome combinatorics. The alphabet considered here has but two letters, *a* and *b*, two fewer than DNA. Panel (A): the number of distinguishable combinations of letters in a 12-letter word as a function of the number of letters *a*. The curve is the famous bell-shaped one of probability and statistics. Panel (B): the logarithm of the number of distinguishable combinations. This number is related to both the information content of the word and its entropy. Note that the potential information content and entropy are greatest when the number of letters *a* equals the number of letters *b*, and that when the word consists of just one type of letter the potential information content is 0.

probability, each nucleotide in the *Haemophilus* genome (or any polynucleotide for that matter) contains 2 'bits' of information. This is because at least two binary 'decisions' are made each time a nucleotide is added: 'Is this nucleotide a purine or a pyrimidine?' and either 'Is this nucleotide a cytosine or a thymine?' or 'Is this nucleotide a guanine or an adenine?' This intuitive view is supported by calculation as follows. The probability, **p**, that the base at a given position is one type and not another is 1/4. This can be written in terms of powers of 2 as

$$\mathbf{p} = 1/4 = 1/2^I = 2^{-I} \tag{9.3}$$

We are interested in powers of two because we wish to relate **p** to some number of binary decisions. The information content in terms of binary decisions, I, is found by taking the base 2 logarithm of both sides of this equation. The result is

$$I = -\log_2 \mathbf{p} = -\log_2(0.25) = -\log_2(2^{-2}) = 2\log_2(2) = 2 \text{ bits per nucleotide} \tag{9.4}$$

Thus, a polynucleotide chain *n* nucleotides long has a sequence information content of 2*n* bits. Similarly, the information content of an arbitrary message in the English language is

$$I = -\log_2(1/27) = 4.76 \text{ bits per character} \tag{9.5}$$

assuming that 26 letters and a space are the only characters allowed and that the probability of occurrence is the same for each type of character

– an obvious oversimplification, but one that will do as a first approximation.

The simplest is not so simple

Let's consider *Escherichia coli*. The genome of this organism is about 4×10^6 bases long, so its genetic information content is 8×10^6 bits. The total number of sequences of this length is not large: at $4^{4\,000\,000} \approx 10^{2\,400\,000}$ it is absolutely astonishingly mind-bogglingly gigantic (for comparison, the temperature at the core of the Sun is $2 \times 10^7\,°C$, the human genome comprises some 2.9×10^9 base pairs, Earth is about 10^{26} *nanoseconds* old, and the universe itself just ten-fold older). Importantly, only a tiny fraction of these sequences ($\leqslant 1\%$) will encode anything **biologically meaningful – enable an organism to maintain its cellular structure by transforming energy from one form to another, grow, adapt, and reproduce.** The conclusion is no different even if we exclude all sequences with a base composition different from that of the *E. coli* genome.

This reasoning can applied to more general questions. The *minimum* number of DNA sequences that are biologically meaningful is the number of different species (with a genome at least as big as that of *E. coli*) and all sequence variations within the species that have ever existed. The *maximum* number of sequences includes all those that have existed plus all others that *could* satisfy all the necessary requirements. All these sequences have the same information content. We have now entered a realm where information theory seems less helpful than we might like it to be, for as yet *we* have no way of making an absolute distinction between sequences that are biologically meaningful and those that are not on the basis of sequence information alone.

Given that all possible sequences are not biologically meaningful, how do the physical, chemical, and biological properties of living organisms select sequences that are biologically meaningful from those that are not? The emphasis on meaning is crucial, as a simple example helps to show. There is often little or no real knowledge communicated in information. Consider, for example, a telephone book. Though it may contain a great amount of information and be *useful* in a variety of ways, a *telephone book* is but a *directory*. A telephone book does not impart knowledge in the way that Shakespeare does, and just as important, it is difficult to imagine how dramatization of a telephone book could be the slightest bit *interesting*. Where do interesting DNA sequences – ones that encode the genomes of living organisms – come from?

Molecular communication

The discussion of biological information can be taken a step further. Nucleic acids are not the only kind of molecular message important to the life of a cell. Another sort is typified by cAMP, which as discussed in a previous chapter is a 'second messenger,' a kind of intracellular signal. cAMP is generated by the cell upon receipt of an appropriate molecular *signal*, for example the binding of the hormone insulin to the extracellular portion of its transmembrane receptor. There are important bioenergetic connections here: insulin plays a key role in regulating the concentration of glucose in the blood, blood delivers the sugar molecules to every cell in the body, and cells use the glucose as the main energy source for production of ATP. In addition, cAMP is the primary intracellular signal in the glycogen phosphorylase kinase and glycogen synthase cascades, important in the metabolism of glucose-storage polymers called glycogen, and cAMP also plays a role in the phosphotransferase system and smooth muscle contraction. cAMP-dependent kinase controls the enzyme acetyl-CoA carboxylase, which catalyzes the first committed step of fatty acid biosynthesis and is one of its rate-controlling steps. It is clear from this that biological information comes in a number of forms.

In *E. coli*, transcription of the β-galactosidase gene occurs when the cell enters an environment in which the glucose level is relatively low and the lactose level relatively high; i.e. when a certain type of signal is received and other conditions are met. The binding of lactose (or, possibly, one of its metabolic products) to the *lac* repressor results in dissociation of the repressor from the operator site, allowing transcription of the β-galactosidase gene to proceed. Once the enzyme β-galactosidase has been synthesized, lactose can be metabolized as an energy source. Transcription of the β-galactosidase gene is inhibited in the absence of lactose in order to conserve energy resources.

Information processing

We can be somewhat more quantitative about the relationship between biological information and thermodynamics. The gain of one bit of information by a cell requires the expenditure of at least $k_B T \ln 2$ units of work: $S = k_B \ln \omega$, and the entropic term of the Gibbs free energy is TS. At 27 °C, one bit of information costs 3×10^{-21} J. The energy required to generate the biological information of an entire cell at this temperature might be 3×10^{-12} J, assuming that the information content of the cell is roughly comparable to that of *Encyclopædia Britannica*. Because 'informational' macromolecules like RNA, DNA, proteins, and polysaccharides constitute at least half of the mass of a bacterial cell, one would be very hard pressed to say how the energetics of information processing can be separated from other aspects of cellular activity. It is also clear that information processing on the cellular level must be a major impetus for the energy consumption of an organism, regardless of its size. It follows that our bodies are processing tremendous quantities of information as long as we're alive, whether we are studying biological thermodynamics, designing biological chemistry experiments, attempting to outfox opponents during a hand of contract bridge, or slumbering away. A nervous system is clearly not necessary for information processing, since organisms like bacteria and yeast do a lot of it without one! Are humans different from other organisms on a basic level? What seems extraordinary about us is that **whereas plants (and digital computers) only carry out the information processing instructions in the program they are running (the plant genome in the one case, a specific piece of software in the other case), human beings can** process information, be aware that this is happening, build models in order to try to understand information/energy flow, search for practical ways of utilizing any new-found knowledge as a means towards a needed or desired end, and consider the meaning of what they are doing.

To summarize this section, information theory provides a semi-objective means of saying what information is and quantifying it. Information theory does not help us to distinguish between biologically meaningful and meaningless DNA sequences. Moreover, information theory does not tell us *where* biologically meaningful information comes from; much less *how* it came to be. Information about a system is similar to the entropy of that system. These measures differ in that the latter relates to the system itself while the former is related to the observer. The distinction, however, must be considered somewhat artificial, since as we have seen thermodynamic quantities are defined only under rather arbitrary circumstances chosen by the observer (experimenter). Information theory in biology is a very large topic, and we have scarcely scratched the surface. Nevertheless, having compared DNA and human language in the context of information theory, we can clearly see that there is something rather earthy about human language, or something unmistakably ethereal about the organization of living matter, or both.

H. Biology and complexity

The situation in the living cell is still far more complex than our mostly qualitative discussion of information theory and biology has so far suggested. For in living things there is not only highly specific encoding of biological information in genetic material, but also continual *interplay* between such information and biological macromolecules. The process of protein biosynthesis is not simply a linear flow of matter, energy, or information, but a highly organized feedback circuit in which proteins and nucleic acids control *each other*. Regulatory proteins bind to DNA and thereby enable or disable the synthesis of mRNA required for making proteins on ribosomes. In most eukaryotes, the nucleic acid message must be spliced in order to remove the non-protein encoding regions called introns. This process, which is carried out by a marvelously complex macromolecular machine, can in some cases lead to a variety of spliced versions of protomessage. The result is that the same gene can be and often is used to produce different forms of the same protein in different tissues.

Each living cell comprises thousands upon thousands of proteins. Each bacterium is made of thousands of different proteins, each encoded by a specific gene and required in a relatively specific amount. The regulation of gene expression and protein metabolism on which the living cell depends is extremely complex. And yet, **all the proteins of cell work together in a co-ordinated way to sustain the highly ordered living state of the cell. Proteins do this by following an**

extremely well organized program encoded in the organism's DNA. The inner workings of a cell are like the automated processes of a factory, only unimaginably more complex. (On a higher hierarchical level, all cells of the body work together in a co-ordinated way to sustain the highly ordered living state of the healthy organism.)

The immune system of vertebrates operates with an astonishing degree of complexity. This marvel of the living world can generate a virtually unlimited variety of antigen-binding sites; it can produce antibodies against almost any antigen it encounters. How is the body able to produce literally *billions* of different antibody structures? There are two basic mechanisms, both of which contribute to **antibody diversity**. They are: **somatic recombination** and **somatic mutation**. Recombination involves just a few gene segments that encode the so-called variable region of the immunoglobulin chain. A limited number of segments can combine in many different ways, in much the same way that a limited number of letters of the Roman alphabet can be joined together different meaningful combinations. Even greater diversity is produced by mutations in the immunoglobulin gene that arise during the differentiation of B cells, the antibody-secreting cells of the immune system. Antibody diversity is an essential means of protection against foreign invaders.

The complexity of living things may make it seem hard to see how they are similar, but there are some common themes of basic importance. A few aspects of the common denominator have been touched on above. Here are a few more. The complexity of a biological organism correlates with its being a far-from-equilibrium open system. It is difficult to imagine that this correlation does not point to a deeper relationship. The signal processing operations carried out by a cell must consume a large amount of energy, whether by dissipating an electrochemical gradient or by the essentially irreversible breaking of chemical bonds, notably the phosphodiester bonds of ATP. The molecules cells use to energize the biochemical reactions necessary to their existence *require* that cells be open thermodynamic systems. (But it does not follow that some aspect of cellular activity must therefore violate either the First or Second Law of Thermodynamics.) Every day in each living cell, thousands of purine nucleotides hydrolyze spontaneously in genomic DNA. To maintain the integrity of the encoded information, the damaged purines must be replaced; an energy consuming process. The chemical structure of DNA plays a key role in this process: the redundancy of genetic information in the form of the complementary strands of the double helix greatly reduces the odds that spontaneous hydrolysis will lead to a permanent change. Other types of mutation can occur in genes, and elaborate DNA repair mechanisms are present to reverse the changes in all organisms. Perhaps most important, the organizational plan of *all* organisms is encoded in DNA, and nearly all organisms use the same code for message transmission to synthesize proteins of the same chirality. The known exceptions are commonly accepted as slight variations on the central theme, not as very distinct competing alternatives. Though it is common knowledge that experimental proof of the triplet nature and degeneracy of the genetic code was worked out by Francis Crick and colleagues, it is less well-known that the triplet code was first proposed by the Ukrainian-American theoretical physicist Georgy (George) Antonovich Gamow (1904–1968), who is also known for developing the Big Bang theory of the origin of the universe with Ralph Alpher and Hans Bethe, popularizing science by means of his highly amusing and helpful Mr Tompkins science books, and collaborating in the design of the hydrogen bomb. **The common features of known living organisms strongly suggest that all of them have a common origin.**

Parts of organisms and indeed entire organisms can usefully be thought of as machines, albeit complex ones, whether the organism has one cell or a lot of them. How far can this analogy be taken? Despite similarities, living organisms and machines are in fact fundamentally different, and as we shall see that difference has to do with complexity. **The physicochemical basis of order in biological organisms remains a major unsolved puzzle**, not least because the Second Law of thermodynamics requires an increase in the entropy of the universe for any real process. To get a better understanding of the machine-like character of organisms, it might be helpful to consider briefly the historical dimension of the concept.

In the 'dualist' philosophy of the French mathematician and philosopher René Descartes (1596–1650), the 'body,' as distinct from the 'mind,' is assigned the properties of a 'machine.' By the middle of the eighteenth century, this extreme mechanistic philosophy had come to be rejected by a number of leading biologists, including George Louis Leclerc de Buffon (1707–1788) and Pierre Louis Maupertuis (1698–1759), both Frenchman, and Albrecht von Haller (1708–1777), a German. These researchers stressed the complexity of life over mechanism, and held that the animate was distinguished from the inanimate by such attributes as 'sensibility,' 'irritability,' and 'formative drive.'

To develop the organism-machine analogy, let the archetypal machine be a personal computer. A PC is made of matter, needs electrical energy to run, and has parts that can wear out. A computer, however, is a relatively static structure, even when boards and chips can swapped about with relative ease and disk drives can fail. That's because the matter a computer is made of doesn't change all that much once the machine and its parts have been produced. A living organism is similar to a PC, but it is also very different. For an organism, be it a bacterium or a bat, utilizes the free energy it has acquired from its environment to carry out a continual process of *self-renewal*, and this is something no machine can do. Each and every cell of an organism simultaneously metabolizes proteins, many of which have no apparent defect, and synthesizes new ones to fill the void. This process requires not merely the expenditure of free energy but a *huge amount* of free energy – on one level to produce the necessary precursors of macromolecules and on another to stitch them together. If the necessary energy requirements are not met, the organism perishes.

We'll come back to death in a moment. First let's be somewhat more quantitative about the role of self-renewal in life. The half-life of a highly purified protein in aqueous solution is on the order of thousands of years. Protein turnover in the living organism is comparatively much more rapid. For example, a 70-kg man (or woman) synthesizes and degrades about 70 g of protein nitrogen – *per day*! The total nitrogen content of a 70-kg man is about 900 g, so the protein turnover rate is roughly 8% – *per day*! The free energy of formation of a peptide bond is about +6 kcal mol^{-1}, not much different in magnitude from the free energy of hydrolysis of ATP, so protein renewal requires a large fraction of an organism's daily energy intake (see Chapter 1). We conclude that **although organisms have machine-like qualities, they are certainly not machines in the way that computers are machines, if organisms can be considered machines at all**.

As we have seen, biological systems are open systems, exchanging matter and energy with their surroundings. Biological systems are most definitely not at equilibrium. In order to maintain themselves, they must have access to a constant throughput of a suitable form of free energy-rich matter. A great deal of research has been done in recent decades on relatively simple non-equilibrium systems and highly ordered structures that can arise within them. An example of this is the swirls that appear in a cup of tea immediately after a spot of milk has been added. The patterns are clearly more ordered than when the milk has been completely stirred in. This sort of order is rather short-lived; it does not persist because maintaining it against dissipative forces would require energy input. There are many other examples. Continuous heating of a fluid can result in the appearance of ordered structures called convection cells and turbulent flows. Under such conditions the entropy of the fluid is not maximal, as it would be if the density were uniform and the same in every direction. A large temperature gradient across a fluid can lead to the formation of Benard cells, highly ordered hexagonal structures (but the temperature gradient must be maintained in order for the Benard cells to persist). And complex weather patterns can result from the combination of solar heating and water vapor. While these non-biological complex systems resemble life in certain ways, there is little doubt that they are a far cry from something as 'simple' as a bacterium.

This discussion raises the question of a possible relationship between biological complexity and the ability to survive. Bacteria, by far the simplest beasts, have been around a very long time, and they are found nearly everywhere on the surface of Earth and many places below. Whatever it is that enables them to adapt to such a broad range of conditions works very well. In view of this, it seems probable that the ability of bacteria to survive all-out nuclear war or some other great calamity would be much greater than that of humans. Assuming that humans are a product of evolution, and evolution is driven by the survival value of genetic changes, how is it that such complex but relatively 'unstable' creatures as humans were able to evolve? Are humans really more fit for survival than bacteria? If humans ever should be able to devise ways of leaving Earth permanently, for example in order for a remnant to save themselves from destruction of the planet or simply to go where no man has gone before, would they be able to depart without taking with them at least a few of the 100 trillion bacteria that inhabit each human intestine? Can evolution be explained in terms of survival value or increasing complexity? Or is the possible or actual long-term survival value of a genetic change not always a relevant question in evolution?

The number of different species in existence today is estimated to be over 1.2 million, of which the animals, from ants to aardvarks, constitute a relatively large proportion. The diverse appearance of animals is largely superficial: the bewildering array of known forms, some of which are downright bizarre, can be sorted into a mere half dozen body plans. Established during embryonic development, a **body plan** limits the size and complexity of an animal. Despite being so similar, different organisms differ strikingly in their ability to adapt to different conditions, e.g. high temperatures. Simple eukaryotes do not adapt to temperatures above about 60 °C. The upper limit for plants and animals is below 50 °C, and the majority of eukaryotes, including humans, are restricted to considerably lower temperatures. Above about 60 °C, the only organisms that have any chance of survival are prokaryotes, unicellular organisms.

How do the complex interactions of organisms known as ecosystems form from individual organisms? Is natural selection acting on random mutations sufficient to generate not only the first living cell but also the totality of the tremendous variety of interacting life forms that have existed on Earth? Does the mutation-selection model predict that an ecosystem will have a *hierarchical* structure, with many interactions between organisms occurring at several different scales of size and complexity? If the employees of a particular company can communicate with each other, and if the communication of one company with another can lead to increased 'productivity,' does such a model provide a sufficient explanation of the origin of the interactions between different organisms in an ecosystem?

What about death? Aging and death may be related to changes in the information processing capability of a cell. One theory assumes that life span is determined by a 'program' encoded in DNA, just as eye color is determined genetically. Indeed, long life often runs in families, and short-lived strains of organisms like flies, rats, and mice can be produced by selective breeding. Nevertheless, there must more to long life than the basic genetic program of aging, to which decades of improvement in human nutrition attest. Another aging theory assumes that cell death results from 'errors' in the synthesis of key proteins like enzymes stemming from faulty messages. Errors in duplication of DNA, incorporation of errors into mRNA, pathological post-transcriptional control, or aberrant post-translational modification are several possible means by which 'impaired' enzymes could be produced in the cell. The 'somatic mutation' theory of aging assumes that aging results from the gradual increase of cells whose DNA has accumulated a significant number of mutations and which no longer function normally. In any case, in some way or other the biologically meaningful information content of a cell declines toward the end of its life.

Before closing this section, let's look briefly at two different situations. The Sun is a massive object, and we know from experience that the Sun's rays can be used to heat a pool of water or, when focused, to start a fire. The Sun transfers heat energy to its surroundings in accordance with the Second Law, and it does so by the transformation of its mass into electromagnetic radiation (at a rate of about 100 million tons per minute). Something similar, but different, happens in our bodies. The biochemical reactions going on within lead to the dissipation of free energy as low frequency electromagnetic radiation, i.e. heat. Because of the various ways in which we use energy to make new proteins or other molecules and dissipate energy as heat, every few hours we have an urge to eat. From the point of view of thermodynamics, we satisfy this urge not to spend time with family or friends, nor to enjoy fine cuisine, but to supply our bodies with a source of free energy so that our bodies will not eat themselves! We need the energy input to maintain body weight and temperature. Without this energy, we would die after a couple of months. Even if we are able to eat well throughout life, we die after just a few score years. In this sense living organisms are like the Sun, which will eventually exhaust its energy resources.

I. | The Second Law and evolution

When the biological world is viewed through a narrow window of time, what one sees is not how much organisms change from to generation to the next, but rather how much they maintain their exquisite order as a species. No wonder species were considered fixed when people thought Earth was relatively young! Nowadays there is no doubt that the order presently exhibited by a particular organism is the result of changes that have occurred over a very long period of time. Moreover, it is clear that in general terms the earliest organisms on Earth were much simpler than the ones in existence today. Exactly how the complex came from the simple is obscure, though in some way or other it must have involved changes in genetic material.

We might ask how there could be such a thing as life at all when the Second Law points to death, annihilation. How can there be a process whereby life forms become increasingly complex wherever the Second Law is required to operate? Has the chain of processes by which the first cell on Earth became the many cells of all the organisms that have ever existed violated the Second Law of Thermodynamics? Some people think so, saying for example that zebras are clearly much more complex and highly ordered organisms than zebra fish and protozoa, protozoa have been around a lot longer than zebra fish and zebras, and the Second Law demands ever-increasing disorder. This view, however, stems from a misunderstanding of what is possible in the context of the First and Second Laws. The Second Law says that the entropy of the universe must increase for any real process, not that order cannot increase anywhere. If order could not increase anywhere, how could rocks be transformed into a pile of cut stones, and stones turned into a majestic feat of architecture like an aqueduct or a cathedral? Or leaving humans out of the argument, how could diamond form from a less ordered array of carbon atoms? The Second Law requires only that any process resulting in a decrease in entropy on a local level must be accompanied by an even larger increase in entropy of the surroundings.

If the degree of order exhibited by an open system increases *spontaneously*, the system is said to be '**self-organizing**;' the environment must exert no control over the system. Under some circumstances, such order can be retained spontaneously, albeit at some energetic cost. Self-organization, which is entirely consistent with the laws of thermodynamics, is usually initiated by internal processes called 'fluctuations' or 'noise,' the kind of particle motion one finds even in systems at equilibrium. Earlier we looked briefly at processes that may have been involved in abiotic synthesis of biological macromolecules. Abiotic formation of macromolecular complexes, assuming it has occurred, has presumably been a matter of self-organization. At some stage, it was necessary for biological macromolecular complexes to catalyze the synthesis of macromolecules capable of storing information. How this could have occurred and where the information to be stored came from are unknown.

Remarkably, the 'collapse' from order into disorder can be *constructive*, at least under limited circumstances. For example, when concentrated guanidine hydrochloride is diluted out of a sample of chemically denatured hen lysozyme, the protein refolds – spontaneously. In other words, the free energy of the disordered state is energetically unfavorable in the absence of chemical denaturant (and at a temperature that favors the native state; see Chapter 2). The situation where the guanidine concentration is low but the protein is unfolded must be one that is far from equilibrium, and the *folding* of the protein must correspond to an *increase* in entropy, since by definition the equilibrium state is the one in which the entropy is a maximum. Because the protein itself is obviously more highly ordered in the folded state than in the unfolded on, the entropy increase must come from the dehydration of the protein surface on structure attainment. The released water molecules become part of the bulk solvent, where their numbers of degrees of freedom are maximal. **In general, an event like the folding of a protein does not occur in isolation from the rest of the world, and it is most important to be aware of the possible consequences of how things do or do not interact.** Certain biochemical reactions, for instance, can take place in cells only because they are chemically coupled to other, energetically favorable ones. The **interaction** between reactions may be crucial to the life of an organism. The Second Law requires that the overall entropy change in the universe be positive for any real process, but just how that the entropy increase is achieved is quite another matter.

According to the Second Law, the entropy of the universe must increase for *any* real process, regardless of the involvement of non-equilibrium states. Non-equilibrium thermodynamics is helpful in thinking about biological processes because it provides a way of rationalizing the local decreases in entropy that are necessary for the formation of such extraordinarily ordered entities as living organisms. There is no violation of the Second Law because the entropy of the universe increases in any non-equilibrium thermodynamic processes. However, non-equilibrium thermodynamics does not explain *how* the first living organism came into existence, *why* it came into being, or why, over time, there has been a marked tendency toward the increase in biological complexity. The First and Second Laws provide restrictions on what processes are *possible*, but we do not say, for example, that the laws of thermodynamics have required a certain biophysicist to choose to investigate the thermodynamic properties of proteins! Nor for that matter could we say that thermodynamics *requires* that the enzyme catalase must have a catalytic efficiency that is limited by physics and by not by chemistry.

How *purposeful* is the coupling of processes that bring about or underpin the order we see in living organisms? This would appear to be a question that thermodynamics cannot answer. While the great local decrease in entropy required for biological reproduction seems very purposeful, the accompanying increase in the entropy of the universe seems to lack purpose, excluding the possibility that its purpose is to enable local increases in order. The individual biochemical events on which brain processes and actions depend may seem purposeless on the level of individual atoms and molecules, but can this be said of the same processes when considered at a less detailed level? After all, a book may make sense on the level of individual words, sentences, paragraphs, sections, chapters, and indeed the entire book. But it would make no sense to think about the purpose of a book on the level of the molecules it is made of. As far as humans are concerned, what could be more purposeful than trying to make sense of the universe and what goes on inside it, being willing to learn, having a keen awareness of the essential and an overall sense of purpose, and testing ideas to see how right they might be?

In thinking about how the Second Law constrains biological processes, a **distinction** must be made between processes that express pre-existing genetic information and ones that involve the appearance of 'new' genetic information. In human growth and differentiation, for example, all the information required to construct the adult body is present in the fertilized ovum. The development of the conceptus into an adult involves the conversion of energy and raw materials into additional order and complexity as prescribed by the genetic material. In the usual case, the 'potential complexity' of the fertilized egg is the same as that of the adult. The 'expressed complexity' of the organism, however, increases dramatically throughout gestation, birth, adolescence, and early adulthood. **Despite the similarities, animal development must differ considerably from speciation, for it is difficult to imagine how the first cell could have contained the genetic information required for the development not only of itself, but also of amoebas, hyacinths, nematodes, koala bears, and humans.**

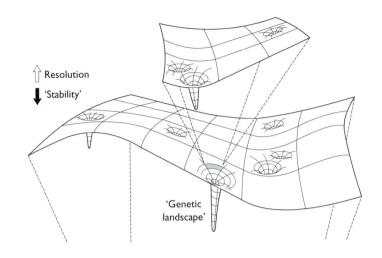

Fig. 9.3 Genetic 'landscape' representing all possible sequences for DNA of a given number of base pairs. 'Pockets' of genomic stability ('wells') correspond to actual species. Though but a relatively small region of the total genetic landscape, each pocket represents substantial genetic 'stability' and allows for considerable sequence variation. There can be a large number of combinations of sequence in different ways: the larger the genome, the greater the number of possible sequences; and the more an organism reproduces itself, the higher the likelihood of variation. Bacteria have small genomes, but there are many of them, on the order of trillions in each human gut. Humans have large genomes, but there have been comparatively few of us (<50 billion). Does this imply that given enough time one could go from a dog, all breeds of which have come about through the selective mating of wolves over a time span of about 100 000 years, to a different *species*, *X*, which, unlike a dog, could not mate with a wolf?

Again assuming the existence of the first cell, it is not hard to see that 'evolution' might not only be compatible with the Second Law of Thermodynamics but indeed driven by it. For replication of the bacterial genome, while good, is not perfect. Moreover, it may be that the fidelity of genome replication has improved with time, not become worse. In other words, given some starting material, the Second Law would seem to demand the generation of new varieties, particularly if the energetic cost of maintaining the genome is high, as it undoubtedly is.

Most mutations resulting from incorrect base pairing, however, are 'corrected' by an elaborate system of repair mechanisms that are found in all organisms. Other mutations are 'tolerated' because they result in no dramatic change in the organism. Biological 'stasis' is energy consuming. Does a model in which random mutation and natural selection drive evolution square with the inclusion in the genetic material of genes that encode enzymes that function to maintain the status quo? The rate of mutation is relatively low, much lower than one would expect if repair enzymes were not present. Is there a selective advantage to having the repair enzymes? Such proteins would appear to slow down evolution. Do only exceptional cases of mutation and natural selection lead to new species? If the basic pattern exhibited by a species is to resist change, how have so many different species arisen in the short time that Earth has existed? And though many different, seemingly unrelated species have existed, because organisms from bacteria to humans have remarkably similar metabolic pathways, it is hard to see how all living (and extinct) organisms (on Earth) could not have had a common origin.

Evolution might be slowed in more ways than one. *C. elegans* is a nematode worm that lives in the soil and feeds on detritus, i.e. decaying organic matter. The adult organism is just under 1 mm long and comprises fewer than 1000 cells. The generation time is a mere three days, and progeny are numerous. Studies of this organism have provided clues about many aspects of eukaryotic life, including for instance the importance of sexual reproduction in higher organisms. It is probable that reproduction of this kind enables an organism to cope with mutations in its genome and thereby to put the brakes on the rate of divergence of the genome. Most mutations have no detectable effect on the phenotype of offspring of sexually reproducing organisms. This works as follows. According to the 'engine and gearbox' view, if you have two worn-out automobiles, one without a gearbox and the other without a driveshaft, you can combine parts and produce a single working vehicle. Something similar appears to happen during sexual reproduction, because the genetic material provided by the male is combined with the DNA of the female. This is particularly important if either dad or mom, but not both, carries of an undesirable genetic trait. In this way, sexual reproduction helps to keep the rate of change of a population low. Organisms that reproduce sexually 'dilute out' the effect of mutations that would reduce either reproductive fitness or number of offspring.

Assuming that mutations and natural selection play an essential role in speciation, it may be that what might be called 'persistent forms' (i.e. the species we know about) are something like local minima in a free energy landscape. On this view, once a species got going it would not change very much; it would be 'stuck' in a 'well' in the 'genetic landscape' (Fig. 9.3). Divergence could still occur via genetic change, but this would entail a comparatively large-scale increase or decrease in, or organization of, genetic information, and most instances of this not yield an organism that could survive long enough to meet another organism with which it could produce viable offspring. In some cases, however, successful reproduction would be possible, and this could be represented diagrammatically as a move from one minimum in the genetic landscape to another. The organisms corresponding to one minimum would be sexually incompatible with organisms of another minimum (the definition of species), if not immediately then after some number of generations. The Second Law of Thermodynamics would not be violated by such a process.

J. | References and further reading

Alberts, B. (1998). The cell as a collection of protein machines: preparing the next generation of molecular biologists. *Cell*, **92**, 291–294.

Altman, S. (1989). Ribonuclease P, an enzyme with a catalytic RNA subunit. *Advances in Enzymology*, **62**, 1–36.

Baldwin, J. E. & Krebs, H. A. (1981). The evolution of metabolic cycles. *Nature*, **291**, 381–382.

Balter, M. (1998). Did life begin in hot water? *Science*, **280**, 31.

Barbieri, M. (1985). *The Semantic Theory of Evolution*. Chur: Harwood Academic.

Behe, M. J. (1996). *Darwin's Black Box: the Biochemical Challenge to Evolution*. New York: Free Press.

Bernstein, M. P., Sandford, S. A. & Allamandola, L. J. (1999). Life's far-flung raw materials. *Scientific American*, **281**, no. 1, 42–49.

Berry, S. (1995). Entropy, irreversibility and evolution. *Journal of Theoretical Biology*, **175**, 197–202.

Birge, R. R. (1995). Protein-based computers. *Scientific American*, **272**, no. 3, 66–71.

Brillouin, L. (1962). *Science and Information Theory*. New York: Academic Press.

Brooks, D. R., Collier, J., Maurer, B., Smith, J. D. H. & Wilson, E. O. (1989). Entropy and information in evolving biological systems. *Biology and Philosophy*, **4**, 407–432.

Bryant, P. J. (1999). *Biodiversity and Conservation: A Hypertext Book*. http://darwin.bio.uci.edu/~sustain/bio65/Titlpage.htm

Casti, J. L. (1997). *Would-be Worlds*. New York: John Wiley.

Cech, T. R. (1986). A model for the RNA-catalyzed replication of RNA. *Proceedings of the National Academy of Sciences of the United States of America*, **83**, 4360–4363.

Cech, T. R. (1986). RNA as an enzyme. *Scientific American*, **255**, no. 5, 76–84.

Chargaff, E. (1978). *Heraclitean Fire*. New York: Columbia University Press.

Cornwell, J. (ed.) (1998). *Consciousness and Human Identity*. Oxford: Oxford University Press.

Crick, F. (1979). Split genes and RNA splicing. *Science*, **204**, 264–271.

Crick, F. (1993). In *The RNA World*, ed. Gesteland, R. F. & Atkins, J. F., pp. xi–xiv. Cold Spring Harbor: Cold Spring Harbor Laboratory Press.

Darnell, J. E. (1985). RNA. *Scientific American*, **253**, no. 4, 54–64.

Davies, P. C. W. (1998). *The Fifth Miracle: the Search for the Origin of Life*. London: Penguin.

Delbrück, M. (1987) *Mind from Matter?* New York: Basil Blackwell.

Dickerson, R. E. (1980). Cytochrome *c* and the evolution of energy metabolism. *Scientific American*, **242**, no. 3, 137–153.

Doolittle, R. (1985). Proteins. *Scientific American*, **253**, no. 4, 88–96.

Drexler, K. E. (1992). *Nanosystems, Molecular Machines and Computation*. New York: John Wiley.

Dyson, F. (1999). *Origins of Life*, 2nd edn. Cambridge: Cambridge University Press.

Eckland, E. H., Szostak, J. W. & Bartel, D. P. (1995). Structurally complex and highly active RNA ligases derived from random RNA sequences. *Science*, **269**, 364–370.

Eigen, M., Gardiner, W., Schuster, P. & Winckler-Oswatitsch, R. (1981). The origin of genetic information. *Scientific American*, **244**, no. 4, 88–118.

Encyclopædia Britannia CD98, 'Aging,' 'Life,' 'Maxwell's demon,' 'Metabolism,' 'The origin of life,' and 'Principles of thermodynamics.'

Ereshefsky, M. (1991). The semantic approach to evolutionary theory. *Biology and Philosophy*, **6**, 59–80.

Felsenfeld, G. (1985). DNA. *Scientific American*, **253**, no. 4, 44–53.

Feynman, R. P., Leighton, R. B. & Sands, M. (1963). *Lectures on Physics*, vol. I, cc. 15 & 16. Reading, Massachusetts: Addison-Wesley.

Flam, F. (1994). Hints of a language in junk DNA. *Science*, **266**, 1320.

Fruton, J. S. (1999). *Proteins, Enzymes, Genes: the Interplay of Chemistry and Biology*. New Haven: Yale University Press.

Galtier, N., Tourasse, N. & Gouy, M. (1999). A nonhyperthermophylic common ancestor to extant life forms. *Science*, **283**, 220–221.

Gesteland, R. F. & Atkins, J. F. (eds) (1993). *The RNA World*. Cold Spring Harbor: Cold Spring Harbor Laboratory Press.

Glansdorff, P. & Prigogine, I. (1974). *Structure, Stability and Fluctuations*. New York: Wiley-Interscience.

Glasser, L. (1989). Order, chaos, and all that! *Journal of Chemical Education*, **66**, 997–1001.

Grene, M. (1987). Hierarchies in biology. *American Scientist*, **75**, 504–510.

Harold, F. M. (1986). *The Vital Force: a Study of Bioenergetics*, cc. 1 & 13. New York: W. H. Freeman.

Heilbronner, E. & Dunitz, J. D. (1993). *Reflections on Symmetry: In Chemistry, and Elsewhere*. Weinheim: VCH.

Helmer, M. (1999). Singular take on molecules. *Nature*, **401**, 225–226.

Hess, B. & Mikhailov, A. (1994). Self-organization in living cells. *Science*, **264**, 223–224.

Hill, T. L. (1963). *Thermodynamics of Small Systems: Parts I and II*. New York: Benjamin.

Huber, C. & Wächtershäuser, G. (1998). Peptides by activation of amino acids with CO on (Ni,Fe)S surfaces: implications for the origin of life. *Science*, **281**, 670–672.

Kasner, E. & Newman, J. (1943). *Mathematics and the Imagination*. New York: Simon and Schuster.

Kaufmann, S. A. (1993) *The Origins of Order: Self-organization and Selection in Evolution*. New York: Oxford University Press.

Keosian, J. (1974). Life's beginnings – origin or evolution? *Origins of Life*, **5**, 285–293.

Kondepudi, D. & Prigogine, I. (1998). *Modern Thermodynamics: from Heat Engines to Dissipative Structures*, cc. 15, 16 & 19.3. Chichester: John Wiley.

Küppers, B. O. (1990). *Information and the Origin of Life*. Cambridge, Massachusetts: Massachusetts Institute of Technology Press.

Lamond, A. I. & Gibson, T. J. (1990). Catalytic RNA and the origin of genetic systems. *Trends in Genetics*, **6**, 145–149.

Lazcano, A. & Miller, S. L. (1994). How long did it take for life to begin and evolve to cyanobacteria? *Journal of Molecular Evolution*, **39**, 546–554.

Lee, D. H., Granja, J. R., Martizez, J. A., Severin, K. & Ghardi, M. R. (1996). A self-replicating peptide. *Nature*, **382**, 525–528.

Lin, S. K. & Gutnov, A. V. (eds). *Entropy: An International and Interdisciplinary Journal of Entropy and Information Studies*. http://www.mdpi.org/entropy/

Löwdin, P.O. (1969). In *Theoretical Physics and Biology*, ed. M. Marois. Amsterdam: North-Holland.

Lwoff, A. (1962). *Biological Order*. Cambridge, Massachusetts: Massachusetts Institute of Technology Press.

MacMahon, J. A., Phillips, D. L., Robinson, J. V. & Schimpf, D. J. (1978). Levels of biological organization: an organism-centered approach. *Bioscience*, **28**, 700–704.

Mahner, M. & Bunge, M. (1998). *Foundations of Biophilosophy*. Berlin: Springer-Verlag.

McClare, C. W. F. (1971). Chemical machines, Maxwell's demon and living organisms, *Journal of Theoretical Biology*, **30**, 1–34.

Mirowski, P. (1989). *More Heat than Light: Economics as Social Physics, Physics as Nature's Economics*, ch. 2. Cambridge: Cambridge University Press.

Moorhead, P. S., Kaplan, M. M. & Brown, P. (1985). *Mathematical Challenges to the Neo-darwinian Interpretation of Evolution: A Symposium Held at the Wistar Institute of Anatomy and Biology*, 2nd printing. New York: Alan R. Liss.

Mojzsis, S. J., Arrhenius, G., McKeegan, K. D., Harrison, T. M., Nutman, A. P. & Friend, R. L. (1996). Evidence of life on earth before 3,800 million years ago. *Nature*, **384**, 55–59.

Morowitz, H. J. (1955). Some order-disorder considerations in living systems, *Bulletin of Mathematical Biophysics*, **17**, 81–86.

Morowitz, H. J. (1967). Biological self-replicating systems. *Progress in Theoretical Biology*, **1**, 35–58.

Morowitz, H. J. (1978). *Foundations of Bioenergetics*, cc. 6 & 14D. New York: Academic Press.

Morowitz, H. J. (1992). *The Beginnings of Cellular Life: Metabolism Recapitulates Biogenesis*. New Haven: Yale University Press.

Oliver, S. G. (1996). From DNA sequences to biological function. *Nature* **379**, 597–600.

Oparin, A. I. (1953). *The Origin of Life*, 2nd edn. New York: Dover.

Orgel, L. E. (1973). *The Origins of Life: Molecules and Natural Selection*. New York: John Wiley.

Orgel, L. E. (1994) The origin of life on the earth. *Scientific American*, **271**, no. 4, 77–83.

Osawa, S., Jukes, T. H., Watanabe, K. & Muto, A. (1992). Recent evidence for evolution of the genetic code. *Microbiological Review*, **56**, 229–264.

Oyama, S. (1985). *The Ontogeny of Information: Developmental Systems and Evolution*. Cambridge: Cambridge University Press.

Pascal, B. (1966). *Pensées*, trans. A. J. Krailsheimer. London: Penguin.

Peusner, L. (1974). *Concepts in Bioenergetics*, cc. 8-11. Englewood Cliffs: Prentice-Hall.

Polanyi, M. (1967). Life transcending physics and chemistry. *Chemical and Engineering News*, **45**, 54–66.

Polanyi, M. (1968). Life's irreducible structure. *Science*, **160**, 1308–1312.

Polanyi, M. & Prosch, H. (1975). *Meaning*. Chicago: University of Chicago Press.

Popper, K. R. & Eccles, J. C. (1977). *The Self and its Brain: An Argument for Interactionism*. London: Routledge.

Prigogine, I., Nicolis, G., & Babloyantz, A. (1972). Thermodynamics of evolution. *Physics Today*, **25**, no. 11, 23–28.

Prigogine, I., Nicolis, G., & Babloyantz, A. (1972). Thermodynamics of evolution. *Physics Today*, **25**, no. 12, 38–44.

Prigogine, I. (1973). In round table with Ilya Prigogine: Can thermodynamics explain biological order? *Impact of Science on Society*, **23**, 159–179.

Prigogine, I. (1978). Time, structure, and fluctuations. *Science*, **201**, 777–785.

Prigogine, I. (1980). *From Being to Becoming*. New York: W. H. Freeman.

Prigogine, I. & Nicolis, G. (1977). *Self Organization in Nonequilibrium Systems*. New York: John Wiley.

Questler, H. (1953). *Information Theory in Biology*. Urbana: University of Illinois Press.

Rebek, J. (1994) Synthetic self-replicating molecules. *Scientific American*, **271**, no. 1, 48–55.

Reid, W. V. & Miller, K. R. (1989). *Keeping Options Alive: The Scientific Basis for Conserving Biodiversity*. Washington, D. C.: World Resources Institute.

Robertson, H. D. (1996). How did replicating and coding RNAs first get together? *Science*, **274**, 66–67.

Rogers, J. & Joyce, G. F. (1999). A ribozyme that lacks cytidine. *Nature*, **402**, 323–325.

Sanchez, J.M. (1995). Order-disorder transitions. *In Encyclopedia of Applied Physics*, ed. G. L. Trigg, vol. 13, pp. 1–16. New York: VCH.

Schrödinger, E. (1945). *What is Life? The Physical Aspect of the Living Cell*. Cambridge: Cambridge University Press.

Schuster, H.G. (1992). Chaotic phenomena. In *Encyclopedia of Applied Physics*, ed. G. L. Trigg, vol. 3, pp. 189–214. New York: VCH.

Senapathy, P. (1995). Introns and the origin of protein-coding genes. *Science*, **269**, 1366–1367.

Shannon, C. E. (1948). The mathematical theory of communication. *Bell System Technical Journal*, **27**, 379–423.

Soltzberg, L. J. (1989). Self-organization in chemistry. *Journal of Chemistry Education*, **66**, 187.

Stevens, T. & McKinley, J. Lithoautotrophic microbial ecosystems in deep basalt aquifers (1995). *Science*, **270**, 450–453.

Sturtevant, J. M. (1993). Effects of mutations on the thermodynamics of proteins. *Pure and Applied Chemistry*, **65**, 991–998.

Voet, D. & Voet, J. G. (1996). 1996 Supplement, *Biochemistry*, 2nd edn, ch. 4. New York: John Wiley.

Waldburger, C. D., Schildbach, J. F. & Sauer, R. T. (1995). Are buried salt bridges important for protein stability and conformational specificity? *Nature Structural Biology*, **2**, 122–128.

Wicken, J. S. (1987). *Evolution, Thermodynamics and Information: Extending the Darwinian Program*. Oxford: Oxford University Press.

Williams, M. B. (1973). Falsifiable predictions of evolutionary theory. *Philosophy of Science*, **40**, 518–537.

Williams, R. J. P. (1993). Are enzymes mechanical devices? *Trends in Biochemical Sciences*, **18**, 115–117.

Wright, M. C. & Joyce, G. F. (1997). Continuous *in vitro* evolution of catalytic function. *Science*, **276**, 614–617.

Yockey, H. P. (1992). *Information Theory and Molecular Biology*. Cambridge: Cambridge University Press.

Zhang, B. L. & Cech, T. R. (1997). Peptide bond formation by *in vitro* selected ribozymes. *Nature*, **390**, 96–100.

Zylstra, U. (1992). Living things as hierarchically organized structures. *Synthese*, **91**, 111–133.

K. | Exercises

1. In order for a physical change to occur spontaneously, the entropy of the universe must increase. Can the increase of the entropy of the universe continue indefinitely? Why or why not? Relate your answer to the constraints on the energy in Eqn. 1.1.

2. Kurt Vonnegut's science fiction novel *The Sirens of Titan* tells the story of a man who has to cope with being converted into pure energy and knowing everything that has already happened or ever will happen. Is this possible in our universe? Why or why not?

3. If Earth was poor in biological information before life began and is now 'biological information-rich,' we should like to know what was the origin of the input of information. Is biological information inherent in the chemical composition of Earth? How did biologically meaningful information come into existence? Does the origin of information reside in the Big Bang?

4. Is a 'DNA world' plausible? Why or why not.

5. Hen egg white lysozyme is 129 amino acid residues long. Calculate the information content of lysozyme on the level of the amino acid sequence. Calculate the information content of the shortest piece of DNA that could encode this protein. Compare the values and comment.

6. According to equation 9.4 each nucleotide in a strand has an information content of 2 bits. Thus a poly-G oligonucleotide 12 bases long has an information content of 24 bits. The logarithm of the

number of distinct ways of arranging the 12 nucleotides, however, is 0. If entropy and information are isomorphous, why are these numbers not identical? (Hint: see Chapter 2.)

7. The human language-DNA analogy must break down at some point. Consider the three basic types of sentence – declarative, imperative, and interrogative – and discuss how they resemble and differ from gene expression.

8. Calculate the information content of the human genome. How many possible different DNA sequences are there with a size identical to that of the human genome?

9. The table below shows the probability of occurrence of letters in English. Note that, unlike the example presented in the text, the probability varies from character to character. When this is the case, the information content of a message is defined as

$$I = -\Sigma p_i \log_2 p_i$$

Calculate the information content of this sentence.

Symbol	Probability, p	Symbol	Probability, p
space	0.2	C	0.023
E	0.105	F, U	0.0225
T	0.072	M	0.021
O	0.0654	P	0.0175
A	0.063	Y, W	0.0175
N	0.059	G	0.012
I	0.055	B	0.011
R	0.054	V	0.0105
S	0.052	K	0.008
H	0.047	X	0.002
D	0.035	J, Q, Z	0.001
L	0.029	—	—

Probability of occurrence of letters in English. From Brillouin (1962).

10. Is the information content of the sickle cell variant of hemoglobin different from or the same as that of the wild-type protein? Is one variant more biologically meaningful than the other? Justify your answer.

11. From the point of view of information theory, entropy measures the observer's lack of knowledge of the microscopic state of a system. Because the information content of a message is only semi-objective, the observer can be said to be a part of any system being studied. Can this subjectivity be circumvented? Why or why not? If yes, how?

12. π, the value of which is 3.1415926..., is a rather remarkable number. Not only does it crop up in areas of mathematics like number theory, but it plays a key role in theoretical physics: π is used to describe the geometry of space (the volume of a sphere is proportional to π).

Moreover, π is a transcendental number and therefore an 'irrational' one, meaning that it cannot be expressed as the ratio of two *integers*; the trail of digits after the decimal place is *infinitely* long. As of late 1999, the value of π had been computed up to over 100 000 million decimal places. Analysis very strongly suggests that there is no organization to this string of numbers. In other words, no distinction can be made between the successive digits of π and numbers between 0 and 9 chosen *at random*. A consequence of this for computing is that there is no way of 'compressing' the digits of π into a shorter string, as one can easily do with a number like 0.321 321 321 . . .. Comment on the relationship between the compressibility of information and entropy. Relate this to the encoding of information in DNA. Speculate on what this might mean for the origin and propagation of life. (The original proof of the irrationality of π was given by Lindemann in 1882. See e.g. Kasner & Newman (1943).)

13. Mathematical transforms and thermodynamics. By means of a Fourier transform, (named after Baron Jean Baptiste Joseph Fourier (1768–1830), a French mathematician and physicist noted for his research on heat diffusion and numerical equations) a signal $h(t)$ in the time domain, for instance a chord played on a piano, may be represented by its spectrum $H(f)$ in the frequency domain, the fundamental vibrational frequencies of the strings corresponding to the keys involved, and *vice versa*. In essence, the Fourier transform decomposes or separates a waveform into sinusoids of different frequency and amplitude, and when these are summed, they give the original waveform. Dr Sylvan C. Bloch, professor of physics at the University of South Florida, has shown how π can be used to generate what is called a spread-spectrum wavelet by the 'direct sequence method.' The digits of π are then encoded on a noise-like wavelet, compressed in time (with a concomitant expansion in frequency), decompressed, and demodulated to recover the information. Explain the similarity between this situation and the First Law of Thermodynamics.

For solutions, see http://chem.nich.edu/homework

Appendix A

General references

Adkins, C. J. (1983). *Equilibrium Thermodynamics*, 3rd edn. Cambridge: Cambridge University Press.

Atkinson, D. E. (1987) *Dynamic Models in Biochemistry*. Menlo Park: Benjamin.

Bederson, B. (1991). Atoms. In *Encyclopedia of Applied Physics*, ed. G. L. Trigg, vol. 2, pp. 245–296, New York: VHC.

Bisswanger, H. (1996). Proteins and Enzymes. In *Encyclopedia of Applied Physics*, ed. G. L. Trigg, vol. 15, pp. 185–214. New York: VCH.

Brey, W. S. (1978). *Physical Chemistry and Its Biological Applications*. New York: Academic Press.

Bridgman, P. W. (1961). *The Nature of Thermodynamics*. New York: Harper & Row.

Brooks, D. R. & Wiley, E. O. (1988). *Evolution as Entropy*, 2nd edn. Chicago: University of Chicago Press.

Cantor, C. R. & Schimmel, P. R. (1980). *Biophysical Chemistry*, vols I, II and III. San Francisco, W. H. Freeman.

Christensen, J. J., Hansen, L. D. & Izatt, R. M. (1976). *Handbook of Proton Ionization Heats and Related Thermodynamic Quantities*. New York: John Wiley.

Colson, S. D. & Dunning, T. H. (1994). The structure of nature's solvent: water. *Science*, **265**, 43–44.

Cooke, R. & Kuntz, I. D. (1974) The properties of water in biological systems. *Annual Review of Biophysics and Bioengineering*, **3**, 95–126.

Dawson, R. M. C., Elliott, D. C., Elliott, W. H. & Jones, K. M. (1986). *Data for Biochemical Research*, 3rd edn. Oxford: Clarendon Press.

Delbrück, M. (1949). A physicist looks at biology. *Transactions of the Connecticut Academy of Arts and Science*, **38**, 173–190.

Edsall, J. T. & Gutfreund, H. (1983). *Biothermodynamics*. New York: John Wiley.

Edsall, J. T. & Wyman, J. (1958). *Biophysical Chemistry*. New York: Academic Press.

Eisenberg, D. & Crothers, D. (1979). *Physical Chemistry with Applications to the Life Sciences*. Menlo Park: Prentice-Hall.

Eisenberg, D. & Kautzmann, W. (1969). *The Structure and Properties of Water*. Oxford: Oxford University Press.

Emsley, J. (1996). *The Elements*. Oxford: Oxford University Press.

Engelking, P. (1994). Molecules. In *Encyclopedia of Applied Physics*, ed. G. L. Trigg, vol. 10, pp. 525–552. New York: VHC.

Franks, F. (ed.) (1973). *Water, a Comprehensive Treatise*, vol. 1–6. New York: Plenum.

Freifelder, D. (1982). *Physical Biochemistry: Applications to Biochemistry and Molecular Biology*, 2nd edn. New York: W. H. Freeman, 1982.

Fruton, J. S. (1999). *Proteins, Enzymes, Genes: the Interplay of Chemistry and Biology*. New Haven: Yale University Press.

Gold, V., Loening, K. L., McNaight, A. D. & Sehmi, P. (1987). *Compendium of Chemical Terminology*. Oxford: Blackwell Scientific.

Goldstein, M. & Goldstein, I. F. (1993). *The Refrigerator and the Universe: Understanding the Laws of Energy*. Cambridge, Massachusetts: Harvard University Press.

Hirst, D. M. (1983). *Mathematics for Chemists*. London: Macmillan.

Hoppe, W., Lohmann, W., Markl, H. & Ziegler, H. (ed.) (1983). *Biophysics*. New York: Springer-Verlag.

Jones, M. N. (ed.) (1979). *Studies in Modern Thermodynamics. 1. Biochemical Thermodynamics*. Oxford: Elsevier.

Kleinauf, H., von Dahren, H. & Jaenicke, L. (ed.) (1988). *The Roots of Modern Biochemistry*. Berlin: de Gruyter.

Klotz, I. M. (1957). *Energetics in Biochemical Reactions*. New York: Academic Press.

Klotz, I. M. (1978). *Energy Changes in Biochemical Reactions*. New York: Academic Press.

Medewar, P. (1984). *The Limits of Science*. Oxford: Oxford University Press.

Mills, I. M. (1989). The choice of names and symbols for quantities in chemistry. *Journal of Chemistry Education*, **66**, 887–889.

Mills, I. M. (ed.) (1993). *Quantities, Units, and Symbols in Physical Chemistry*. Oxford: Blackwell Scientific.

Morris, J. G. (1974). *A Biologist's Physical Chemistry*. Reading, Massachusetts: Addison-Wesley.

National Institute for Standards and Technology Web site: http://webbook.nist.gov

Phoenix, D. (1997) *Introductory Mathematics for the Life Sciences*. London: Taylor & Francis.

Rose, A. H. (ed.) (1967). *Thermobiology*. New York: Academic Press.

Segal, I. H. & Segal, L. D. (1993). Energetics of biological processes. In *Encyclopedia of Applied Physics*, ed. G. L. Trigg. New York: VCH.

Selected Values of Chemical Thermodynamic Properties (Washington, D.C.: U.S. National Bureau of Standards, 1952).

Sober, H. A. (ed.) (1970). *Handbook of Biochemistry*, 2nd edn. Cleveland, OH: Chemical Rubber Company.

Steiner, E. (1996). *The Chemistry Maths Book*. Oxford: Oxford University Press.

Stillinger, F. H. (1980). Water revisited, *Science*, **209**, 451–457.

Stull, D.R. & Sinke, G. C. (1956). *Thermodynamic Properties of the Elements*. Washington, D.C.: American Chemical Society.

Tanford, C. (1961). *Physical Chemistry of Macromolecules*. New York: John Wiley.

Tanford, C. (1980). *The Hydrophobic Effect*. New York: John Wiley.

von Baeyer, H. C. (1998). *Maxwell's Demon: Why Warmth Disperses and Time Passes*, New York: Random House.

Weast, R. C. (ed.) (1997). *CRC Handbook of Chemistry and Physics*, 78th edn. Boca Raton: Chemical Rubber Company.

Weinberg, R. A. (1985). The molecules of life. *Scientific American*, **253**, no. 4, 34–43.

Weiss, T. F. (1995). *Cellular Biophysics*, vol. 1 & 2. Cambridge, Massachusetts: Massachusetts Institute of Technology Press.

Westhof, E. (1993). *Water and Biological Macromolecules*. Boca Raton: Chemical Rubber Company.

Wilkie, D. R. (1960). Thermodynamics and the interpretation of biological heat measurements. *Progress in Biophysics and Biophysical Chemistry*, **10**, 259–298.

Appendix B

Biocalorimetry

A. | Introduction

Calorimetry is the only means by which one can make direct, model-independent measurements of thermodynamic quantities. Spectroscopic techniques, despite being extremely sensitive and able to provide high-resolution structure information, can give but an indirect, model-dependent determination of the thermodynamics of a system. Calorimetric analysis therefore complements spectroscopic studies, giving a more complete description of the biological system of interest. Modern microcalorimeters are both accurate and sensitive, so that measurements require relatively small amounts of material (as little as 1 nmol) and can yield data of a relatively low uncertainty.

Diffuse heat effects are associated with almost all physico-chemical processes. Because of this, microcalorimetry provides a way of studying the energetics of biomolecular processes at the cellular and molecular level. Microcalorimetry can be used to determine thermodynamic quantities of a wide range of biological processes. Among these are: conformational change in a biological macromolecule, ligand binding, ion binding, protonation, protein–DNA interaction, protein–lipid interaction, protein–protein interaction, protein–carbohydrate interaction, enzyme–substrate interaction, enzyme–drug interaction, receptor–hormone interaction, and macromolecular assembly. Calorimetry is also useful in the analysis of the thermodynamics of very complex processes, for example enzyme kinetics and cell growth and metabolism. That is, calorimetry is not narrowly applicable to processes occurring at equilibrium.

There are three broad classes of biological calorimetry: bomb calorimetry, differential scanning calorimetry (DSC) and isothermal titration calorimetry. The choice of instrumentation depends mainly on the type of process being studied. Bomb calorimetry is used to measure the energy content of foods and other materials; discussion of the technique can be found in Chapters 1 and 2. Here we focus on DSC and ITC. An appealing and important feature of these two methods is that, because they are used to study the energetics of non-covalent or reversible interactions, they are at least in principle nondestructive. In fact, in practice most of the sample can often be recovered from a DSC or ITC experiment and put to further use.

B. | Differential scanning calorimetry

DSC (Fig. 2.10A) measures the heat capacity of a process over a range of temperatures. The technique has been employed in the thermodynamic characterization of different kinds of biopolymers, but it has primarily been used to study the heat-induced unfolding of proteins and polynucleotides. Other applications of DSC include measurement of the stability of protein-nucleic complexes (e.g. ribosomes) and determination of the melting temperature of the gel–liquid crystal phase transition of lipid vesicles (Chapter 4).

Temperature is the principal independent variable in a DSC experiment. Cosolvent concentration, pH, or ion concentration can be a useful second independent variable, as for example in studies of protein stability (Chapter 5). DSC experiments take place at constant pressure, so the heat effect corresponds to the enthalpy of the reaction (Chapter 2). The shape of the heat capacity function ($<C_p>$ versus T) provides information on the thermodynamics of the order-disorder transition. DSC is the only model-independent means of determining ΔH, ΔC_p, T_m, and cooperativity of a structural change in a biological macromolecule. Measurement of these thermodynamic quantities provides a 'complete' thermodynamic description of an order-disorder transition, because these quantities are sufficient to simulate a heat capacity function ($<C_p>$ versus T). (See Chapter 6 for further details.) Determination of ΔH and T_m permits evaluation of ΔS (Chapter 2).

To rationalize the results of a DSC experiment, one often needs information on the thermodynamics of interactions between specific chemical groups. That is, although a single DSC experiment can in principle provide a 'complete' description of the thermodynamics of an order/disorder transition, such a description is consistent with many possible molecular mechanisms, and independently acquired data are needed to exclude some of the various possibilities.

The usual scanning rate in a DSC experiment is 1 °C per minute. This will be appropriate, however, only if the system will come to equilibrium relatively quickly throughout the temperature range of the experiment. A process is reversible if it proceeds through a succession of equilibrium or near-equilibrium states (Chapter 2), and if the reverse process yields the starting material *as it was before heating*. A large number of small proteins exhibit nearly complete reversibility after unfolding. The degree of reversibility often depends on solution conditions and the duration of heating in the unfolded state. When equilibrium is approached on a longer time scale than the scan rate, kinetic effects must be taken into account in interpreting the results of an experiment. Of course, any changes in protein association must also be accounted for, as for instance when the degree of oligomerization changes with temperature.

Analysis of the properties of point mutants has revealed that proteins are extremely complicated thermodynamic systems. Single amino acid replacement probably causes numerous small effects, and these are likely to be distributed throughout the molecule. This makes an observed thermodynamic quantity difficult to rationalize in terms of molecular structure. One of the most interesting findings of such studies, which is often though not always observed, is 'enthalpy–entropy' compensation (Chapter 5). This describes the situation where a point mutation alters the enthalpy of unfolding but the calculated free energy at, say, 25 °C, is relatively unchanged. The molecular origin of this effect is often not entirely clear.

DSC can also be used to study interactions between molecules. For instance, if a protein has a high-affinity binding site for an ion, say calcium, the concentration of calcium will have a marked effect on protein stability. The transition temperature will increase with ion concentration until all the sites remain filled as long as the protein is folded. DSC can thus be used to determine affinity constants of ligands (Chapter 7). The same approach can be used to measure the stability of polynucleotides, as well as how this varies with the concentration of ions or DNA-binding proteins.

C. | Isothermal titration calorimetry

ITC (Fig. 2.8) is used to characterize the binding interactions of a macromolecule to a ligand at constant temperature. The ligand can be another macromolecule, a peptide, an oligonucleotide, a small chemical compound, an ion or even just a proton. (See Chapter 7 for an in-depth discussion of binding.) ITC permits analysis of the binding properties of native molecules without modification or immobilization. In a well-designed experiment, an ITC instrument can be used to acquire a complete profile of the binding interaction and to measure the binding constant, stoichiometry, and other thermodynamic functions. Modern instruments are accurate and easy to use. Specific applications of ITC include: disaggregation of cationic micelles, enzyme–substrate interactions, antibody–antigen recognition, peptide–antibiotic interactions, and protein–DNA binding. The approach can also be used to study whole organism metabolism. ITC is a versatile technique.

ITC measures the heat of a reaction. The experiment takes place at constant pressure, so the heat absorbed or evolved is the enthalpy of reaction (Chapter 2). In an ITC experiment, small aliquots of a titrant solution containing the ligand are added sequentially to a macromolecule in a reaction cell, and the instrument records changes in enthalpy as the binding sites become saturated. The shape of the titration curve provides information on the strength of the binding interaction and the number of ligands recognized by the macromolecule. Determination of ΔH, K_{eq}, and n, the number of ligands, permits evaluation of ΔS (Chapter 4). ITC can also be used to measure ΔC_p of binding if the titration experiment is carried out at several different temperatures ($\Delta C_p = \Delta \Delta H / \Delta T$, Chapter 2).

Results from ITC studies can throw light on the nature of macromolecular interactions on the molecular level. This is particularly true if high-resolution structural information on the macromolecule or ligand is available. Knowledge of the binding thermodynamics and structural details of the macromolecule in the free and liganded states can lead to insights on the rules governing such interactions and enable the manipulation of biomolecular recognition processes at the molecular level.

In antibody binding, for example, a solution of the antibody is titrated with small aliquots of the antigen, and the heat evolved in the formation of the antigen–antibody complex is measured with a sensitivity as high as 0.1 μcal. The free energy of binding (ΔG), the binding enthalpy (ΔH), and the binding entropy (ΔS) can often be measured in a single experiment. Moreover, no spectroscopic or radioactive label must be attached to the antigen or antibody, simplifying both experimental design and interpretation of results. The change in heat capacity accompanying the formation of the complex can be determined by measuring ΔH over a range of temperatures. ΔC_p measured in this way is often found to be large and negative (why?). A binding experiment can give several other types of thermodynamic data: protein–ligand interaction in the binding pocket, structural changes that might occur on binding, reduction of the translational degrees of freedom of the antigen. One must design experiments cleverly to measure individual contributions to an overall heat effect. For instance, suppose that the hydroxyl group of a Tyr residue of a macromolecule forms a hydrogen bond with a hydrogen bond acceptor in a ligand. Replacement of the Tyr by Phe would permit determination of the contribution of a single hydrogen bond to the binding thermodynamics, as long as no other changes occurred. In practice, however, even such small changes as Tyr→Phe are often accompanied by the rearrangement of water molecules in the binding site, making it difficult to be certain of the proportion of the measured effect that is attributable to the chemical modification.

ITC is also useful for drug discovery in the pharmaceutical industry. The data provided by the technique complement rational drug design. It can reduce the time or money needed to take a lead compound into the marketplace, because sample through-put for the instrument is high. For instance, one could use ITC to screen potential inhibitors of an enzyme, using the heat effect as an indication of binding. Additional experiments could then be done to investigate whether or not the ligands affect enzymatic activity. Direct measurement of the heat of reaction is one of the best ways to characterize the thermodynamics of binding, and ITC enables rapid determination of binding affinity.

Equilibrium binding constants are inherently difficult to measure when the affinity is high. Biological processes in this category include cell surface receptor binding and protein–DNA interactions. Technique-independent difficulties stem from loss of signal intensity at low concentrations and from slow off-rate kinetics. The largest binding constant that can be measured reliably by titration microcalorimetry is about 10^9 M^{-1}. This poses a problem if the binding constant is large under physiological conditions. One way of dealing with this situation is to try to determine the binding constant under other conditions, and then to make the necessary corrections. This can be done because the Gibbs free energy is a state function (see Chapter 2 for a discussion of the characteristics of state functions).

D. | The role of biological techniques

Certain biological techniques play an important role in present-day biological calorimetry. For instance, site-directed mutagenesis enables one to study the thermodynamic consequences of changing single amino acids, so one is no longer limited to chemical modifications, useful as they can be. Large-scale production of recombinant proteins is also extremely useful. Although DSC and ITC are nondestructive techniques, protein folding/unfolding can be considerably less reversible than one might like, and recovery of a macromolecule or ligand from an ITC experiment might not be worth the effort if large quantities of pure materials are relatively easy to prepare.

E. | References and further reading

Blandamer, M. J. (1998). Thermodynamic background to isothermal titration calorimetry. In *Biocalorimetry: Applications of Calorimetry in the Biological Sciences*, ed. J. E. Ladbury & B. Z. Chowdhry. Chichester: John Wiley.

Chellani, M. (1999). Isothermal titration calorimetry: biological applications. *Biotechnology Laboratory*, **17**, 14–18.

Cooper, A. & Johnson, C. M. (1994a). Differential scanning calorimetry. In *Methods in Molecular Biology, Vol. 22: Microscopy, Optical Spectroscopy, and Macroscopic Techniques*, ed. C. Jones, B. Mulloy & A. H. Thomas, ch. 10, pp. 125–136. Totowa, NJ: Humana Press.

Cooper, A. & Johnson, C. M. (1994b). Isothermal titration microcalorimetry. In *Methods in Molecular Biology, Vol. 22: Microscopy, Optical Spectroscopy, and Macroscopic Techniques*, ed. C. Jones, B. Mulloy & A. H. Thomas, ch. 11, pp. 137–150. Totowa, NJ: Humana Press.

Freire, E., Mayorga, O. L. & Straume, M. (1990). Isothermal titration calorimetry. *Analytical Chemistry*, **62**, 950A–959A.

MicroCal, Inc., Web site http://www.microcalorimetry.com.

Plotnikov, V. V., Brandts, J. M., Lin, L. N. & Brandts, J. F. (1997). A new ultrasensitive scanning calorimeter. *Analytical Biochemistry*, **250**, 237–244.

Straume, M. (1994). Analysis of two-dimensional differential scanning calorimetry data: elucidation of complex biomolecular energetics. *Methods in Enzymology*, **240**, 530–568.

Sturtevant, J. (1987) Biochemical applications of differential scanning calorimetry. *Annual Review of Physical Chemistry*, **38**, 463–488.

Appendix C

Useful tables

A. Energy value and nutrient content of foods

	Energy (kJ)	Water (g)	Carbohydrate (g)	Protein (g)	Fat (g)	Alcohol (g)
Whole wheat flour	1318	14	63.9	12.7	2.2	—
White bread	1002	37.3	49.3	8.4	1.9	—
White rice, boiled	587	68	30.9	2.6	1.3	—
Milk, fresh, whole	275	87.8	4.8	3.2	3.9	—
Butter, salted	3031	15.6	trace	0.5	81.7	—
Cheese, Cheddar	1708	36	0.1	25.5	34.4	—
Steak, grilled	912	59.3	0	27.3	12.1	—
Tuna, canned in oil, drained	794	63.3	0	27.1	9	—
New potatoes, boiled in unsalted water	321	80.5	17.8	1.5	0.3	—
Peas, frozen, boiled in unsalted water	291	78.3	9.7	6	0.9	—
Cabbage, boiled in salted water	75	92.5	2.5	0.8	0.6	—
Orange	158	86.1	8.5	1.1	0.1	—
Apple, raw	199	84.5	11.8	0.4	0.1	—
White sugar	1680	trace	105	Trace	0	—
Beer*, canned	132	—	2.3	0.3	Trace	3.1
Spirits* (brandy, gin, rum, whiskey)	919	—	trace	Trace	0	31.7

All values per 100 g edible portion, except those indicated with an asterisk. *Values per 100 ml. Data are from *The Composition of Foods*, 5th ed. (1991), The Royal Society of Chemistry and the Controller of Her Majesty's Stationery Office.

B. Physical properties of amino acids

	Volume properties of amino acid residues			Accessible surface areas of amino acids in a Gly-X-Gly tripeptide in an extended conformation		Side-chain atoms			Packing of residues in the interior of proteins	
	Van der Waals volume (Å³)	Partial volume in solution (Å³)	Partial specific volume (cm³)⁻¹	Total (Å²)	Main-chain atoms (Å²)	Total (Å²)	Non-polar (Å²)	Polar (Å²)	Fraction of residues at least 95% buried	Relative free energy of residue in interior relative to surface (kcal mol²¹)[a]
Alanine, Ala (A)	67	86.4	0.732	113	46	67	67		0.38	−0.14
Arginine, Arg (R)	148	197.4	0.756	241	45	196	89	107	0.01	1.40
Asparagine, Asn (N)	96	115.6	0.610	158	45	113	44	69	0.12	0.75
Aspartic acid, Asp (D)	91	108.6	0.573	151	45	106	48	58	0.15	0.78
Cysteine, Cys (C)	86	107.9	0.630	140	36	104	35	69	0.40[b] 0.50[c]	−0.61[bc]
Glutamine, Gln (Q)	114	142.0	0.667	189	45	144	53	91	0.07	0.80
Glutamic Acid, Glu (E)	109	128.7	0.605	183	45	138	61	77	0.18	1.15
Glycine, Gly (G)	48	57.8	0.610	85	85				0.36	0
Histidine, His (H)	118	150.1	0.659	194	43	151	102	49	0.17	0.02
Isoleucine, Ile (I)	124	164.6	0.876	182	42	140	140		0.60	−0.68
Leucine, Leu (L)	124	164.6	0.876	180	43	137	137		0.45	−0.59
Lysine, Lys (K)	135	166.2	0.775	211	44	167	119	48	0.03	2.06
Methionine, Met (M)	124	160.9	0.739	204	44	160	117	43	0.40	−0.65
Phenylalanine, Phe (F)	135	187.3	0.766	218	43	175	175		0.50	−0.61
Proline, Pro (P)	90	120.6	0.748	143	38	105	105		0.18	0.50
Serine, Ser (S)	73	86.2	0.596	122	42	80	44	36	0.22	0.40
Threonine, Thr (T)	93	113.6	0.676	146	44	102	74	28	0.23	0.32
Tryptophan, Trp (W)	163	225.0	0.728	259	42	217	190	27	0.27	−0.39
Tyrosine, Tyr (Y)	141	190.5	0.703	229	42	187	144	43	0.15	0.28
Valine, Val (V)	105	136.8	0.831	160	43	117	117		0.54	−0.55
weighted average			0.703							

[a] Calculated as −RT ln(fraction in interior / fraction on surface), with the relative free energy of Gly set to zero.
[b] When in disulfide form. [c] When in thiol form. Data from Tables 4.3, 4.4 and 6.3 of Creighton and references therein.

C. Protonation energetics at 298 K

Buffer	ΔH (kJ mol^{-1})	ΔC_p (J K^{-1}mol^{-1})	Buffer	ΔH (kJ mol^{-1})	ΔC_p (J K^{-1}mol^{-1})
Acetate	0.49	−128	Imidazole	36.59	−16
Cacodylate	−1.96	−78	TES	32.74	−33
MES	15.53	16	HEPES	21.01	49
Glycerol-2-phosphate	−0.72	−179	EPPS	21.55	56
ACES	31.41	−27	Triethanolamine	33.59	48
PIPES	11.45	19	Tricine	31.97	−45
Phosphate	5.12	−187	Tris	47.77	−73
BES	25.17	2	TAPS	41.49	−23
MOPS	21.82	39	CAPS	48.54	33

Data were prepared and revised by H. Fukuda & K. Takahashi, Laboratory of Biophysical Chemistry, College of Agriculture, University of Osaka Prefecture, Sakai, Osaka 591, Japan.

D. Buffer ionization constants

Acid	pK
Oxalic acid	1.27 (pK_1)
H_3PO_4	2.15 (pK_1)
Citric acid	3.13 (pK_1)
Formic acid	3.75
Succinic acid	4.21
Oxalate$^-$	4.27 (pK_2)
Acetic acid	4.76
Citrate$^-$	4.76 (pK_2)
Succinate	5.64 (pK_2)
MES[a]	6.09
Cacodylic acid	6.27
H_2CO_3	6.35 (pK_1)
Citrate^{2-}	6.40 (pK_3)
ADA[b]	6.57
PIPES	6.76
ACES	6.80
$H_2PO_4^{-}$	6.82 (pK_2)
MOPS[c]	7.15
HEPES[d]	7.47
HEPPS[e]	7.96
Tricine[f]	8.05
TRIS[g]	8.08
Glycylglycine	8.25
Bicine[h]	8.26
Boric acid	9.24
Glycine	9.78
HCO_3^{-}	10.33 (pK_2)
Piperidine	11.12
HPO_4^{2-}	12.38 (pK_3)

Data from Dawson *et al.*, *Data for Biochemical Research*, 3rd edn (Oxford: Clarendon, 1986) or Good *et al.* (1966) *Biochemistry*, 5,467.

Abbreviations: [a]morpholinoethanesulfonic acid, [b]acetamidoiminodiacetic acid, [c]morpholinopropanesulfonic acid, [d]hydroxytethylpiperazine-ethanesulfonic acid, [e]hydroxyethylpiperazone-propanesulfonic acid, [f]trishydroxymethylmethylglycine, [g]trishydroxymethylaminomethane, [h]bishydroxymethylglycine.

E. Energetics of the reactions of the citric acid cycle

Reaction	Enzyme	Comments	$\Delta G^{\circ\prime}$ (kJ mol^{-1})	ΔG (kJ mol^{-1})
Acetyl-CoA + H$_2$O oxaloacetate → citrate + CoASH	Citrate synthetase	An acetyl group is added to oxaloacetate, changing the carbonyl carbon at C-3 from +2 to a +1 oxidation state. Water is used for release of free CoA. This favorable energy change makes the first step in the cycle essentially irreversible.	−31.5	<0
Citrate → cis-aconitate + H$_2$O → isocitrate	Aconitase	A hydroxyl group is transferred from C-3 to C-2 on citrate by successive dehydration and hydration reactions.	~5	~0
Isocitrate + NAD$_{ox}$ → NAD$_{red}$ + CO$_2$ + 2-oxoglutarate (α-ketoglutarate)	Isocitrate dehydrogenase	Isocitrate is decarboxylated via NAD-linked oxidation.	−21	<0
2-Oxoglutarate + CoASH + NAD$_{ox}$ → NAD$_{red}$ + CO$_2$ + succinyl-CoA	2-Oxoglutarate dehydrogenase multienzyme complex	2-oxoglutarate is decarboxylated via NAD-linked oxidation. Succinyl is attached to CoA, forming succinyl-CoA.	−33	<0
Succinyl-CoA + P$_i$ + GDP → succinate + GTP + CoASH	Succinyl-CoA synthase	Substrate-level phosphorylation is driven by the redox reaction in the previous step. The terminal phosphate group of GTP can be transferred to ATP by nucleoside diphosphate kinase.	−2.1	~0
Succinate + FAD$_{ox}$ → Fumarate + FAD$_{red}$	Succinate dehydrogenase	The FAD is covalently linked to the enzyme, which is bound to the inner mitochondrial membrane.	+6	~0
Fumarate + H$_2$O → malate	Fumarase	Water is added to fumarate to form malate.	−3.4	~0
L-Malate + NAD$_{ox}$ → oxaloacetate + NAD$_{red}$	Malate dehydrogenase	This energy barrier is overcome by maintaining a low concentration of oxaloacetate in the mitochondrial matrix.	+29.7	~0

Data are from Table 19-2 of Voet & Voet (1996)

Appendix D

BASIC program for computing the intrinsic rate of amide hydrogen exchange from the backbone of a polypeptide

The computer program IRATE calculates intrinsic rates of hydrogen exchange for all backbone amide hydrogens in a given amino acid sequence, using the method of Bai *et al.* (1993) *Proteins*, **17**, 75–86. The input file must be an ASCII file and the amino acid sequence must be in single letter code. The program accepts both upper case or lower case letters, but the input file must not contain spaces.

```
' ****************************************************
' * Program: IRATE (Intrinsic RATe of Exchange)      *
' * Author: Donald T. Haynie, Ph.D.                  *
' * Date of Original Program: May 1996               *
' * Previous Modification: Th.23.v.96                *
' * Most recent Modification: Th.23.ix.99            *
' * Acknowledgement: Dr Christina Redfield           *
' ****************************************************
'
CLS
PRINT " "
'  Read input file containing amino acid sequence.
'
INPUT 'Name of ASCII file that contains amino acid sequence'; infile$
OPEN infile$ FOR INPUT AS #1
INPUT #1, sequence$
CLOSE #1
'
'  Determine length of the sequence.
'
length = LEN(sequence$)
'
'  Create an array of the length of the sequence.
'
DIM seq(length - 1) AS INTEGER
PRINT 'The length of the sequence is:'; length
PRINT ' '
INPUT 'Name of protein/peptide'; protein$
PRINT ' '
'
'  Create an output file.
'
```

```
INPUT 'Output filename'; outfile$
OPEN outfile$ FOR OUTPUT AS #1

'   Convert the characters of amino acid sequence input string to integers and store
'   in sequential elements of array 'seq.' The input string must be in single letter
'   code.
'
position=0
DO WHILE position<length
    aa$=MID$(sequence$, position+1, 1)
    IF ASC(aa$)>96 THEN
            seq(position)=ASC(aa$) - 97
    ELSE
            seq(position)=ASC(aa$) - 65
    END IF
    position=position+1
LOOP
'
'   Prompt for temperature and pD
'
PRINT " "
INPUT "Temperature (deg. C)"; temp
PRINT " "
INPUT "Exchange medium: H2O (0) or D2O (1)"; solvent
PRINT " "
INPUT "Is protein/peptide deuterated (y/n)"; deuterated$
'
'   Set the reference exchange rates in accordance with experimental conditions.
'
flag=0
DO WHILE flag<1
    IF LEFT$(deuterated$, 1)="y" OR LEFT$(deuterated$, 1)="Y" THEN
        IF solvent=0 THEN
```

```
' See Connelly et al. (1993) Proteins 17:87-92.
'
                    arefrate=1.4
                    brefrate=9.87
                    wrefrate=-1.6
                    flag=1
            ELSE
                    PRINT "IRATE does NOT calculate rates of exchange of deuterons into D2O"
            END IF
    ELSE
        SELECT CASE solvent
            CASE 0
' See Connelly et al. (1993) Proteins 17:87-92.
'
                    arefrate=1.39
                    brefrate=9.95
                    wrefrate=0!
            CASE 1
' See Bai et al. (1993) Proteins 17:75-86.
'
                    arefrate=1.62
                    brefrate=10.05
                    wrefrate=-1.5
        END SELECT
        flag=1
    END IF
LOOP
PRINT " "
PRINT arefrate, brefrate, wrefrate
PRINT " "
INPUT "Measured pH or pD (uncorrected) at 20 deg. C"; pd
'
```

```
' If solvent is D20, a correction must be made to electrode reading.
' Ionization constants are for 20 deg. C and are from p. D-166 of CRC
'
IF solvent=1 THEN
        pd=pd+.4
        pod=pd - 15.049
ELSE
        pod=pd - 14.1669
END IF
'
'   Convert temperature to degrees K
'
temp=temp+273.15
PRINT " "
'
' Create table of the measured effects of amino acid side chains on HX rates of
' neighboring peptides. See Bai et al. (1993).

DIM table(27, 3) AS SINGLE
table(0, 0)=0!                     'Ala acid catalysis left
table(0, 1)=0!                     'Ala acid catalysis right
table(0, 2)=0!                     'Ala base catalysis left
table(0, 3)=0!                     'Ala base catalysis right
table(17, 0)=-.59                  'Arg acid left
table(17, 1)=-.32                  'Arg acid right
table(17, 2)=.08                   'Arg base left
table(17, 3)=.22                   'Arg base right
table(13, 0)=-.58                  'Asn
table(13, 1)=-.13
table(13, 2)=.49
table(13, 3)=.32
table(3, 0)=.9                     'Asp base
table(3, 1)=.58
table(3, 2)=-.3
```

```
table(3, 3) = .18
table(1, 0) = -.9
table(1, 1) = -.12
table(1, 2) = .69
table(1, 3) = .6          'Asp acid
table(2, 0) = -.54
table(2, 1) = -.46
table(2, 2) = .62
table(2, 3) = .55         'Cysteine
table(9, 0) = -.74
table(9, 1) = -.58
table(9, 2) = .55
table(9, 3) = .46         'Cystine
table(6, 0) = -.22
table(6, 1) = .22
table(6, 2) = .27
table(6, 3) = .17         'Gly
table(16, 0) = -.47
table(16, 1) = -.27
table(16, 2) = .06
table(16, 3) = .2         'Gln
table(4, 0) = -.9
table(4, 1) = .31
table(4, 2) = -.51
table(4, 3) = -.15        'Glu base
table(14, 0) = -.6
table(14, 1) = -.27
table(14, 2) = .24
table(14, 3) = .39        'Glu acid
table(7, 0) = 0!
table(7, 1) = 0!
table(7, 2) = -.1
table(7, 3) = .14         'His base
table(20, 0) = -.8        'His acid
```

```
table(20, 1) = -.51
table(20, 2) = .8
table(20, 3) = .83
table(8, 0) = -.91                'Ile
table(8, 1) = -.59
table(8, 2) = -.73
table(8, 3) = -.23
table(11, 0) = -.57               'Leu
table(11, 1) = -.13
table(11, 2) = -.58
table(11, 3) = -.21
table(10, 0) = -.56               'Lys
table(10, 1) = -.29
table(10, 2) = -.04
table(10, 3) = .12
table(12, 0) = -.64               'Met
table(12, 1) = -.28
table(12, 2) = -.01
table(12, 3) = .11
table(5, 0) = -.52                'Phe
table(5, 1) = -.43
table(5, 2) = -.24
table(5, 3) = .06
table(15, 0) = 0!                 'Pro (trans)
table(15, 1) = -.19
table(15, 2) = 0!
table(15, 3) = -.24
table(23, 0) = 0!                 'Pro (cis)
table(23, 1) = -.85
table(23, 2) = 0!
table(23, 3) = .6
table(18, 0) = -.44               'Ser
table(18, 1) = -.39
table(18, 2) = .37
```

```
table(18, 3) = .3
table(19, 0) = -.79          'Thr
table(19, 1) = -.47
table(19, 2) = -.07
table(19, 3) = .2
table(22, 0) = -.4           'Trp
table(22, 1) = -.44
table(22, 2) = -.41
table(22, 3) = -.11
table(24, 0) = -.41          'Tyr
table(24, 1) = -.37
table(24, 2) = -.27
table(24, 3) = .05
table(21, 0) = -.74          'Val
table(21, 1) = -.3
table(21, 2) = -.7
table(21, 3) = -.14
table(25, 0) = 0!            'N-term
table(25, 1) = -1.32
table(25, 2) = 0!
table(25, 3) = 1.62
table(26, 0) = .96           'C-term base
table(26, 1) = 0!
table(26, 2) = -1.8
table(26, 3) = 0!
table(27, 0) = .05           'C-term acid
table(27, 1) = 0!
table(27, 2) = 0!
table(27, 3) = 0!
'
'     Write information about protein and HDX conditions to output file.
'
PRINT #1, "Calculated intrinsic amide exchange rates for ** "; protein$; " **"
PRINT #1, "at ** pD (corrected)"; pd; "** and **"; temp; "K **"
```

```
PRINT #1, " "
IF solvent = 1 THEN
        PRINT #1, "Solvent is D2O"
ELSE
        PRINT #1, "Solvent is H2O"
END IF
PRINT #1, " "
PRINT #1, "All rates are in units of inverse minutes"
PRINT #1, " "
PRINT #1, "res R L"; TAB(12); "fracar"; TAB(40); "fracal"; TAB(70); "acid rate"; TAB(90); "base rate";
    TAB(110); "water rate"; TAB(130); "intrinsic rate"; TAB(160); "dG++"
PRINT #1, " "
'
'   Set the activation energies.
'
PRINT " "
'
'   Acid
aae = 14000!
'
'   Base
'
bae = 17000!
'
'   Water
wae = 19000!
PRINT " "
CLS
DEFDBL C, F, H, K, Z
'
'   Factor for converting log to ln
'
```

```
constant = LOG(10)
' See Eq. (3) of Bai et al. (1993).
corr = (1 / temp - 1 / 293!) / 1.9872
' Fraction protonated to the right
fracar = 0
' Fraction protonated to the left
fracal = 0
position = 0
' Calculate percentage of protons bound for acidic side chains.
DO WHILE position<length
    ' What is this residue?
    SELECT CASE seq(position)
        ' If this is a Cys, is it in an S-S bridge?
        CASE 2
            PRINT "Residue"; position+1; "is a Cys"
            INPUT "Is it involved in an S-S bridge"; reply$
            PRINT " "
            IF LEFT$(reply$, 1) = "y" OR LEFT$(reply$, 1) = "Y" THEN
            ' If yes, change the array.
                seq(position) = 9
```

```
            END IF
            logarl = table(9, 0)
            logbrl = table(9, 2)

'   If this residue is an Asp, do this:
'   '

        CASE 3

            PRINT "The default pKa of Asp"; position+1; "is 3.95"
            INPUT "Do you wish to change this value (y/n)"; reply$
            PRINT " "
            IF LEFT$(reply$, 1) = "y" OR LEFT$(reply$, 1) = "Y" THEN
                INPUT "What is the pKa for this residue"; pKa
            ELSE
                pKa = 3.95
            END IF
            PRINT "pKa is "; pKa
            PRINT " "
            fracal = (10 ^ (pKa - pd)) / (1 + 10 ^ (pKa - pd))
            logarl = fracal * table(1, 0) + (1 - fracal) * table(3, 0)
            logbrl = fracal * table(1, 2) + (1 - fracal) * table(3, 2)

'   If this residue is a Glu, do this:
'   '

        CASE 4

            PRINT "The default pKa of Glu"; position+1; "is 4.4"
            INPUT "Do you wish to change this value (y/n)"; reply$
            PRINT " "
            IF LEFT$(reply$, 1) = "y" OR LEFT$(reply$, 1) = "Y" THEN
                INPUT "What is the pKa for this residue"; pKa
            ELSE
                pKa = 4.4
            END IF
            PRINT "pKa is "; pKa
            PRINT " "
```

```
            fracal = (10 ^ (pKa - pd)) / (1 + 10 ^ (pKa - pd))
            logarl = fracal * table(14, 0) + (1 - fracal) * table(4, 0)
            logbrl = fracal * table(14, 2) + (1 - fracal) * table(4, 2)

' If this residue is a His, do this:
'
        CASE 7
            PRINT "The default pKa of His"; position+1; "is 6.5"
            INPUT "Do you wish to change this value (y/n)"; reply$
            PRINT " "
            IF LEFT$(reply$, 1) = "y" OR LEFT$(reply$, 1) = "Y" THEN
                INPUT "What is the pka for this residue"; pKa
            ELSE
                pKa = 6.5
            END IF
            PRINT "pKa is "; pKa
            PRINT " "
            fracal = (10 ^ (pKa - pd)) / (1 + 10 ^ (pKa - pd))
            logarl = fracal * table(20, 0) + (1 - fracal) * table(7, 0)
            logbrl = fracal * table(20, 2) + (1 - fracal) * table(7, 2)

' If this residue is none of the above . . . .
'
        CASE IS < 2
            logarl = table(seq(position), 0)
            logbrl = table(seq(position), 2)

' then just read values straight . . . .
'
        CASE 5 TO 6
            logarl = table(seq(position), 0)
            logbrl = table(seq(position), 2)

' from the array "table" . . . .
```

```
            CASE IS>7
                logar1 = table(seq(position), 0)
                logbr1 = table(seq(position), 2)

        END SELECT

' If the second residue has been reached:

'       IF position>0 THEN

' Check the identity of the previous residue.

'           SELECT CASE seq(position - 1)
'               CASE 3

' Compute weighted average rate if this is an Asp:

'               logar = fracar * table(1, 1) + (1 - fracar) * table(3, 1) + logar1
'               logbr = fracar * table(1, 3) + (1 - fracar) * table(3, 3) + logbr1

'               CASE 4

' Compute weighted average rate if this is a Glu:

'               logar = fracar * table(14, 1) + (1 - fracar) * table(4, 1) + logar1
'               logbr = fracar * table(14, 3) + (1 - fracar) * table(4, 3) + logbr1

'               CASE 7

' Compute weighted average rate if this is a His:

'               logar = fracar * table(20, 1) + (1 - fracar) * table(7, 1) + logar1
'               logbr = fracar * table(20, 3) + (1 - fracar) * table(7, 3) + logbr1

'               CASE IS<3

' Or just read the values . . .
```

```
                logar=table(seq(position-1), 1)+logarl
                logbr=table(seq(position-1), 3)+logbrl
        CASE 5 TO 6
'
'  . . . from the array "table" . . .
'
                logar=table(seq(position-1), 1)+logarl
                logbr=table(seq(position-1), 3)+logbrl
        CASE IS>7
'
'  . . . if this is neither Asp nor Glu nor His.
'
                logar=table(seq(position-1), 1)+logarl
                logbr=table(seq(position-1), 3)+logbrl
        END SELECT
'
'  Take polypeptide chain end effects into account as follows:
'
        IF position=1 THEN
            logar=logar+(10 ^ (7.4-pd)) / (1+10 ^ (7.4-pd)) * table(25, 1)
            logbr=logbr=(10 ^ (7.42pd)) / (1+10 ^ (7.4-pd)) * table(25, 3)
        END IF
        IF position=length-1 THEN
            logar=logar+(10 ^ (3.9-pd)) / (1+10 ^ (3.9-pd)) * table(27, 0) + (1 - (10
 ^ (3.9-pd)) / (1+10 ^ (3.9-pd)) * table(26, 0)
            logbr=logbr+(1 - (10 ^ (3.9-pd)) / (1+10 ^ (3.9-pd))) * table(26, 2)
        END IF
'
'  Compute rates:
'
'  Acid rate:
'
        acidrate=EXP((arefrate+logar-pd) * constant-aae * corr)
```

```
'  Base rate:
'
                baserate = EXP((brefrate+logbr+pod) * constant-bae * corr)
'
'  Water rate:
'
                waterrate = EXP((wrefrate+logbr) * constant-wae * corr)
'
'  Sum rates to give total rate:
'
                k = acidrate+baserate+waterrate
                PRINT #1, position+1; TAB(6); CHR$(seq(position-1)+65); SPC(1); CHR$(seq(position)+65);
TAB(12); fracar; TAB(40); fracal; TAB(70); acidrate; TAB(90); baserate; TAB(110); waterrate; TAB(130); k;
TAB(160); USING "##.###"; -1.9872 * temp * LOG(6.6254E-27 * k / 1.38046E-16 / temp / 60!) / 1000!
                END IF
'
'  Switch left to right, etc., and go to the next residue:
'
                fracar = fracal
                fracal = 0
                position = position+1
LOOP
CLOSE
```

Glossary

abiotic synthesis – non-biological synthesis of a biochemical, usually a macro-molecule.

acid – proton donor. Compare *base*.

acidity constant – pH at which dissociation of protons from a specific titratable site is half complete; a measure of the *free energy* of protonation.

activated complex – structure of enzyme–substrate complex in the transition state.

activation barrier – schematic representation of the *energy* that must be added to reactants to convert them to products.

activation energy – minimum *energy* input required to initiate a chemical reaction under given conditions.

active site – region on the surface of an enzyme where the *substrate* binds and catalysis occurs. See *catalyst*.

active transport – transport of ions or metabolites across a biological membrane against a concentration gradient at the expense of *energy* resources of the cell (*ATP* hydrolysis). Compare *diffusion* and *facilitated diffusion*.

activity – effective concentration of a chemical species.

activity coefficient – factor by which the concentration of a chemical species is multiplied to give the *activity*.

actual information – calculated information content of a message. Compare *potential information*.

Adair equation – general ligand binding equation first proposed by Gilbert Adair.

adjustable parameter – component of a mathematical model (of a *process*) the value of which is determined by fitting the model to experimental data.

allosteric regulation – modulation of enzyme function through the binding of small molecules or ions to sites on the enzyme other than where catalysis occurs.

amino acid composition – percentage of each amino acid type for a given polypeptide.

anaerobic autotrophs – organisms that synthesize all their cellular constituents from simple molecules, some inorganic, in the absence of oxygen.

antibody diversity – vast repertoire of antibodies produced in individual mammals by means of genetic recombination (combinatorics) and mutation.

association constant – binding constant for association of ligand and macromolecule. Compare *dissociation constant*.

ATP – small molecule compound that is the main *energy* 'currency' of all known organisms. *ATP* is also utilized in the communication of biological information: it is directly involved in the synthesis of *second messengers*, *mRNA*, and *DNA*, and in the propagation of chemical signals by *phosphorylation* of amino acid side chains.

barbed end – the end of an actin filament where *ATP*-bound actin monomers associate preferentially. So named from appearance of myosin S1 fragment-bound actin filaments by scanning electron microscopy. Synonym of *plus end*.

base – proton acceptor. Compare *acid*. *Base* is also used to describe a hydrogen bond-forming information storage unit in DNA or RNA.

Big Bang – cataclysmic explosion about 20 billion years ago by which the universe is thought to have come into existence.

binding capacity – number of binding sites per macromolecule.

binding site – precise location on a macromolecule where a ligand binds.

biochemist's standard state – defined reference state of greatest use to biochemists, as it accounts for pH and assumes that reactions occur in aqueous solvent.

bioenergetics – the study of *energy* changes in living organisms, particularly as these concern glucose metabolism and *ATP* production.

biological information – the one-dimensional information content of genetic material and the three-dimensional information content of proteins and other biological macromolecules.

bit of information – information content of one binary decision.

body plan – one of but several different basic organization schemes into which all known organisms can be classified.

Bohr effect – effect of pH on oxygen-binding properties of hemoglobin, first described by Christian Bohr.

Boltzmann distribution – the most probable distribution of a system at equilibrium if the system contains a large number of molecules; first described by Ludwig Boltzmann.

Boltzmann factor – relative contribution of a *state* to the magnitude of the *partition function*; named after Ludwig Boltzmann. Synonym of *statistical weight*.

boundary – conceptual barrier where the system meets the surroundings which may or may not be permeable to heat or matter.

breathing motions – stochastic fluctuations in the structure of proteins and other biological macromolecules. See *Le Châtelier's principle*.

Brønsted–Lowry definitions – see *acid* and *base*.

buffering capacity – quantitative ability of a buffered solution to resist changes in pH upon addition of *acid* or *base*.

calorimetric enthalpy – heat absorbed or evolved during a *process*, usually occurring at constant pressure.

calorimetry – science of measuring heat transfer from system to surroundings and vice versa.

carbon – extraordinary element whose ability to form up to four relatively stable covalent bonds per atom is essential for life as we know it.

catalyst – substance whose presence increases the rate of a chemical reaction but is not consumed by the reaction.

chemical potential – the *free energy* of a compound in solution.

chemosynthesis – biochemical *process* by which *ATP* is synthesized by reduction of inorganic compounds and not by absorption of photons. Compare *photosynthesis*.

chirality – molecular handedness.

chlorophyll – major antenna for absorption of sunlight in plants.

citric acid cycle – set of coupled reactions in the mitochondrial matrix that oxidize acetyl groups and generate CO_2 and reduced intermediates used to make *ATP*; also known as Krebs cycle and tricarboxylic acid (TCA) cycle.

closed system – one that is permeable to heat but not matter. See *system*.

cold-denaturation – unfolding of a protein molecule by cooling rather than heating.

collision theory – one explanation of chemical reactivity. Compare *transition-state theory*.

competitive inhibition – blocking of a biochemical interaction by an inhibitor through direct competition with the ligand for the ligand binding site.

configuration – arrangement of particles in a *system*.

conformational change – alteration of the three-dimensional structure of a molecule but not of its covalent bonds.

conformational entropy – entropy of fixing the three-dimensional structure of a molecule.

conservation of energy – apparently fundamental principle of physics that *energy* is neither created nor destroyed in any physical, chemical, or biological *process*. See *First Law of Thermodynamics*.

cooperativity – degree of 'concertedness' of a change in conformation or arrangement of particles in a system.

coupled reaction – an overall spontaneous reaction made so by the product of an unfavorable reaction being a reactant in a more favorable one.

critical concentration – concentration of actin monomers below which polymerization will not occur under specified conditions.

Dalton's law – overall pressure of a gas is the sum of the partial pressures of the constituent gases.

degeneracy – number of distinguishable states of the same *energy* level.

differential scanning calorimetry (DSC) – device for measuring the heat exchanged at constant pressure as a function of temperature.

diffusion – random movement of particles at a given *temperature*. It is also called *passive diffusion*, in the context of net movement of a chemical species across a membrane at a *rate* proportional to the concentration gradient. Compare *facilitated diffusion* and *active transport*.

dissipation – expenditure of *free energy* or increase of entropy in which no work is done. See also *substrate cycling*.

dissociation constant – concentration of ligand at which half of all sites are occupied.

distribution – real or conceptual dispersal of something in space and time.

disulfide bond – covalent linkage between two sulfur atoms, each of which is donated by the amino acid cysteine.

Donnan equilibrium – equilibrium involving a semi-permeable membrane, permeant ions, and impermeant ions (usually a biological macromolecule).

dynamic equilibrium – state of no net change, not no change, as in any chemical equilibrium.

effector – ion or molecule involved in allosteric regulation of enzymatic activity.

efficiency – ratio of work done by a system to heat added to the system.

electromotive force – synonym of *voltage*.

electroneutrality – condition of a net charge of zero.

electroweak interactions – interparticle interactions mediated by, the electroweak *force*, one of the fundamental forces of nature; electromagnetic and weak nuclear interactions.

endergonic reaction – one which does not occur spontaneously at constant *temperature* and *pressure* unless *work* is done on the *system*. Antonym of *exergonic reaction*.

endothermic reaction – one which involves the absorption of heat. Antonym of *exothermic reaction*.

energy – the most fundamental concept of science; the capacity have an effect, the capacity to do work.

energy transducer – common intermediate in chemically coupled reactions.

energy well – local minimum in the *free energy* surface.

enthalpy – thermodynamic state function usually measured as heat transferred to or from a system at constant pressure.

enthalpy of binding – enthalpy difference between the bound and unbound states of a ligand–macromolecule system.

enthalpy of denaturation – enthalpy change of protein unfolding at a given temperature.

enthalpy–entropy compensation – phenomenon observed in weakly stable systems in which changes in enthalpy are attended by changes in entropy but little or no change in *free energy*.

enthalpy of hydration – enthalpy change on solvation of an ion or molecular compound.

entropy – thermodynamic state function that is a measure of disorder.

equilibrium – condition of no further net change in a close system which is not to be confused with *steady state*.

equilibrium constant – provides a means of calculating the standard *free energy* change for a reaction by measuring the amounts of products and reactants in a *system*.

ergodic hypothesis – assumption that the short term behavior of a large collection of identical objects is equivalent to the long term behavior of a small collection of such objects.

evolution – gradual change in genetic material with time.

EX1 mechanism – rate of hydrogen exchange limited by the intrinsic rate.

EX2 mechanism – rate of hydrogen exchange limited by the rate of exposure of the labile hydrogen to the exchange medium (solvent).

exergonic reaction – one which does occur spontaneously at constant temperature and pressure in the absence of work being done on the system. Antonym of *endergonic reaction*.

exothermic reaction – one which involves the release of heat. Antonym of *endothermic reaction*.

extremophile – bacterium that thrives in harsh physical or chemical environments.

extrinsic property – quantity that does depend on amount of substance present, for example *energy*. Compare *intrinsic property*.

facilitated diffusion – membrane protein-aided transport of an ion or molecule across a membrane and down its concentration gradient. Compare *diffusion* and *active transport*.

feedback inhibition – in metabolism, down-regulation of a metabolic pathway by interaction between a product of the pathway and one of its enzymes.

First Law of Thermodynamics – statement of the conservation of *energy*.

first-order reaction – one in which reaction rate is proportional to the first power of the concentration of reactant.

force – in mechanics, physical agency that changes the velocity of an object.

fractional saturation – quantitative measure of partial saturation of available binding sites.

frequency factor – parameter of reaction rate at a given temperature.

function – variable quantity related to one or more other variables in terms of which it may be expressed or on the value of which its own value depends.

Gibbs free energy – thermodynamic potential for a system under the constraints of constant temperature and constant pressure.

glucose – predominant source of chemical *energy* in cells.

glycolysis – anaerobic conversion of sugar to lactate or pyruvate with the production of *ATP*.

group-transfer potential – driving force for the chemical transfer of a given type of chemical group, e.g. phosphoryl group.

half-life – time required for half of a given amount of reactant to be converted into product.

half reaction – conceptual reduction reaction showing the transfer of electrons explicitly.

heat – *energy* transfer by random motion. Compare *work*.

heat capacity – change in enthalpy per unit change in temperature.

heat capacity at constant pressure – change in enthalpy per unit change in temperature at constant pressure. The heat capacity specifies the temperature dependence of the enthalpy and entropy functions.

heat engine – system that uses heat transfer to do work.

heat sink – thing which absorbs thermal *energy*.

heat source – thing which radiates thermal *energy*.

helix propensity – one of several definitions is the relative probability that an amino acid type is found in helical structure in the folded states of proteins.

Henderson–Hasselbalch equation – mathematical relationship between pH and *acidity constant*.

Hess's law – additivity of independently determined enthalpies; a statement of the First Law.

Hill equation – mathematical relationship between free ligand concentration, binding constant, number of cooperative subunits, and fractional saturation of bindings sites; named after Archibald Hill.

Hill plot – popular but non-ideal graphical representation of binding data which can be used to determine the cooperativity of binding, named after Archibald Hill.

homoallostery – allosteric regulation in which the *effector* is the same chemical species as the *substrate*.

hydrogen electrode – standard for measurement of redox potential.

hydrogen exchange – *acid*- and *base*-catalyzed chemical *process* in which one labile hydrogen atom is exchanged for another, generally donated by the solvent.

hydrophobic interaction – in biological macromolecules, particularly proteins, favorable intermolecular interaction between apolar moieties, e.g. aliphatic amino acid side chains.

ideal gas law – quantitative relationship between pressure, volume, temperature, the number of moles of ideal gas in a closed system.

information content – minimum number of binary decisions required to construct a message.

information theory – science of data communication.

inherent meaning – meaning of a message that is completely independent of an observer.

interaction – the effect of one object on another.

internal energy – thermodynamic *state function* that measures the *energy* within the *system*.

intrinsic property – quantity that does not depend on amount of substance present, for example pressure. Compare *extrinsic property*.

intrinsic rate of exchange – rate of exchange of a specific labile hydrogen atom with solvent, usually in a completely unstructured polypeptide.

irreversibility – *process* not reversible. Dissipation or destruction.

isoelectric point – pH at which the net charge on a macromolecule (usually a polypeptide) is zero.

isosbestic point – wavelength at which the value of a spectroscopic variable is independent of the structure of system (usually a macromolecule in solution).

isothermal system – one at constant temperature.

isolated system – one permeable neither to heat nor matter.

isothermal titration calorimetry (ITC) – measures heat exchanged at constant temperature and pressure in binding experiment.

kinetic barrier – synonym of *activation energy*.

Kirchoff's enthalpy law – mathematical relationship between reference state enthalpy change, heat capacity change, temperature change, and overall enthalpy change.

KNF model – one of the two most popular models of allosteric regulation, named after Koshland, Némethy, and Filmer. See *MWC model*.

Langmuir adsorption isotherm – mathematical relationship between the free ligand concentration, the association constant, and degree of saturation of binding sites at constant temperature; named after Irving Langmuir.

latent heat – the enthalpy of a phase change, i.e. of the reorganization of the state of matter.

law – in science and philosophy, a theoretical principle stating that a particular phenomenon always occurs if certain conditions are met.

Le Châtelier's principle – system at equilibrium responds to a disturbance of a system by minimizing the effect of the disturbance.

life – qualitative property of highly-organized matter, essential features of which are: growth, development, metabolism, and reproduction under the control of a genetic program, and reproduction by means of transmission of genetic material.

Lifson–Roig model – popular model of helix–coil transition theory.

ligand – ion or molecule (other than an enzyme substrate) that binds (usually specifically) to a macromolecule (usually a protein).

linking number – parameter describing the number of complete turns of the DNA backbone within defined boundaries.

local unfolding – fluctuation of structure, not complete unfolding. May result in some amide protons becoming available for exchange but not others.

machine – structure of any kind; an apparatus for applying power.

macroscopic system – comprises such a large number of particles that measured properties at equilibrium are approximately constant in time and fluctuations are relatively small.

Marcus theory – widely accepted mathematical description of the energetics of electron transfer.

mass action – in a system at equilibrium, a change in the amount of reactants (products) results in a compensating change in the amount of products (reactants), so that the relative proportion of reactants and products is minimized.

mass action ratio – ratio of the product of the activities (concentrations) of products to the product of the activities (concentrations) of reactants.

mean chemical potential – chemical potential of a chemical species whose activity is calculated as a *mean ionic activity*.

mean free path – average distance between collisions of particles.

melting temperature – temperature at which a solid undergoes a phase transition to the liquid state; in the case of proteins, temperature at which denaturation occurs.

metabolism – the biological enzymatic breakdown of molecules.

methanogens – bacteria that live only in oxygen-free milieus and generate methane by the reduction of carbon dioxide.

Michaelis–Menten equation – important mathematical relationship of enzyme kinetics, named after the biochemists who described it.

microscopic system – one member of a large ensemble of identical objects, for example one protein molecule in a concentrated solution of identical protein molecules.

minus end – synonym of *pointed end*.

mole fraction – a type of concentration scale.

molecular motor – *energy*-consuming protein molecule involved in force generation.

molecular switch – protein molecule whose conformation and biological function is controlled by binding (e.g. of protons, dissolved gas molecules, inorganic ions, small organic compounds, proteins, nucleic acids. . .).

molten globule – partly-ordered state of proteins, characterized by compactness, intact secondary structure, and fluctuating tertiary structure.

momentum – mass times velocity.

MWC model – one of the two most popular models of allosteric regulation, named after Monod, Wyman, and Changeux. See *KNF model*.

Nernst equation – mathematical relationship between electrical potential across a membrane and the ratio of the concentrations and valences of ions on either side of the membrane, named after Walther Nernst.

non-competitive inhibition – inhibition of enzyme activity resulting from the binding of an inhibitor to a location on the surface of the enzyme other than the site where the substrate binds (active site).

non-equilibrium thermodynamics – thermodynamic concepts and relationships pertinent to systems that are not at *equilibrium*.

nucleation of α-helix – in helix–coil transition theory, the formation of the first $i, i + 4$ hydrogen bond of an α-helix.

number of ways – number of distinct arrangements of particles in a system.

observable quantity – measurable property of a system.

observer – key component of any scientific experiment, from deciding what the aim of the experiment will be, to how the experiment is designed, to how the data are analyzed, and what the data mean. The role of the observer is highlighted in information theory.

open system – permeable to heat and matter.

optimization – process whereby an observable quantity of a system is made as large or as small as possible within given constraints.

order – non-arbitrary arrangement of things. Compare *random*.

order of reaction – sum of the powers of the molar concentrations of the reactants in the *rate law* of a reaction.

osmosis – movement of water across a semi-permeable membrane from a region of high to a region of low impermeant solute concentration.

osmotic work – (mechanical) *work* done by *osmosis*.

osmotic pressure – *pressure* arising from *osmosis*, e.g. by the displacement of a quantity of solvent against the force of gravity.

oxidant – synonym of 'oxidizing agent,' electron acceptor. Compare *reductant*.

panspermia – hypothesis that life on Earth originated at a remote location.

partition function – sum of all relevant statistical weights of a system.

passive transport – diffusive movement of an ion or molecule across a membrane, often through a protein pore.

path function – thermodynamic properties that relate to the preparation of the state.

phase – a state of matter that is uniform throughout, both in chemical composition and physical state.

phosphates – important constituent of many biological molecules, e.g. *ATP*, DNA, ...

phosphoanhydride bond – type of bond that is cleaved when *ATP* is hydrolyzed to ADP.

photosynthesis – biological *process* by which photosynthetic bacteria and plants convert the *free energy* of photons into chemical energy.

pointed end – minus end of actin filament, where ADP-bound actin monomers dissociate from polymer. So named from appearance of myosin S1 fragment-bound actin filaments by electron microscopy.

plus end – synonym of *barbed end*.

polymerization assays – means of testing the effect of a potential actin-binding protein on the polymerization of actin.

potential information – in information theory, maximum possible information content of a message. Compare *actual information*.

precursors – subunits of which a polymer is made, e.g. free amino acids in the case of a polypeptide.

pressure – force per unit area.

process – course of action or succession of actions, taking place or carried on in a definite matter.

propagation – in helix–coil transition theory, lengthening of the helix following nucleation.

protection factor – ratio of intrinsic rate of exchange to measured rate of exchange of polypeptide backbone amide proton.

proton motive force – proton concentration gradient across a membrane and the membrane electrical potential.

quenched-flow pulse labeling – technique for measuring the rate of stabilization of structure in proteins.

random – not sent, guided, or arranged in a discernibly special way. Compare *order*.

rate constant – proportionality constant between *rate of reaction* and molar concentrations of reactants.

rate-determining step – slowest step of a multi-step chemical reaction.

rate law – experimentally determined mathematical relationship between molar concentrations of reactants and *rate of reaction*. See *rate of reaction*.

rate of reaction – mathematical expression in term of molar concentrations of reactants and *rate constant*.

rectangular hyperbola – shape of ligand-binding curve when saturation is plotted as a function of free ligand concentration and there is no binding *cooperativity*, as in the case of myoglobin.

redox – chemical *process* in which electrons are transferred from the reductant to the oxidant. Important in metabolic reactions. See *redox couple* and *standard redox potential*.

redox couple – electron donor and acceptor.

reductant – synonym of 'reducing agent,' electron donor. Compare *oxidant*.

reference state – the most stable state of an element under defined conditions; alternatively, any set of defined conditions.

respiration – biological *process* by which oxygen is used an electron acceptor in the metabolism of food, principally glucose.

reversibility – *process* runs backwards or forwards. Near equilibrium states throughout.

RNA world – hypothesis that living organisms were preceded on earth by abiotic synthesis of RNA.

salt bridge – energetically favorable electrostatic interaction between an ionized acid and an ionized base.

salting in – increased solubility of protein in low ionic strength aqueous solution relative to pure water. Compare *salting out*.

salting out – decreased solubility of protein in high ionic strength aqueous solution relative to low ionic strength solution. Compare *salting in*.

saturation – complete filling of available binding sites by *ligand*.

Scatchard plot – popular but non-ideal graphical representation of binding data which can be used to determine the binding affinity and number of ligand binding sites.

Second Law of Thermodynamics – increase of entropy. In symbols, $\Delta S = q/T$.

second messenger – intracellular signaling molecule the concentration of which rises or falls in response to *binding* of an extracellular *ligand* to a receptor.

second-order reaction – one in which reactant rate is proportional to the second power of the concentration of reaction rate.

self-organization – spontaneous appearance of order on a local level.

semantic level – levels of meaning.

sliding filament model – model of interaction between actin and myosin underlying force generation in skeletal muscle.

solubility – the extent to which a chemical species, for instance a metabolite or protein, will dissolve in a solvent, usually water.

somatic mutation – major mechanism by which *antibody diversity* is generated; involves point mutations in B cells.

somatic recombination – major mechanism by which *antibody diversity* is generated; involves genetic recombination in B cells.

specific heat – the heat capacity per unit mass of material.

spontaneity – tendency of a chemical reaction to proceed in a certain direction without the addition of *energy*.

stability – difference in *free energy* between states, usually between the unfolded and folded states of a protein.

stability curve – variation of stability with some independent variable (usually temperature but often denaturant concentration).

standard state – unambiguous reference state.

standard state enthalpy change – enthalpy change for a *process* under standard conditions.

standard redox potential – measured voltage difference between a half cell consisting of both members of a *redox couple* (an electron donor and acceptor) in their standard states and a standard reference half cell (usually a *hydrogen electrode*).

state – thermodynamic state of a system, for example the folded conformation or unfolded conformation of a protein.

state of the system – specified by values of state variables.

state function – thermodynamic quantity whose value depends only on the current *state of the system* and is independent of how the *system* has been prepared, e.g. internal *energy*, *enthalpy*, *entropy*.

state variable – thermodynamic quantity under the control of the observer which, when fixed, determines the state of the system, e.g. pressure, volume, temperature.

statistical factors – binding constant coefficients related to the number of ways in which ligands can associate with and dissociate from a macromolecule.

statistical weight – synonym of *Boltzmann factor*.

steady state – condition of an open system in which the rate of flow of *energy* or a substance into the system is identical to the rate of flow out of the system. Compare *equilibrium*.

steady state assumption – in enzyme kinetics, assumption that the time rate of change of concentration of enzyme–*substrate* is zero.

substrate – chemical compound on which an enzyme acts.

substrate cycling – in metabolism, the formation and breakdown of a certain molecular compound that results in a net change in the concentration of *ATP* but not of the molecular compound itself.

sulfate-reducing bacteria – prokaryotic organisms that can grow and reproduce at temperatures as high as 100 °C and under very high pressures.

supercoiling – coiling of circular double-stranded DNA.

surroundings – the part of the universe that is not the system.

symbol – abstract representation of a thing. For instance, 'G' can represent the base guanine or the amino acid glycine. The meaning of the sign will of course depend on the context.

system – part of universe chosen for study.

temperature – measure of *thermal energy*, or how fast molecules are moving.

temperature of maximum stability – temperature at which the free energy difference between the folded and unfolded states of a protein at a given pressure is a maximum.

thermal energy – the average energy of a particle at a given temperature.

thermodynamic potential function – measures the *free energy* difference between states and is therefore an indicator of whether a reaction will be spontaneous.

thermodynamics – the study of the nature of heat and its relationship to other forms of *energy*.

thermostability – the *free energy* difference between states, usually of a macromolecule, under specified conditions, especially *temperature*.

Third Law of Thermodynamics – the *entropy* of a *system* approaches zero at the *temperature* goes to absolute zero (0 K).

titration – gradual filling up or removal of a *ligand* from a binding site. See *dissociation constant*.

transfer free energy – *free energy* change on transfer of a compound from one medium to another, for example from an organic solvent phase to aqueous solution.

transition – structural change of a system, e.g. the unfolding of a protein from conformation (e.g. the folded state) to another (e.g. a denatured state).

transition state – the crucial configuration of atoms at which the potential energy of the reactants is a maximum.

transition state theory – a means of identifying the main features of governing the size of a rate constant with a model description of a chemical reaction.

transition temperature – temperature at which a phase change occurs: in the context of protein denaturation, the temperature of the midpoint of a folding/unfolding transition.

transmission coefficient – the proportionality constant when the rate of passage of the activated complex through the transition state is assumed to be proportional to the vibrational frequency along the reaction coordinate.

treadmilling – simultaneous polymerization and depolymerization of a filamentous structure, e.g. an actin microfilament.

turnover number – the number of catalytic reactions per enzyme molecule per unit time.

twist – parameter describing the frequency of turns of the DNA double helix.

two-state approximation – in order–disorder transitions, characterization of a structural change in terms of two states (e.g. folded and unfolded, native and denatured, etc.), and in ligand binding, characterization of the equilibrium in terms of just the bound and unbound states of the ligand.

van der Waals interaction – type of intermolecular interaction named after van der Waals.

van't Hoff enthalpy – the *enthalpy* change calculated from the *temperature* dependence of the *equilibrium constant*.

van't Hoff graph – a plot of the equilibrium constant versus temperature, from which the van't Hoff enthalpy can be obtained.

velocity – speed of motion in a certain direction (mechanics) or rate of enzyme catalysis (enzyme kinetics).

voltage – synonym of *electrical potential*, named after Alessandro Volta.

work – *energy* transfer by organized motion. Compare *heat*.

writhe – *parameter* describing the pathway of the DNA backbone in space in the supercoiling of circular DNA.

Zeroth Law – simple argument by which the concept of *temperature* is justified.

Zimm–Bragg model – one of the two most popular models of helix–coil transition theory, named after the persons who were the first to describe it.

Index of names

Subject index